IN CASE OF ACCIDENT[1]

*In case of accident notify the laboratory instructor **immediately**.*

FIRE

Burning Clothing. Prevent the person from running and fanning the flames. Rolling the person on the floor will help extinguish the flames and prevent inhalation of the flames. If a safety shower is nearby hold the person under the shower until flames are extinguished and chemicals washed away. Do not use a fire blanket if a shower is nearby. The blanket does not cool and smouldering continues. Remove contaminated clothing. Wrap the person in a blanket to avoid shock. Get prompt medical attention.

Do not, under any circumstances, use a carbon tetrachloride (toxic) fire extinguisher and be very careful using a CO_2 extinguisher (the person may smother).

Burning Reagents. Extinguish all nearby burners and remove combustible material and solvents. Small fires in flasks and beakers can be extinguished by covering the container with an asbestos-wire gauze square, a big beaker, or a watch glass. Use a dry chemical or carbon dioxide fire extinguisher directed at the base of the flames. **Do not use water.**

Burns, either Thermal or Chemical. Flush the burned area with cold water for at least 15 min. Resume if pain returns. Wash off chemicals with a mild detergent and water. Current practice recommends that no neutralizing chemicals, unguents, creams, lotions, or salves be applied. If chemicals are spilled on a person over a large area quickly remove the contaminated clothing while under the safety shower. Seconds count and time should not be wasted because of modesty. Get prompt medical attention.

CHEMICALS IN THE EYE: Flush the eye with copious amounts of water for 15 min using an eye-wash fountain or bottle, or by placing the injured person face up on the floor and pouring water in the open eye. Hold the eye open to wash behind the eyelids. After 15 min of washing obtain prompt medical attention, regardless of the severity of the injury.

CUTS: Minor Cuts. This type of cut is most common in the organic laboratory and usually arises from broken glass. Wash the cut, remove any pieces of glass, and apply pressure to stop the bleeding. Get medical attention.

Major Cuts. If blood is spurting place a pad directly on the wound, apply firm pressure, wrap the injured to avoid shock, and get **immediate** medical attention. Never use a tourniquet.

---

[1] Adapted from *Safety in Academic Chemistry Laboratories,* prepared by the American Chemical Society Committee on Chemical Safety, March 1974.

ORGANIC
EXPERIMENTS

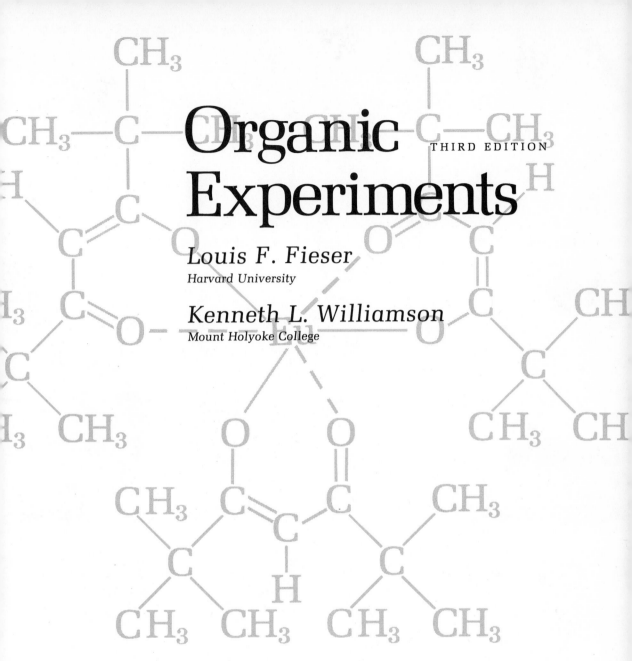

Organic
THIRD EDITION
Experiments

Louis F. Fieser
Harvard University

Kenneth L. Williamson
Mount Holyoke College

D. C. HEATH AND COMPANY
Lexington, Massachusetts / Toronto

$$C_{10}H_{12}N_2O$$

Foreword

This is the seventh version of a book published in 1935 under the title *Experiments in Organic Chemistry*. The name was retained in a 2nd edition of 1941, a 3rd edition of 1955, and a 3rd edition, Revised, of 1957. The next revisions were called *Organic Experiments,* 1st Ed. (1964), and 2nd Ed. (1968). With the present edition, the book has gained from the participation of Professor Kenneth L. Williamson as joint author. It is a pleasure to welcome him to a task which I know will be pleasurable, and which I hope will be important.

LOUIS F. FIESER

Preface

The approach to this, the latest edition of *Organic Experiments*, does not differ materially from that stated in the preface to the 1st edition, published under the title *Experiments in Organic Chemistry*, and used four decades ago: "... to provide the beginning student with carefully standardized and detailed direction, in order that a good technique may be acquired with the greatest possible economy of time and materials."

Many of the experiments included in the first and subsequent editions of this book have become laboratory classics: The Martius Yellow experiment, the isolation of cholesterol from human gallstones, the synthesis of chemiluminescent luminol, and others. All of the experiments reflect a common objective of experimental organic chemistry: rapid, efficient laboratory work that results in high yields of pure products.

Since the last edition of this book many new terms and concepts have been introduced into organic chemistry, e.g., nuclear magnetic resonance, phase transfer catalysis, shift reagents, and crown ethers. We have incorporated a number of the newest and most important of these concepts in meaningful experiments in this edition of *Organic Experiments*.

The first three chapters introduce apparatus and elementary techniques of the organic laboratory; the next six chapters take up the standard laboratory operations (distillation, crystallization, extraction, etc.) with a series of appropriate experiments. Chapters 10–15 deal with the organic applications of spectroscopy (IR, UV, and nmr) and gas chromatography. The remaining 49 chapters include a variety of experiments, e.g., synthesis, analysis, structure determination, and natural product isolation, all of which involve the student in the important areas of modern organic chemistry.

The majority of experiments in this edition are unchanged from the previous editions; however, many experiments have been modernized to take advantage of newer reagents and reactions. For example, the preparation of dichlorocarbene is accomplished using chloroform and a concentrated base in the presence of one of the remarkable phase-transfer catalysts. A biochemical experiment— the familiar biosynthesis of ethanol—has been introduced in conjunction with a distillation experiment. In the simple, elegant synthesis of the rather complex natural product carpanone, no less than four asymmetric centers are introduced in one reaction. Other similar experiments reflecting current student interest in biochemistry and natural products are the isolation of cholesterol from gallstones, of lycopene from tomatoes, of carotene from carrots, and of oleic acid from olive oil. The enzymic resolution of DL-alanine illustrates the use of enzymes in a very practical way. The now widely used Jones reagent and Collins reagent are employed for the oxidation of cholesterol. A new and very straightforward kinetics experiment involving nucleophilic displacement on methyl iodide by t-butoxide ion has been added, including an example of a typical computer program for analysis of the data. Qualitative organic analysis has been given significant coverage; data are supplied for the systematic classification and identification of several hundreds of unknowns, which have been selected for low cost and are listed with pertinent data in the chapter. The classical procedures of analysis can be employed with or without the aid of spectroscopy.

Spectroscopy plays an important part in modern organic chemistry. With the advent of relatively inexpensive infrared, visible-ultraviolet, and nuclear magnetic resonance spectrometers, these instruments have become part of the equipment of the modern organic laboratory. Not every undergraduate in a first course in organic chemistry will have the opportunity to operate a spectrometer personally, but the student should at least be well acquainted with how these instruments work, with how to prepare samples for them, and with the interpretation of the spectra. Accordingly, we have included chapters on infrared, ultraviolet, and nuclear magnetic resonance spectroscopy, with an emphasis on the techniques used for preparing samples and running the spectra. Throughout the book spectra of starting materials and products are included, and the salient points of many spectra are drawn to the student's attention to illustrate what to look for in a spectrum. Lanthanide shift reagents are introduced into the nmr chapter as a means of simplifying otherwise complex spectra. A discussion of gas chromatography is followed by an experiment utilizing this technique for separation of the mixture of olefins resulting from dehydration of an alcohol.

Throughout this edition we have employed the best of current techniques and apparatus. We assume that standard-taper glassware is in universal use, although organic chemists still need to know how to bore corks and bend glass. Standard-taper glassware of 14/20 or 19/22 size is prescribed. Consequently, most synthetic experiments are carried out on a 10-g scale, with a consequent decrease in reagent cost and an increase in the safety factor compared to larger-scale experiments. A number of illustrations of research apparatus have been included so that beginning students will be aware of commercially available equipment.

An instructor's guide,* which lists the experiments according to type and cites the time required for each experiment, is available. The guide also contains answers to questions and problems in the text, a list of required equipment, and notes on all of the experiments, which point out the observations students can be expected to make.

As stated previously, the emphasis of *Organic Experiments, 3rd Edition* is on the experimental aspects of organic chemistry. Hopefully, in this area, the text will contribute to the education of future organic chemists and biochemists. We deliberately have not included long discussions of the theoretical background for each experiment. We expect students to relate the experiments to their knowledge of chemistry in general.

---

* Available without cost from the publisher upon request.

We are indebted to Dr. Mary Fieser for excellent advice in the preparation of the material for this book. We also gratefully acknowledge the assistance of Professor Paul Kropp, University of North Carolina, for many valuable suggestions; of Dr. Paul P. Bryant, D. C. Heath and Co., for editorial comments; and Louise Williamson for help at all stages of the revision. Our thanks too, to Perkins-Elmer Corporation for the ultraviolet spectra reproduced in Chapter 15.

LOUIS F. FIESER
KENNETH L. WILLIAMSON

Contents

1 *Apparatus*

KEYWORDS   Standard taper   Claisen distilling head
Separatory funnel   Vacuum adapter
Condenser   Cold finger condenser
Distilling column

Welcome to the organic chemistry laboratory. One of the first things you will do is to check over your equipment. By referring to the drawings in this and the next chapter, you can identify the pieces of equipment that are not now familiar to you.

We strongly recommend the use of standard-taper ground glass equipment of the types shown in the drawings, either the $^{19}/_{22}$ or the $^{14}/_{20}$ size (Fig. 1.1). (The numbers refer to the diameter and length of the ground joint in mm.) This equipment is easy to put together and just as easy to take apart; when carefully greased it is vacuum tight. Do not leave it joined together for long periods of time as the joints may become stuck. If this should happen, consult your laboratory instructor for suggestions.

Check to see that your thermometer is reading correctly (20°C = 72°F) and that there are no little star-shaped cracks in your flasks. Replace all flasks in this condition. Remember that porcelain apparatus and equipment with graduations and ground glass joints are expensive; Erlenmeyer flasks, beakers, and test tubes are, by comparison, fairly cheap. If ground glass equipment is not available in your laboratory, follow local instructions for making similar assemblies, as they will be adequate for your work.

Figure 1.2 shows an assembly for fractional distillation (the separation of two miscible liquids of similar boiling points) and Fig. 1.3, an assembly for refluxing a small amount of a reaction mixture under a cold finger condenser. The water-cooled tube is thrust through an inverted No. 3 neoprene adapter and rested in the test tube at a convenient height. Figure 1.4 shows a typical apparatus for determination of the melting point of an organic compound.

A separatory funnel with a Teflon stopcock is illustrated in Fig. 1.5. Teflon has a low coefficient of friction and is chemically inert, but it deforms under tension and the stopcock will stick in the glass bore of the funnel if not cared for. Remember to loosen the stopcock after each laboratory period. If the stopcock should become stuck, cool the end of the funnel in ice or Dry Ice. Usually the stopcock will shrink enough to be loosened.

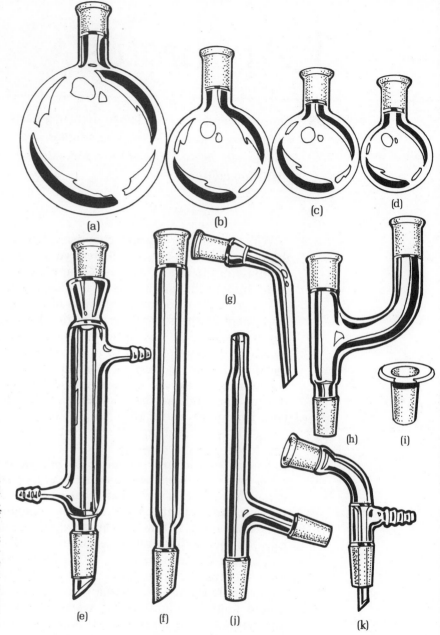

Figure 1.1
Standard taper ground
glass apparatus (19/22).
Round-bottomed flasks of
(a) 250, (b) 100, (c) 50,
and (d) 25 ml capacity;
(e) condenser; (f) dis-
tilling column; (g) simple
bent adapter; (h) Claisen
distilling head;
(i) stopper; (j) distilling
head; and (k) vacuum
adapter.

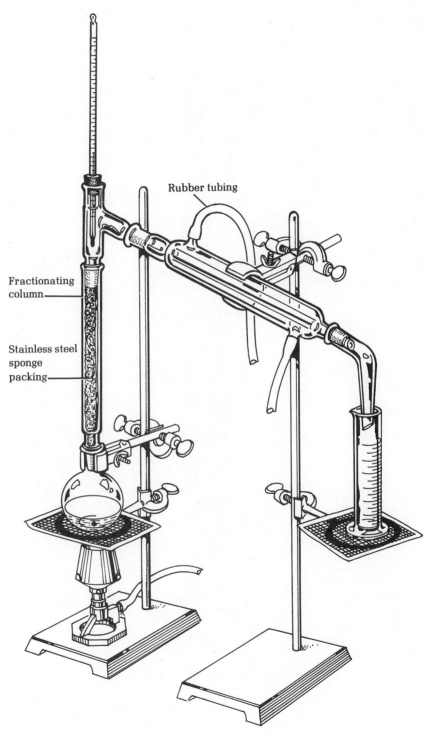

Rubber tubing

Fractionating
column

Stainless steel
sponge
packing

Figure 1.2
Fractional distillation
apparatus.

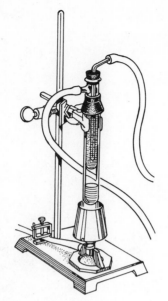

Figure 1.3
Cold finger reflux
condenser.

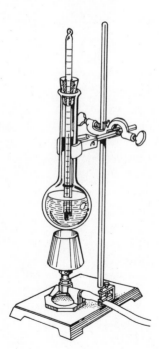

Figure 1.4
Melting point apparatus.

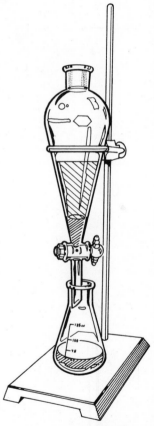

Figure 1.5
Separatory funnel with
Teflon stopcock.

2 Techniques, Important Laboratory Operations

The synthesis, separation, and purification of organic compounds involves a number of processes such as distillation, crystallization, and extraction, which we shall detail in the following chapters. Here we present some miscellaneous but important operations which you will employ frequently. Make mental notes of these as you inspect the drawings for identification of your apparatus.

1. Washing and Drying Laboratory Equipment

Considerable time can be saved by cleaning each piece of equipment soon after it has been used, for you will know then what contaminant is present and be able to select the proper method for removing it. It will be easier to remove before it has dried and hardened. A small amount of organic tar usually can be removed with a few milliliters of an appropriate organic solvent. If the amount of tar is large, it may be best to first try warm water and a detergent; let the vessel soak for a time and then see if the material can be dislodged with a test-tube brush. Acetone (bp 56.1°) has great solvent power and is often effective. It is miscible with water and vaporizes readily, and so is easy to remove from the vessel. Cleaning up after one operation often can be carried out during the period in which a second one is in process.

A polyethylene bottle (shown in Fig. 2.1) serves as a convenient wash bottle for acetone. The Nalgene® 400-ml wash bottles are

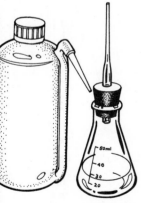

Figure 2.1
Polyethylene wash bottle and a Calcutta wash bottle.

Wash bottles

Clean up as you go

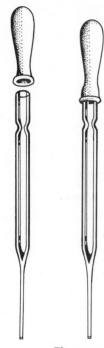

**Figure 2.2
Capillary dropping
tubes (Pasteur
pipettes).**

Use caution!

recommended containers for other common solvents. The name, symbol, or formula of a solvent can be written on a bottle with a Magic Marker or wax pencil. A volume up to 25 ml per squeeze can be delivered from the spout into a graduate; a large volume of solvent can be poured from the top. The dispensing spout gives an ultrafine stream; it can be cut back to increase the size of the stream. For crystallizations and for the quick cleaning of apparatus, it is convenient to have available one bottle for each frequently-used solvent. For dispensing ether, the Calcutta wash bottle is invaluable (Fig. 2.1).[1] It is operated by pointing the tip downward and allowing the base of the flask to be warmed by the heat of the hand. The Calcutta wash bottle is made by cutting a capillary dropping tube to the appropriate length and inserting it in a cork. You will find this little wash bottle very useful in washing down flasks used in ether extractions and filtrations.

Capillary dropping tubes (Pasteur pipettes), Fig. 2.2, are useful for transferring small quantities of liquid, adding reagents dropwise, and washing down the insides of flasks.

Sometimes a reaction flask will not be clean after washing with a detergent and with acetone. If so, try an abrasive household cleaner. Do not hesitate to bend the handle of your test-tube brush to reach the curved sides of the flask. As a last resort use a powerful oxidizing agent. To carry out this process, rinse the flask with water, let it drain, and *carefully* add about 5 ml of concentrated sulfuric acid and 1 ml of concentrated nitric acid. Let the mixture remain in the flask for a time if there is a vigorous reaction, then heat on the steam bath. After the reaction is over, you will find that, on decantation of the acids and washing with water, the flask will be clean.

To dry a piece of apparatus rapidly, rinse with a few milliliters of acetone and invert over a beaker to drain. *Do not use compressed air.* You will find if you hold a piece of clean filter paper in front of the compressed air jet that the air contains droplets of oil, water, and particles of rust. It is much better practice to draw a slow stream of air through the apparatus using the suction of your water aspirator (see Section 5 of this chapter).

2. Breaking and Bending Glass Tubes

To cut a glass tube, first make a fine straight scratch, extending about a quarter of the way around the tube, with a glass scorer. This is done by applying firm pressure on the scorer and rotating the glass tube slightly (Fig. 2.3). Only one scratch should be made; in no case should you try to saw a groove in the tube. The tube is then grasped with the scratch away from the body and the thumbs pressed together at the near side of the tube just opposite the scratch, with the arms pressed tightly against the body (Fig. 2.4). A straight,

Break by pulling

[1] The device was suggested by Dr. Bidyut K. Bhattacharyya, a co-worker from Calcutta, hence the name.

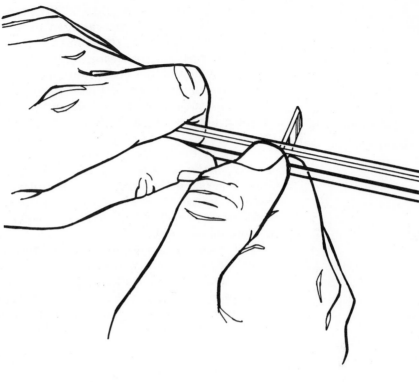

Figure 2.3
Scratching glass tubing
with glass scorer prior to
breaking. The scratch is
about one-fourth the
tube's circumference in
length.

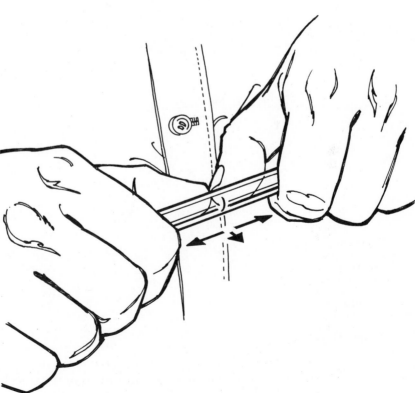

Figure 2.4
Breaking glass tubing.
Thumbs are opposite the
scratch. Pull, about 90%;
bend, about 10%.

clean break will result when *slight* pressure is exerted with the thumbs and a strong force applied to pull the tube apart. It is a matter of 90% pull and 10% bend.

The sharp edges that result from breaking a glass rod or tube will cut both you and the corks and rubber stoppers and tubing being fitted over them. Remove these sharp edges by holding the end of the rod (or tube) in a Bunsen burner flame and rotating the rod until the sharp edges melt and disappear. This fire polishing process can be done even for Pyrex glass if the flame is hot enough. Open the air inlet at the bottom of the burner to its maximum; the hottest part of the flame is about 7 mm above the inner blue cone. A stirring rod with a flattened head, useful for crushing lumps of solid against the bottom of a flask, is made by heating a glass rod until a short section at the end is soft and quickly pressing the end onto a smooth metal surface.

Fire polishing

The secret to successful glass working is to have the glass thoroughly and uniformly heated before an operation. Since Pyrex glass softens at 820° and soft glass at 425°, the best way to work Pyrex is with a gas-oxygen torch, but with patience it can be satisfactorily heated over an ordinary Bunsen burner with a wing top attached (Fig. 2.5). Stopper the tube at the left-hand end, grasp in

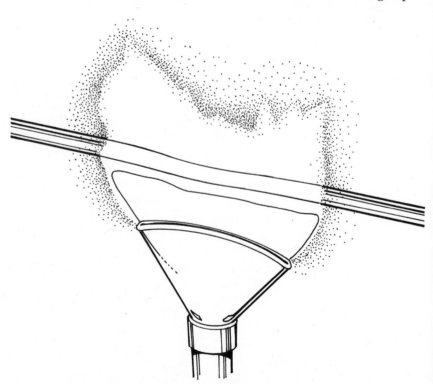

Figure 2.5 Heating glass tubing prior to bending. The wing top produces a broad flame that heats enough of the tubing to allow a good bend to be made. The tubing is held about 7 mm above the inner blue flame.

the left hand with palm down and in the right hand with palm up so you can swing the open end of the tube into position for blowing without interruption of the synchronous rotation of the two ends. Adjust the air intake of the burner for the maximum amount of air possible (too much will blow out the flame) and rotate the tube constantly, holding it about 7 mm above the inner blue cone. A bit of coordination is needed to rotate both ends at the same speed once the glass begins to soften; when the flame is thoroughly tinged with yellow (from sodium ions escaping from the hot glass) and the tube begins to sag, remove the tube from the flame and bend it in the vertical plane with the ends upward and the bend at the bottom. Should the tube become constricted at the bend, blow into the open end immediately upon completion of the bend to expand the glass to its full size.

Rotate constantly

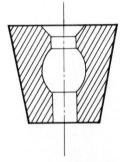

Figure 2.6 Cross section of the fusiform type of rubber stopper. (Arthur H. Thomas Co.)

3. Using Rubber Stoppers and Boring Corks

Insertion of a glass tube into a rubber stopper is easy if the glass is lubricated with a small drop of glycerol. Grasp the tube very close to the end to be inserted; if it is grasped at a distance, especially at the bend, the pressure applied for insertion may break the tube and result in a serious cut. Stoppers with the cross section shown in Fig. 2.6 are safer and easier to use than those with a cylindrical bore. If a glass tube or thermometer should become stuck to a stopper, it can be removed by painting on glycerol and forcing the pointed tip of an 18-cm spatula between the rubber and glass. Another method is to select a cork borer that fits snugly over the glass tube, moisten it with glycerol, and slowly work it through the stopper. When the stuck object is valuable, such as a thermometer, the best policy is to cut the stopper with a sharp knife.

In the construction of some apparatus, the necessity will arise for insertion of a large diameter glass tube into a flask or condenser. Since corks are much easier to bore than rubber stoppers, they are used in this case. Select a high quality cork that fits only 4–5 mm into the opening of the flask and soften it in a cork roller (Fig. 2.7a) or by rolling underfoot. The cork should then be flexible enough to go about half its length into the opening of the flask and will not split when being bored.

Boring corks

Select a sharp cork borer slightly smaller than the size of the hole required, hold it in the right hand and work it into the cork which is held in the left hand. After each twist, grasp the cork at a new place and check the alignment. When the cork has been cut halfway through, withdraw the borer by twisting, push out the plug if present, and finish boring the hole from the other end of the cork (Fig. 2.7b). If these operations are done carefully, it will not be necessary to ream out the hole with a rat-tail file. The secret to good cork-boring technique is a sharp borer and 90% twisting motion and 10% pushing motion on the borer.

Mostly a twisting motion

To sharpen the cork borer, insert the sharpener as far as it will go and turn the borer while pressing on the blade with your thumb

Figure 2.7(a)
Cork roller used to
soften cork stoppers.

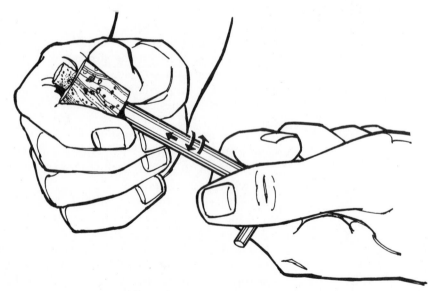

Figure 2.7(b)
Boring a hole through a
softened cork.

(Fig. 2.8). A fine shaving of brass will come off, leaving a razor-sharp edge on the borer.

4. Wash Bottles

A wash bottle of distilled water is indispensable in the laboratory,

Figure 2.8
Sharpening a cork borer.

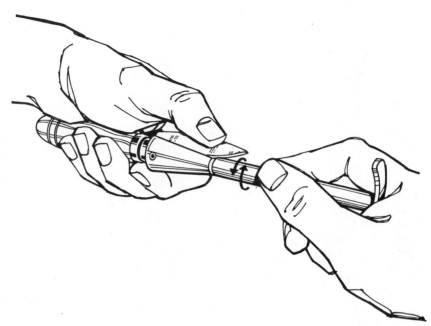

Figure 2.9
Water wash bottle
(Florence flask).

and the classical design shown in Fig. 2.9 is recommended over a variety of others. It is easily constructed from a 500-ml Pyrex Florence flask, 7-mm tubing, and a 2-hole, No. 5 rubber stopper—preferably of the fusiform type. One can direct the stream of water from the wash bottle by guiding the flexible tip with the forefinger. After an initial delivery, the few drops of water remaining in the tip can be forced out separately by a quick breath of air. A large volume of water, for example 25, 50, or 100 ml, can be measured rapidly into a graduate by inverting the wash bottle and delivering the liquid into the container through the mouthpiece (Fig. 2.10). Cut off the flow a little short of the required amount and then add water until the mark is reached, using the wash bottle in the normal fashion.

Since methanol is required frequently for crystallization, a methanol wash bottle made from a 125-ml Florence flask is very useful.

5. The Aspirator and Filter Trap

You will make frequent use of a vacuum to promote sublimation, to reduce the boiling points of liquids, to hasten drying of solids, to remove the liquid from a crystalline product, and to speed up slow filtrations. Water flowing through the aspirator (which works on Bernoulli's principle) will produce a vacuum equal to the vapor pressure of the water flowing through it (17 torr at 20°, 5 torr at 4°). Polypropylene aspirators are inexpensive and will not corrode as do brass ones. A check valve is built into all aspirators, but when the water is turned off it may back up into the vacuum system. For

Figure 2.10
Delivering large amounts
of water from wash bottle.

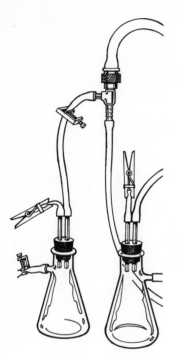

Figure 2.11
Filter traps functioning
as pressure gauges, (*left*)
under vacuum, and
(*right*) at atmospheric
pressure.

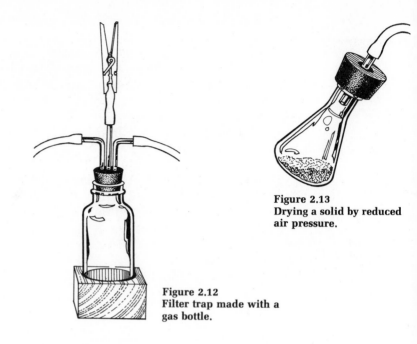

Figure 2.13
Drying a solid by reduced
air pressure.

Figure 2.12
Filter trap made with a
gas bottle.

A convenient
pressure gauge

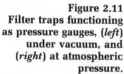

The wooden base
of the filter trap has
an opening 62 mm in
diameter which serves
as a convenient rest-
ing place for round-
bottomed flasks of 50,
100, 250, or 500-ml
capacity and takes the
place of a set of
cork rings.

this reason a trap is always installed in the line (Fig. 2.11). A valve for ready release of vacuum is made by inserting a 6-cm long piece of thin-walled rubber tubing ($^3/_{16}$ dia hole $\times$ $^1/_8$ in. wall thickness) between the jaws of a wooden or plastic clothes clip and slipping the tubing a short distance over the upright glass tube of the flask.

The water passing through the aspirator should always be turned on full force. With proper adjustment of the clothes clip and with an aspirator that is working properly, the thin-walled rubber tubing will collapse and the clothes clip will bend over, acting as an indicator (see Fig. 2.11). Thus, you will be able to see at a glance if all connections in the system are tight. You will, in time, learn to hear the difference in the sound of an aspirator working on an open system and one working on a closed system. Figure 2.12 shows a filter trap made from a 250-ml gas bottle with a three-hole rubber stopper, set in a wooden block to provide stability.

6. Evacuation of a Solid to Constant Weight; The Aspirator Tube

A solid that has been collected by suction filtration may dry slowly at atmospheric conditions, particularly if the solvent is water and the solid noncrystalline. For rapid drying, place the moist solid in a tared flask of appropriate size (a tared flask is one that has been weighed and the weight recorded) and attach a large one-hole rubber stopper by means of a 6-mm dia glass tube to the suction tubing leading to the aspirator. Atmospheric pressure will hold the flask to the stopper when the aspirator starts working (Fig. 2.13). Flasks from 10 ml to 125 ml can be evacuated in this manner and used to dry a solid. Turn on the aspirator to full force, apply steam (see Section 7), and observe the pressure gauge. When the pointer reaches a position of maximum declination, kink the vacuum tube and see whether the pointer moves any farther; if so, all connections are not tight. After about 15 min of evacuation on the steam bath, disconnect the flask, wipe it, and weigh. Repeat the operation until the weight is constant.

A reaction product is often isolated by extraction (Chapter 8) with ether or another organic solvent. The extract is then dried (water removed) by filtration through a cone of anhydrous sodium sulfate held on a filter paper fitted into a stemless conical funnel. The reaction flask and funnel then are rinsed with ether, delivered from a Calcutta wash bottle, and the combined extracts evaporated on the steam bath under an aspirator tube in a tared flask. Ether (bp 34.6°) is volatile and highly flammable, but a *small* quantity can be evaporated on the steam bath with safety by removal of the flammable vapor with the aspirator tube (Fig. 2.14). Since you have the

Extinguish all nearby flames when working with ether!

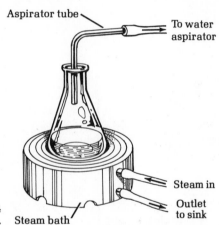

Aspirator tube

To water aspirator

Steam in

Outlet to sink

Figure 2.14
Aspirator tube in use.   Steam bath

tare weight of the flask, you can then evacuate it to constant weight after all the solvent seems to have evaporated and been carried away through the aspirator tube. It is convenient to scratch an identifying mark on the outside of flasks frequently used and to keep a list of their tare weights in your notebook.

7. Steam Baths and Hot Plates

Many operations in the organic laboratory are conducted on the steam bath because the solvents used often boil below 100° and are flammable. Figure 2.15 shows the correct way to heat solvents in a round-bottomed flask on a steam bath; an annular space of a few mm is left between the flask and the support ring (three wood matchsticks serve as spacers), and a *slow* trickle of steam is allowed to come up around the flask. When heating solvents in an Erlenmeyer flask, again let a slow trickle of steam rise around the flask by hav-

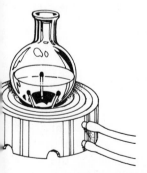

Figure 2.15
Heating a round-bottomed flask on the steam bath. Matchstick spacers allow steam to come up around flask for more efficient heating.

Figure 2.16
Heating an Erlenmeyer
flask on the steam bath.
Placing flask slightly off
center allows steam to
rise around flask.

ing the flask slightly off center on the rings (Fig. 2.16).

The 3-in., 70-watt hot plate shown in Fig. 2.17 is ideal as a heat source for carrying out crystallizations in small Erlenmeyer flasks. Since a single hot plate does not draw much current, several plates can be plugged in on the same bench without risk of blowing a fuse. The hot plate is free from fire hazard; at full heat the element does not glow, and ether dropped on it does not ignite. Because it is slow to heat you should plug in the hot plate at the beginning of the period, if you anticipate using it.

The steam bath is useful for *controlled* heating. After prolonged heating, a thermometer in 100 ml of water heated on a steam bath will register 90° as the maximum temperature, significantly below the boiling point of water. The hot plate will heat 150 ml of Dow silicone fluid to the much higher temperature of 172° after prolonged heating. [Silicone fluid (bp > 300°) gives excellent service in melting-point baths but is more expensive than di-n-butyl phthalate (bp 340°), the fluid recommended for student use.]

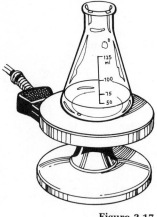

Figure 2.17
70-Watt hot plate
(Wilkens-Anderson
40635).

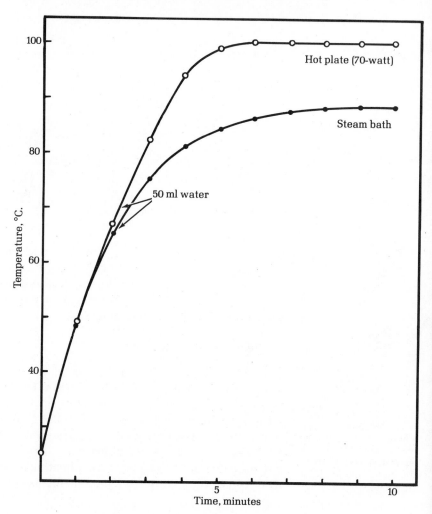

Figure 2.18
Time-temperature curve
for heating 50 ml of
water on the hot plate
and on a steam bath.

The time-temperature curve in Fig. 2.18 shows that the hot plate will heat 50 ml of water to boiling in 5 min; the steam bath does not heat as rapidly. Ten ml of methanol can be brought to the boiling point in 36 seconds on the hot plate, and, as seen in Fig. 2.19, there

Figure 2.19
Hot plate being used in crystallization.

is enough space on it for two or three 25-ml Erlenmeyer flasks.

8. Swirling and Dissolving Solids

Heat alone will not suffice to bring a solid into solution rapidly. Lumps of solid should first be crushed with the flattened end of a stirring rod, and then the hot mixture swirled by grasping the top of the flask and moving the bottom in a circular motion (Fig. 2.20). Alternatively, the flask can be rotated while inclined in the steam bath. A motor-driven stirrer is essential when stirring must be done over a long period of time or in large-scale operations (Fig. 2.21).

9. Transfer of a Solid

The proper technique for transfer of a solid to a reaction vessel or a crystallization flask is shown in Fig. 2.22. The sample is placed on a creased filter paper (or, better, smooth glazed weighing paper) and the near end of the crease rested over the mouth of the flask. A stainless steel spatula is used to scrape the product into the flask. Toward the end of the operation, the creased paper should be raised

Figure 2.20
Swirling a liquid in a flask. The bottom of the flask is given a circular motion, the top remains stationary.

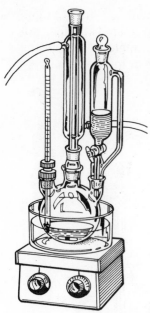

Figure 2.21
A reaction mixture being
heated in an oil bath on
a hot plate and stirred
with a Teflon-covered
stirring bar. Apparatus
of this type is used by
research chemists in
carrying out large-scale
reactions.

until nearly vertical and tapped on both sides with the spatula to
dislodge adhering solid. The same technique serves for transfer of
a small sample of a solid to a specimen vial, except that the vial
should be placed on a piece of clean paper in case of spillage.

10. Electric Drier

The air-drying of products can be speeded considerably by application of the gentle heat from the 70-watt hot plate. A filter paper,
carrying the substance to be dried, is placed on a tray of adjustable
height made from a flat piece of 5 × 8-in. galvanized steel-mesh
fencing[1] mounted as shown in Fig. 2.23. The wire tray is held on a
Fisher clamp as seen in Fig. 2.24.

The temperature of the solid, registered on a thermometer with
the bulb imbedded in the solid being dried, depends mainly upon
the height of the drying tray above the hot plate and also upon the
depth and density of the crystals. Figure 2.23 shows that a 5-g
sample held 9 cm above the plate is heated to 49°. Lowering the
sample to 7 cm raises the temperature to 60°; at 6 cm it is 75–80°.
With this apparatus a substance of sufficiently high melting point
and stability that is damp with water can be dried about three times
faster than at room temperature.

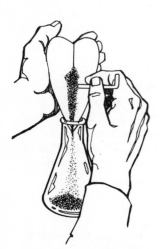

Figure 2.22
A technique of trans-
ferring solids.

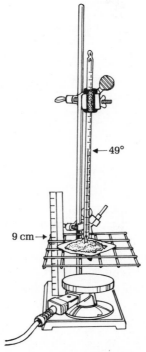

Figure 2.23
Drying a solid on a
drying tray.

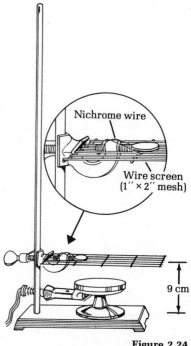

Figure 2.24
Drying tray mounted on a ring
stand (tray available from
Wilkens-Anderson, No. 40638).

[1] Welded galvanized steel-mesh fencing (1 × 1 in. or 1 × 2 in. 14-gauge) is available from J. C.
Penney and from Sears.

3 Weights and Measures and the Laboratory Notebook

KEYWORDS   Estimation of volumes and weights   Calibrated dropping tubes
Measuring tube

Weight vs. volume

Figure 3.1 shows that equal weights of solid materials occupy considerably different volumes; estimation of weight by inspection can be very misleading. Since some of the experiments of this book require accurate estimation of quantities too small to be weighed on available balances, a procedure is given below for measuring the weight of a sample from the volume it occupies in a calibrated melting point capillary.

Estimation of volume

Occasions arise where a rough estimate of volume is adequate and measurement in a graduate inconvenient, for example, when a solution that half fills a 125-ml Erlenmeyer flask is to be evaporated to a volume of 15 ml and the concentrated solution let stand for crystallization. A simple solution is to measure 15 ml of meth-

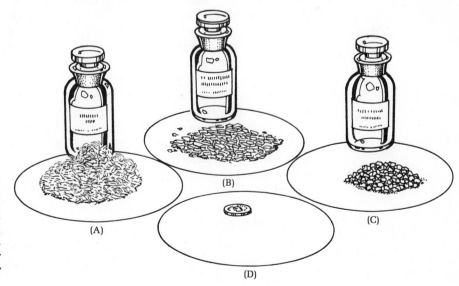

Figure 3.1
Equal weights (6.1 g):
(A) benzoic acid,
(B) 2,3-dimethyl-
naphthalene,
(C) *dl*-**hydrobenzoin,**
and (D) a quarter-
dollar coin.

(A)

(B)

(C)

(D)

17

anol or acetone into a second 125-ml Erlenmeyer flask and to use this as a guage. A low-boiling organic solvent of low surface tension is preferable to water, because the emptied flask drains and dries faster. Most new Erlenmeyer flasks have graduations showing approximate volumes (Fig. 3.2).

EXPERIMENTS

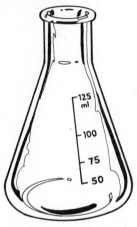

Figure 3.2
Erlenmeyer flask with approximate volumes graduations.

1. Estimation of Weights Using Measuring Tubes

To prepare the measuring tubes, make marks with a wax pencil 22 cm from a cleanly cut end of a piece of 7-mm (o.d.) glass tubing. Cut the tube at the 22-cm mark following the procedure previously described for breaking glass tubing. Then heat the tube with a micro-burner at the midpoint and draw it down to the stages shown in Fig. 3.3a and b. Then, in turn, seal off each half as shown (Fig. 3.3c), heat each seal until soft, and press the tube down onto a flat, heat-resistant surface (a coin) to form two flat-ended cylindrical tubes a little over 10 cm long. Do not fire polish the open end, but rather blunt the outer edge with a file. Prepare a second pair of measuring tubes of 5-mm tubing. Select the better tube of each size for calibration and put the others aside as substitutes that can be assumed to be of the same inside diameter as those calibrated. Fill the 7-mm measuring tube with sodium bicarbonate by thrusting the open end into a pile of solid, backed by a spatula, and shaking it down by the agitation provided by light stroking with a file. Do not pack the solid in as tightly as possible, but rather develop your own technique for packing by stroking with a file, in a way that you can reproduce. When the tube is filled, measure the length of the column of solid and then, by stroking with a file, shake out the bicarbonate onto a tared paper and weigh. Calculate the weight of bicarbonate per unit of length of column, *i.e.*, mg per mm. Now fill the 5-mm tube with bicarbonate packed to the same degree of density as the 7-mm tube, measure the length of the column and, to determine the weight, invert this tube over the 7-mm tube and empty its contents into the larger tube by stroking with a file. Pack the column using your packing technique, measure the length of the column of solid, calculate the weight that filled the 5-mm tube, and so calibrate the 5-mm tube in mg of bicarbonate per mm of length.

Shake the solid out of the measuring tubes, blow out any adhering particles with an air blast, and store the tubes for use whenever you wish to measure small amounts of bicarbonate. For measurement of another solid, you can either repeat the calibration process just described or determine the density of the solid relative to that of sodium bicarbonate by filling the 7-mm tube and weighing the contents.

2. Volumes

Small volumes can be measured conveniently with the cali-brated capillary dropping tube shown in Fig. 3.4; these have many

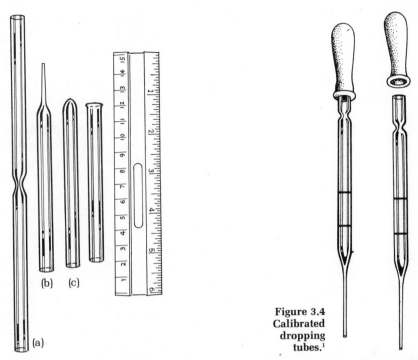

Figure 3.3
Preparation of a 7-mm
(o.d.) measuring tube.

Figure 3.4
Calibrated
dropping
tubes.[1]

uses for the transfer of liquids. In case commercial dropping tubes are not available, a pair of tubes can be made at a time by marking a length of 8-mm glass tubing at distances of 11, 15, and 26 cm, cutting the tube at the 26-cm mark, and heating it at the middle and drawing it down. The capillary tip can be cut off cleanly to a length of 5–6 cm after scratching it with a scorer; the other end should be fire polished.

Adjust the bulb to the extreme end of the capillary dropping tube, insert the tip into water, squeeze out all the air possible, and draw as much water as you can into the tube. Eliminate air bubbles so that the tube is filled solidly with liquid from tip to meniscus. Weigh a 10-ml Erlenmeyer flask on a single pan balance or a triple beam balance to 0.01 g, introduce the water into it by micro drops, and count the drops. From the gain in weight of the flask and the specific gravity of water, calculate the ml delivered by your tube per micro drop. Then, by addition or removal of an appropriate number of micro drops, adjust the volume in the 10-ml flask to exactly 1.0 ml, draw this amount of water completely into the tube and mark the level of the meniscus with a rubber band. To check your first measurement, draw in concentrated hydrochloric acid to the 1-ml mark, count the drops delivered, and see if the weight is close to the expected value (see table, inside back cover of this book). If so, you have a calibrated tube with which you can measure 1 ml of any liquid by volume, or measure fractional volumes of water or

[1] Uncalibrated pipets with bulbs can be calibrated at the $\frac{1}{2}$ and 1 ml marks as described in Section 2.

of concd hydrochloric acid by micro drops. A permanent calibration mark on the tube can be made by making a light scratch with a file or scorer, daubing the scratch with a colored marker (Carter Ink Co.), and rubbing the ink off the smooth glass surface with a moistened towel. Whereas the weight per drop of an aqueous solution as concentrated as the 36% acid is slightly different from that for water, the calibration by drops established for water can be used for measurement of dilute (5–10%) aqueous solutions of acids and bases. To determine if the water calibration applies also to organic liquids, count the number of drops delivered by 1.0 ml of methanol.

3. Titration

Approximation: 50-mm column in standard capillary

To check the calibration of your measuring tube with that of the dropping tube, use the measuring tube and place 1 millimole ($\frac{1}{1000}$ mole) of sodium bicarbonate into a 10-ml Erlenmeyer flask, dissolve it in 1 ml of water, and determine the pH by dipping a strip of Hydrion paper into the solution and at once comparing the color with the color charts on the two sides of the plastic container. With your dropping tube measure 1.0 ml of concd hydrochloric acid into a 10-ml graduate, dilute with water to a volume of 10 ml, and calculate the number of drops of the dilute solution required to give 1 millimole of acid (your instructor can help). Run about 80% of the theoretical amount of acid into the bicarbonate solution, check the pH, and then complete the titration by dropwise addition until you reach the end point of pH 4. Compare calculated and found values. Before discarding the solution, let a piece of used Hydrion paper soak in it for a time to determine if the pigment is water soluble. If it is, the paper should never be left long in a solution being tested.

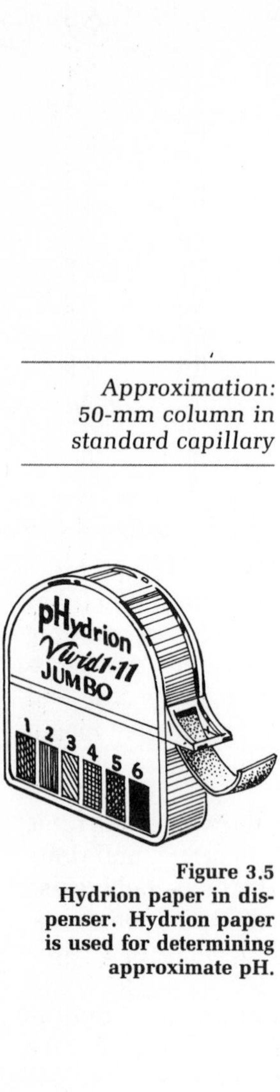

Figure 3.5
Hydrion paper in dispenser. Hydrion paper is used for determining approximate pH.

4. Applications

The above techniques are useful in the investigation of unknowns. The density of a liquid unknown can be determined in the course of calibrating a dropping tube already calibrated for water. A capillary tube for measurement of a solid unknown can be calibrated by the simple method described in the last paragraph of Section 3.1. A test of an aqueous solution of a liquid or solid unknown with Hydrion paper will reveal the presence of acidic or basic groups and indicate the strength, and a titration will establish the number of ionic groups if the molecular weight is known. Even if the substance is nonionic, knowledge of the approximate molecular weight should facilitate identification.

THE LABORATORY NOTEBOOK

A complete, accurate record is an essential part of laboratory work. Failure to keep such a record means laboratory labor lost. An adequate record includes the procedure (what was done), observations (what happened)? and conclusions (what the results mean).

Use a lined, paperbound, $8\frac{1}{2} \times 11$ in. notebook and record all data in ink. Allow space at the front for a table of contents, number the pages throughout, and date each page as you use it. Reserve the left-hand page for calculations and numerical data, and use the right-hand page for notes. Never record *anything* on scraps of paper to be recorded later in the notebook. It is not good practice to erase, remove, or obliterate notes; simply draw a single line through incorrect entries.

The notebook should contain a statement or title for each experiment followed by balanced equations for all principal and side reactions, and where relevant, the mechanisms of the reactions. Consult your textbook for supplementary information on the class of compounds or type of reaction being dealt with. Give a reference to the procedure being used; do not copy verbatim the procedure in the laboratory manual.

Before coming to the lab to do preparative experiments prepare a table (in your notebook) of reagents to be used and the products expected, with their physical properties. (An illustrative table appears with the first preparative experiment, the preparation of n-butyl bromide, Chapter 17.) From your table, use the molar ratios of reactants and determine the limiting reagent and calculate the theoretical yield (in grams) of the desired product. Enter all data in your notebook (left-hand page).

Include an outline of the method of purification of the product. This is most easily done by means of a flow sheet, which lists all possible products, by-products, unused reagents, solvents, etc., that appear in the crude reaction mixture. On the flow sheet diagram indicate how each of these is removed, e.g., by extraction, various washing procedures, distillation, or crystallization. With this information entered in the notebook before coming to the laboratory you will be ready to carry out the experiments with the utmost efficiency. Plan your time before coming to the laboratory. Often two and three experiments can be run simultaneously.

When working in the laboratory, record everything you do and everything you observe *as it happens*. The recorded observations constitute the most important part of the laboratory record as they form the basis for the conclusions you will draw at the end of each experiment. Record the physical properties of the product, the yield in grams, and the percentage yield. When your record of an experiment is complete, another chemist should be able to read the account with complete understanding and to determine what you did, how you did it, and what conclusions you reached. In other words, from the information in your notebook a chemist should be able to repeat your work.

4 Biosynthesis of Ethanol

KEYWORDS Emden-Meyerhof-Parnas scheme   Fermentation
Sucrose   Filter aid (celite)
Enzymes, substrates   Büchner funnel

$$\text{Sucrose} \xrightarrow[\text{Na}_2\text{HPO}_4]{\text{Enzymes}} 4\ \text{CH}_3\text{CH}_2\text{OH} + 4\ \text{CO}_2$$

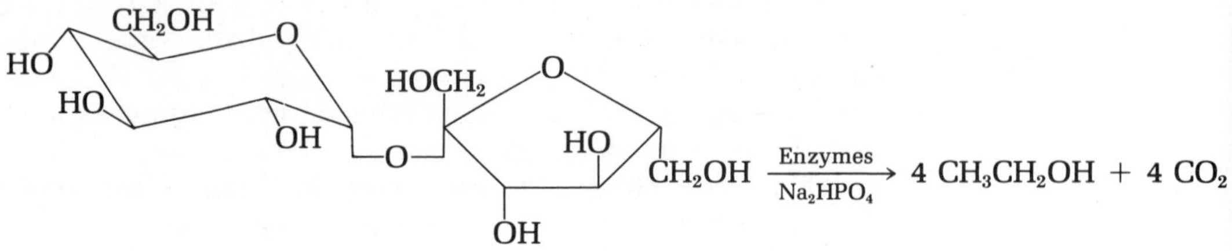

Sucrose

Sucrose, ordinary table sugar, is converted into ethanol and carbon dioxide with the aid of some twenty enzymes as catalysts, in addition to adenosine triphosphate (ATP), phosphate ion, thiamine pyrophosphate, magnesium ion, and reduced nicotinamide adenine dinucleotide (NADH), all present in yeast. Since the process of fermentation has been of interest to man since prehistoric times, the chemistry of this process has been thoroughly studied; however, the exact details of interaction of enzymes (large globular polypeptides of MW 60,000 to 360,000) with the substrate (the small molecule) in each step are not known.

The fermentation process—known as the Emden-Meyerhof-Parnas scheme—involves the hydrolysis of sucrose to glucose and fructose which, as their phosphates, are cleaved to two three-carbon fragments. These fragments, as their phosphates, eventually are converted to pyruvic acid, which is decarboxylated to give acetaldehyde. Acetaldehyde, in turn, is reduced to ethanol in the final step. Each step requires a specific enzyme as a catalyst and often inorganic ions, such as magnesium and, of course, phosphate. Thirty-one kilocalories of heat are released per mole of glucose consumed in this sequence of anerobic reactions.

Enzymes are remarkably efficient catalysts, but they are also labile (sensitive) to such factors as heat and cold, changes in pH, and various specific inhibitors. In the experiments of this chapter

22

you will have an opportunity to observe the biosynthesis of ethanol and to test the effects of various agents on the enzyme system. In the next experiment (Chapter 5) you will isolate the ethanol in pure form by simple and fractional distillation.

$$
\begin{array}{ccccc}
\text{CHO} & & \text{CH}_2\text{OH} & & \text{CH}_2\text{OPO}_3{}^{2-} \\
\text{H}-\text{C}-\text{OH} & & \text{C}=\text{O} & & \text{C}=\text{O} \\
\text{HO}-\text{C}-\text{H} & \rightleftharpoons & \text{HO}-\text{C}-\text{H} & \rightarrow & \text{HO}-\text{C}-\text{H} \\
\text{H}-\text{C}-\text{OH} & & \text{H}-\text{C}-\text{OH} & & \text{H}-\text{C}-\text{OH} \\
\text{H}-\text{C}-\text{OH} & & \text{H}-\text{C}-\text{OH} & & \text{H}-\text{C}-\text{OH} \\
\text{CH}_2\text{OPO}_3{}^{2-} & & \text{CH}_2\text{OPO}_3{}^{2-} & & \text{CH}_2\text{OPO}_3{}^{2-}
\end{array}
$$

Glucose-6-phosphate   **Fructose-6-phosphate**   **Fructose-1,6-diphosphate**

Dihydroxyacetone phosphate
$$
\begin{array}{c}
\text{CH}_2\text{OPO}_3{}^{2-} \\
\text{C}=\text{O} \\
\text{CH}_2\text{OH}
\end{array}
$$

$\updownarrow$

Glyceraldehyde-3-phosphate
$$
\begin{array}{c}
\text{CHO} \\
\text{H}-\text{C}-\text{OH} \\
\text{CH}_2\text{OPO}_3{}^{2-}
\end{array}
$$

$$
\begin{array}{cccccc}
\begin{array}{c}\text{O}\diagdown \\ \text{C}-\text{OPO}_3{}^{2-} \\ \text{H}-\text{C}-\text{OH} \\ \text{CH}_2\text{OPO}_3{}^{2-}\end{array} & \rightarrow &
\begin{array}{c}\text{COOH} \\ \text{H}-\text{C}-\text{OH} \\ \text{CH}_2\text{OPO}_3{}^{2-}\end{array} & \rightarrow &
\begin{array}{c}\text{COOH} \\ \text{H}-\text{C}-\text{OPO}_3{}^{2-} \\ \text{CH}_2\text{OH}\end{array} & \rightarrow &
\begin{array}{c}\text{COOH} \\ \text{C}-\text{OPO}_3{}^{2-} \\ \text{CH}_2\end{array}
\end{array}
$$

1,3-Diphospho-glyceric acid   **3-Phospho-glyceric acid**   **2-Phospho-glyceric acid**   **Phosphoenolpyruvic acid**

$$
\begin{array}{ccccc}
\begin{array}{c}\text{COOH} \\ \text{C}=\text{O} \\ \text{CH}_3\end{array} & \xrightarrow{\text{CO}_2} &
\begin{array}{c}\text{CHO} \\ \text{CH}_3\end{array} & \rightarrow &
\begin{array}{c}\text{CH}_2\text{OH} \\ \text{CH}_3\end{array}
\end{array}
$$

Pyruvic acid   **Acetaldehyde**   **Ethanol**

EXPERIMENTS   **1. Fermentation of Sucrose**

Macerate (grind) one-half cake of yeast or half an envelope of dry yeast in 50 ml of water in a beaker, add 0.35 g of disodium hydrogen phosphate, and transfer this slurry to a 500-ml round-bottomed flask. Add a solution of 51.5 g of sucrose in 150 ml of water, and shake to ensure complete mixing. Fit the flask with a one-hole rubber stopper containing a bent glass tube that dips below the surface of a saturated aqueous solution of calcium hydroxide (limewater) in a 6-in. test tube (Fig. 4.1). The tube in limewater will act as a seal to prevent air and unwanted enzymes from entering

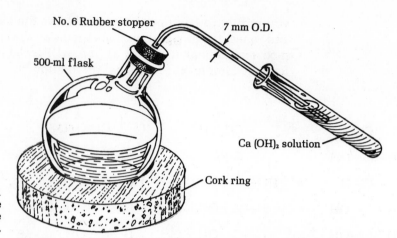

No. 6 Rubber stopper

7 mm O.D.

500-ml flask

Ca (OH)₂ solution

Cork ring

Figure 4.1
Apparatus used in the
fermentation of sucrose
experiment.

the flask but will allow gas to escape. Place the assembly in a warm spot in your desk (the optimum temperature for the reaction is 35°) for a week, at which time the evolution of carbon dioxide will have ceased.

Upon completion of fermentation add 10 g of filter aid (diatomaceous earth or face powder) to the flask, shake vigorously, and filter. Use a 5.5-cm Büchner funnel placed on a neoprene adapter or Filtervac atop a 500-ml filter flask that is attached to the water aspirator through a trap by vacuum tubing (Fig. 4.2). Since the

Reaction time,
1 week

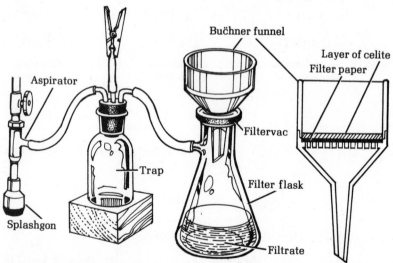

Büchner funnel

Layer of celite
Filter paper

Aspirator

Filtervac

Trap

Filter flask

Splashgon

Filtrate

Figure 4.2
Vacuum filtration
apparatus.

apparatus is top-heavy, clamp the flask to a ring stand. Moisten the filter paper with water and apply gentle suction (water supply to aspirator turned on full force, clothespin on trap partially closed), and slowly pour the reaction mixture onto the filter. Wash out the flask with a few ml of water from your wash bottle and rinse the filter cake with this water. The filter aid is used to prevent the pores of the filter paper from becoming clogged with cellular debris from the yeast.

Filter aid

The filtrate, which is a dilute solution of ethanol contaminated with bits of cellular material and other organic compounds (acetic acid if you are not careful), is saved in a stoppered flask for the next experiment (Chapter 5).

2. Effect of Various Reagents and Conditions on Enzymatic Reactions

About 15 min after mixing the yeast, sucrose, and phosphate, remove 20 ml of the mixture and place 4 ml in each of five test tubes. To one tube add 1.0 ml of water, to the next add 1.0 ml of 95% ethanol, to the next add 1.0 ml of 0.5 M sodium fluoride. Heat the next tube for 5 min in a steam bath and cool the next tube for 5 min in ice. Add 10 to 15 drops of mineral oil on top of the reaction mixture in each tube (to exclude air, since the process is anerobic). Place tubes in a beaker of water at room temperature for 15 min, then take them one at a time and connect each to the manometer as shown in Fig. 4.3. Allow about 30 sec for temperature equilibration,

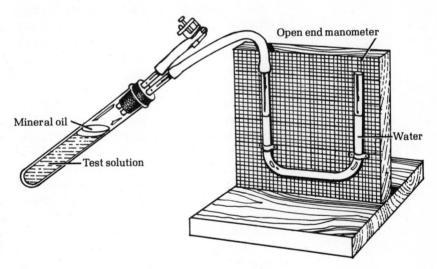

Figure 4.3
Manometric gasometer.

then clamp the vent tube and read the manometer. Record the height of the manometer fluid in the open arm of the U-tube every minute for 5 minutes or until the fluid reaches the top of the manometer. Plot a graph of the height of the manometer fluid against time for each of the four reactions. What conclusions can you draw from the results for these five reactions?

5 Simple and Fractional Distillation

KEYWORDS   Liquid-vapor equilibrium   Thermometer calibration
Fractionating column, packing   Boiling stone
Azeotrope

The distillation process

When water is heated with a flame in a simple distillation apparatus (Fig. 5.1), the vapor pressure of the liquid, or the tendency of molecules to escape from the surface, increases until it becomes equal to the atmospheric pressure, at which point the liquid begins to boil. Addition of more heat will supply the heat of vaporization required for conversion of the liquid water to gas (steam), which rises in the apparatus, warms the distillation head and thermometer, and flows down the condenser. The cool walls of the condenser remove heat from the vapor and the vapor condenses to the liquid form. Distillation should be conducted slowly and steadily and at a rate such that the thermometer bulb always carries a drop of condensate and is bathed in a flow of vapor. Liquid and vapor are then in equilibrium, and the temperature recorded is the true boiling point. If excessive heat is applied to the walls of the distilling flask above the liquid level, the vapor can become superheated, the drop will disappear from the thermometer, the liquid-vapor equilibrium is upset, and the temperature of the vapor rises above the boiling point.

Ethanol boils at 78.4°, water at 100°. When distilled in a simple distilling apparatus a mixture of these two miscible liquids starts to distil a little above the boiling point of ethanol and stops distilling a little below the boiling point of water. All fractions of the distillate are mixtures and little separation of the two components is achieved. A better separation could be obtained by redistillation of each fraction. If this process of redistillation is repeated often enough, separation of the two components of the mixture will eventually be achieved. Fortunately this series of condensations and redistillations is done automatically in a fractionating column.

Fractionating column packed with a stainless steel sponge

The fractionating column shown in Fig. 5.2 contains a stainless steel scouring sponge (Fig. 5.3), which forms a porous packing for equilibration of vapor and condensate. At the start of the distilla-

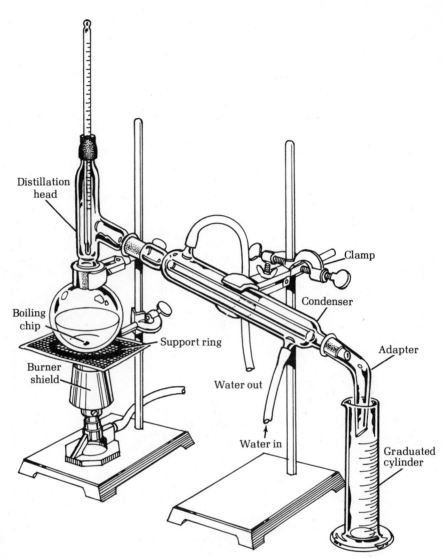

Figure 5.1
Apparatus for simple
distillation.

tion of a mixture of ethanol and water the mixture boils and the
vapors condense in the lowest part of the fractionating column.
The composition of this condensate is similar to that of the first
fraction collected in a simple distillation—richer in ethanol than in
water, but by no means pure ethanol. Then, as more ethanol and
water boil, the temperature of these vapors is higher than that of
the first portion of the mixture, because this portion contains less
ethanol and more water. These hot vapors contact the liquid already
in the fractionating column from the first part of the distillation and
a heat transfer takes place, which causes the less volatile component (ethanol) to boil from that liquid. A succession of these condensations and redistillations occur up and down the column. The
efficiency of a column is rated by the number of simple distillations

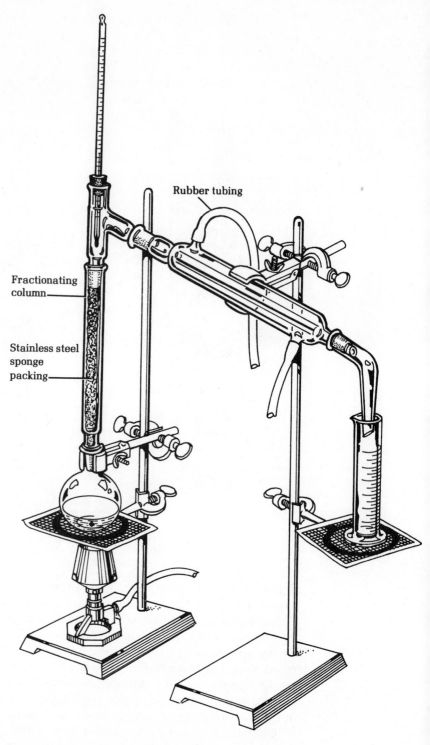

Rubber tubing

Fractionating
column

Stainless steel
sponge
packing

**Figure 5.2
Apparatus for fractional
distillation. The position
of the thermometer bulb
is critical.**

that take place within the column. After ethanol-water vapor has
warmed the entire length of the column, the less volatile part con-

denses and trickles down over the surface of the packing, while fresh vapor from the flask forces its way through the descending condensate with attendant heat interchange. A number of equilibrations between ascending vapor and descending condensate take place throughout the column. The vapor that eventually passes into the receiver is highly enriched in the more volatile ethanol, whereas the condensate that continually drops back into the flask is depleted of the volatile component and enriched with the less volatile water. The packing is used in the column to increase the vapor-liquid contact area. Since equilibration is fairly slow, the slower the distillation the better the separation.

Equilibration between ascending vapor and descending condensate

In research, fractionating heads are frequently used at the top of a distillation column (Fig. 5.4). These heads are designed in such a

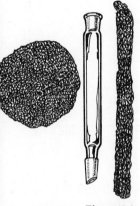

Figure 5.3
Fractionating column and packing.

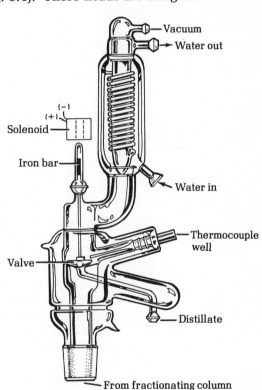

Figure 5.4
Total reflux fractionating head, vacuum jacketed.

way that all the vapor boiled from the flask can be condensed and returned to the column (total reflux). A solenoid lifts the small iron bar and opens the valve about once a minute to release one drop at a time to a receiver; thus, there is time for complete equilibration to take place in the column between drops. With this type apparatus, liquids with boiling points differing by only one or two degrees can be separated. The stainless steel sponge-packed column illustrated and specified for these experiments achieves a fairly sharp separation (see the distillation curve of Fig. 5.5), since the boiling points of ethanol and water differ by 22°.

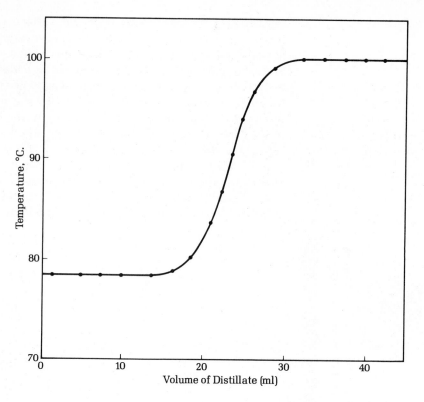

Figure 5.5
Fractionation graph of a
mixture of ethanol and
water.

In the graph of Fig. 5.6 the boiling point of an ethanol-water mixture is plotted as a function of the composition (heavy solid curve). The dashed curved line indicates the composition of the distillate obtained at the boiling point of the mixture. For example, if a mixture of 25% ethanol and 75% water were distilled, it would boil at the temperature indicated by (A). If a bit of that vapor were condensed and analyzed, it would be found to be a mixture of 50% ethanol and 50% water (B). If that bit of vapor were condensed to a tiny drop of liquid, the liquid would, of course, still be half ethanol and half water, and it would have a boiling point represented by point (C). The vapor in equilibrium with that 50-50 liquid has, when analyzed, the composition (D), and so on. Thus, distillation of a mixture of ethanol and water with a fractionating column should give a first fraction considerably enriched in ethanol. But note there is a minimum in the curve. Ethanol and water do not form an ideal mixture; they do not conform to Raoult's law. Because of molecular interaction, a mixture of 95.5% (by weight) of ethanol and 4.5% of water boils *below* (78.15°) the boiling point of pure ethanol (78.3°). Thus, no matter how efficient the distilling apparatus we can never obtain 100% ethanol by distillation of a mixture of 75% water and 25% ethanol. A mixture of liquids of a certain definite composition that distils at a constant temperature without change in composition is called an **azeotrope;** 95% ethanol is such

95% Ethanol, a low-
boiling azeotrope

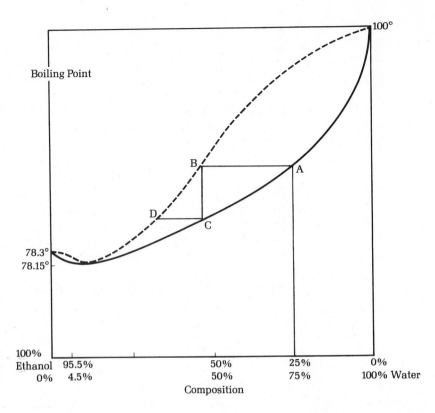

an azeotrope. To prepare 100% ethanol the water can be removed chemically (reaction with calcium oxide) or by removal of the water as an azeotrope with still another liquid. An azeotrope mixture of 32.4% ethanol and 67.6% benzene (bp 80.1°) boils at 68.2°. A ternary azeotrope, bp 64.9°, contains 74.1% benzene, 18.5% ethanol, and 7.4% water. Absolute alcohol (100% ethanol) is made by addition of benzene to 95% alcohol and removal of the water in the volatile benzene-water-alcohol azeotrope.

A pure liquid has a constant boiling point. A change in boiling point during distillation is an indication that the liquid is not pure. The converse proposition, however, is not always true, and constancy of a boiling point does not necessarily mean that the liquid consists of only one compound. For instance, two miscible liquids of similar chemical structure and which boil at the same temperature individually will have nearly the same boiling point as a mixture. And, as noted previously, azeotropes have constant boiling points which can be either above or below the boiling points of the individual components.

Azeotropes

When a solution of sugar in water is distilled, the boiling point recorded on a thermometer located in the vapor phase is 100° (at 760 torr) throughout the distillation, whereas the temperature of the boiling sugar solution itself is initially somewhat above 100° and continues to rise as the solution becomes more concentrated. The

Nonvolatile solute

vapor pressure of the solution is dependent upon the number of water molecules present in a given volume, and hence with increasing concentration of nonvolatile sugar molecules and decreasing concentration of water, the vapor pressure at a given temperature decreases and a higher temperature is required for boiling. However, sugar molecules do not leave the solution, and the drop clinging to the thermometer bulb is pure water in equilibrium with pure water vapor.

When a distillation is carried out in a system open to the air and the boiling point is thus dependent on existing air pressure, the prevailing barometric pressure should be noted and allowance made for appreciable deviations from the accepted bp temperature (see Table 5.1). Distillation can also be done at the lower pressures that can be achieved by an oil pump or an aspirator with substantial reduction of boiling point (see Chapter 62).

Table 5.1   Variation in Boiling Point with Pressure

Pressure, mm	Water, °C	Benzene, °C
780	100.7	81.2
770	100.4	80.8
760	100.0	80.1
750	99.6	79.9
740	99.2	79.5
584*	92.8	71.2

* Instituto de Quimica, Mexico City, altitude 7700 ft (2310 meters).

EXPERIMENTS   ## 1. Calibration of Thermometer

Test the 0° point of your thermometer with a well-stirred mixture of crushed ice and distilled water. To check the 100° point put 10 ml of water in a 25 × 150-mm test tube, clamp the tube in a vertical position, add one carborundum boiling stone or porcelain chip to prevent bumping, and boil gently holding the thermometer in the vapor from the boiling water. Then immerse the bulb in the liquid and see if you can observe superheating. Check the atmospheric pressure to determine the true boiling point of the water.

2. Apparatus for Simple Distillation

The 500-ml round-bottomed flask is clamped over a Bunsen burner fitted with a shield (chimney) to exclude drafts (see Fig. 5.1). The flask is fitted with a distillation head (also called a stillhead) carrying a thermometer and connected to a short condenser supported by a clamp. The condenser is positioned so that the lip at the delivery end directs the drip into the bent adapter. A bent adapter leads the distillate into a 50-ml graduated cylinder. Note that the bulb of the thermometer is below the opening into the side arm of the distillation head. Each ground glass joint is greased. This operation is done by putting 3 or 4 stripes of grease lengthwise around the male joint and pressing the joint firmly into the other without twisting. The air is thus eliminated and the joint will appear almost transparent. (Do not use excess grease as it will contaminate the product.) Water enters the condenser at the tublature nearest the receiver. Because of the large heat capacity of water only a very small stream ($1/8''$ dia) is needed; too much water pressure will cause the tubing to pop off.

3. Simple Distillation

Caution: 95% Ethanol
is flammable

Add the filtrate from the previous experiment (Section 1, Chapter 4), which consists of a dilute solution of ethanol in water, to the 500-ml flask and add a boiling stone. After making sure all connections are tight, heat the flask directly with a small flame until boiling starts. Then adjust the flame until the distillate drops at a regular rate of about one drop per second. Record both the temperature and the volume of distillate at regular intervals. After 50 ml of distillate are collected, discontinue the distillation. Record the barometric pressure, make any thermometer correction necessary, and plot boiling point (bp) *vs.* volume of distillate. Save the distillate for fractional distillation (Section 5, of this chapter).

4. Apparatus for Fractional Distillation

A 100-ml round-bottomed flask is supported with a clamp on a wire gauze and ring over a Bunsen burner fitted with a draft shield (Fig. 5.2). A Hempel distillation column is packed with one third or one fourth of a stainless steel sponge and mounted in the flask, together with a stillhead and thermometer. The stillhead delivers into a short condenser fitted with a bent adapter leading into a 10-ml graduated cylinder.

5. Fractional Distillation of Ethanol-Water Mixture

Add your ethanol-water mixture from the simple distillation (Section 3, preceding) to the flask, add a boiling chip, and quickly bring the mixture to a boil. As soon as boiling starts, turn the flame to the smallest, just nonluminous, flame that will stay lit (cut the air supply to the burner). Heat slowly at first and do not hurry. A ring of condensate will rise slowly through the column; if you cannot at first see this ring, locate it by touching the column with the fingers. The rise should be very gradual, in order that the column can acquire a steady temperature gradient. Do not apply more heat until you are sure that the ring of condensate has stopped rising, then increase the heat gradually. In a properly conducted operation, the vapor-condensate mixture reaches the top of the column only after several minutes. Once distillation has commenced, it should continue steadily without any drop in temperature at a rate not greater than 1 ml in 1.5–2 min. Observe the flow and keep it steady by slight increases in heat as required. Protect the column from drafts.

Best fractionation by
slow and steady
heating

Record the temperature as each ml of distillate collects, and make more frequent readings when the temperature starts to rise abruptly. Each time the graduated cylinder fills, quickly empty it into a series of labeled 25-ml Erlenmeyer flasks. Stop the distillation when a second constant temperature is reached. Plot a distillation curve and record what you observed inside the column in the course of the fractionation. Combine the fractions which you think are pure 95% ethanol and turn in half of the product in a bottle labeled with your name, desk number, the name of the product, the bp range, weight, and percentage yield (calculated on the basis of the amount of sucrose used). Save the other half for the azeotropic distillation experiment.

Dry apparatus

Your apparatus must be dry for both of the next two experiments; therefore, extinguish the flame and, while the column is cooling, clamp the condenser in a vertical position over a beaker and rinse the inner walls with a few drops of acetone from your plastic wash bottle; rinse the stillhead and thermometer and flask in the same way, and let the pieces dry. Rinse the cooled column with acetone, allow the solvent to drain out, and then connect the column to the suction pump and draw a slow stream of air through the packing to remove the acetone as it evaporates.

6. Distillation of the Benzene-Ethanol-Water Ternary Azeotrope[1]

Add half of the 95% ethanol you isolated by fractional distillation to a 50-ml round-bottomed flask, along with a boiling chip. Now add a volume of benzene equal to one half the volume of the 95% ethanol and set up the apparatus for fractional distillation. Distil slowly and steadily, and record boiling points and milliliters distilled. Distil until about 3–4 ml is left in the boiling flask. *Never distil to dryness!* Change receivers each time you think you have reached the boiling point of a pure component. Add water to each fraction except the last, and note how much benzene is present.

Absolute ethanol

The last fraction should be 100% (200 proof)[2] ethanol. Inevitably, this last fraction will contain traces of benzene, which is toxic. Bottle the product and turn it in with the usual information on the label. From your observations see if you can identify the ternary and binary azeotropes mentioned in the introduction to this chapter.

7. Fractional Distillation of Carbon Tetrachloride-Toluene Mixture

Fractionate a mixture of 10 ml each of carbon tetrachloride and toluene and plot a distillation curve. If, at the start, water collects in droplets inside the stillhead, hold a folded towel around

[1] Several students may wish to pool their 95% ethanol for this experiment; otherwise the last fraction will be too small to work with.
[2] An old method of testing the alcoholic content of whisky involved pouring it on gunpowder and setting it afire. Ignition of the gunpowder after the alcohol burned away was proof the whisky did not contain too much water.

this piece until the moisture is gone. Ideally the process within the fractionating column should be an adiabatic one (no heat exchange with the surroundings). Try insulating the column with asbestos tape or with glass wool covered by aluminum foil with the shiny side in.

8. Unknowns[3]

You will be supplied with an unknown, prepared by the instructor, that is a mixture of two solvents from those listed in Table 5.2. The solvents are mutually soluble and differ in boiling point by more than 20°. Fractionate the unknown, identify the components from the boiling points, and estimate the composition of the original mixture from the distillation curve.

Table 5.2   Some Properties of Common Solvents

Solvent	Bp	Latent heat of vaporization[a]	Surface tension[b] (20°)
Acetone	56.5°	125.3	23.7
Methanol	64.7°	261.7	22.6
Hexane	68.7°	79.2	18.4
Carbon tetrachloride	76.7°	46.4	26.8
Benzene	80.1°	93.5	29.0
Water	100.0°	536.6	72.7
Toluene	110.6°	86.8	28.4

[a] Calories per gram at the bp
[b] Dynes per centimeter

QUESTIONS

1. In the simple distillation of Section 1 can you account for the boiling point of your product in terms of the known boiling points of pure ethanol and pure water?

2. From your plot of bp vs. volume of distillate in the simple distillation, what can you conclude about the purity of your product?

3. From the temperature/volume plot in the fractionation of ethanol-water mixture, what conclusion can you draw about the homogeneity of the distillate? Does it have a constant boiling point? Is it a pure substance since it has a constant boiling point?

4. How could you prove you have carried out a biosynthesis of ethanol? Could you prove, for instance, that you have isolated ethanol and not methanol? How could you show that your initial distillate from the fractional distillation contains 5% water?

5. Why is it necessary to dry the apparatus before carrying out either of the last two experiments?

[3] Because of the large holdup of the column relative to the small volume distilled, the unknown should not contain less than 10 ml of the less volatile component. The pair benzene-methanol is eliminated by the specification of a 20° difference in boiling point, but note also that these liquids form an azeotrope bp 58.3°. Methanol and toluene form an azeotrope bp 63.8° (69% methanol).

6 Melting Points

KEYWORDS Melting point capillary tube
Depression of melting point
Mixed melting point
Stem correction
Rast molecular weight determination
Molality
Evacuating and sealing a capillary tube

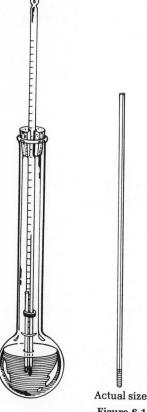

Actual size

Figure 6.1
Melting point flask and capillary tube. The Pyrex flask is 22 cm long, the neck 18 mm o.d., the bulb 55 mm o.d. (Wilkens-Anderson, 5712).

The melting point of an organic solid can be determined by introducing a tiny amount of the substance into a small capillary tube (Fig. 6.1), attaching this to the stem of a thermometer centered in a heating bath, heating the bath slowly, and observing the temperatures at which melting begins and is complete. Pure samples usually have sharp melting points, for example 149.5–150° or 189–190°; impure samples of the same compounds melt at lower temperatures and over a wider range, for example 145–148° or 187–189°. The contaminant that depresses the melting point and extends the melting range may be an indefinitely characterized resinous material, or it may be a trace of a second chemical entity of melting point either higher or lower than that of the major component. Under equilibrium conditions (no super-cooling) the temperature at which a pure solid melts is identical with that at which the molten substance solidifies or freezes. Just as salt lowers the freezing point of water, so one compound (A) depresses the melting point of another (B) with which it is mixed. If pure A melts at 150–151° and pure B at 120–121°, mixtures of A with small amounts of B will melt unsharply at temperatures below 150° and mixtures of B containing A will melt below 120°. Both the temperature and sharpness of melting are useful criteria of purity.

A third substance C may have exactly the same melting point as A, namely 150–151°, but if a mixed melting point determination is made, that is, if A and C are mixed and the melting point of the mixture is observed, the one substance will be found to depress the melting point of the other. Depression, or nondepression, of melting point is invaluable in the identification of unknowns. An unknown D found to melt at 150–151° can be suspected of being identical with one or the other known substances A and C; observation that the mixture AD shows a melting point depression would exclude identity with A, and failure of C to depress the melting point of D would prove C and D identical.

If a substance melts at 150° or higher, the thread of mercury in

Mixed melting point

Stem correction

the upper part of the thermometer will be cooler than that in the bulb, and hence the temperature recorded will be a little lower than the actual bath temperature. The extent of the error due to stem exposure depends upon the design of the thermometer and the type of heating bath used; it may amount to 2–5° at 200°, 3–10° at 250°. An approximate correction for stem exposure can be calculated from the formula:

$$\text{Stem correction (°C)} = 0.000154 \, (t - t')N$$

where the fraction represents the difference in the coefficients of expansion of glass and of mercury, t is the temperature read, t' is the average temperature of the exposed column of mercury (determined, approximately, by reading the temperature of a second thermometer whose bulb is placed midway between the bath and the point corresponding to t), and N represents the length, measured in degrees, of the thread exposed between the top of the heating liquid and the point t. The convention for reporting a corrected melting point is: mp 283.5–284.5° (corr). The error can be eliminated by use of a heating bath designed to accommodate a series of short thermometers that can be totally immersed.

EXPERIMENT   1. Apparatus

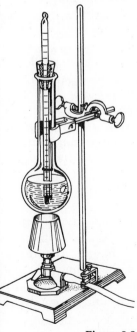

Figure 6.2
Apparatus for melting point determination.

The thermometer is fitted through a cork, a section of which is cut away for pressure release and so that the scale is visible (Fig. 6.1). A single-edge razor blade is convenient for cutting, and the cut can be smoothed or deepened with a triangular file. The curvature of the walls of the flask causes convection currents in the heating liquid to rise evenly along the walls and then descend and converge at the center; hence, care must be taken to center the thermometer in the flask. The long neck prevents spilling and fuming and minimizes error due to stem exposure. The bulb of the flask (dry!) is three-quarters filled with di-n-butyl phthalate; if it darkens in use the liquid should be replaced. The heating flask is mounted as shown in Fig. 6.2, with the bulb close to the chimney of a microburner. Careful control of heat input required in taking a melting point is accomplished both by regulating the gas supply with the screw pinchclamp and by raising or lowering the heating bath.

Capillary melting point tubes can be obtained commercially or they may be made by drawing out 12-mm soft-glass tubing. The 12-mm tubing is rotated in the hottest part of the Bunsen burner flame until it is very soft and begins to sag. It should not be drawn out at all during the heating. Remove from the flame, hesitate for a second or two, and then draw it out steadily but not too rapidly to arm's length. With some practice it is possible to produce 10–15 good tubes in a single drawing. The long capillary tube can be cut into sections with the sharp edge of a scorer. Alternatively, bring

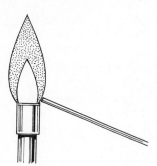

Figure 6.3
Sealing a melting point
capillary tube.

a sharp file resting on your forefinger underneath the tube, put your thumb over the place to be cut, and gently draw in the file so as to produce a slight scratch. On applying gentle pressure with the thumb the tube will now break cleanly. The tube is sealed by rotating the end in the edge of a small flame (Fig. 6.3). The actual size of a satisfactory capillary tube is shown in Fig. 6.1.

2. Operation

Put a small pile of the sample, about 3–4 mm in diameter, on a filter paper and crush the sample to a powder with a small spatula. Scrape the powder into a mound and use the spatula as a stop while pushing the open end of a capillary tube into the sample to scoop it up. The sample may be shaken down to the bottom of the tube by gentle stroking with a file, or it may be tamped down by dropping the tube through a long glass tube held vertically and resting on a hard surface. The column of solid should be no more than 2–3 mm in length and it should be tightly packed. If the sample is in the form of a compact cake, it is often convenient to fill the capillary by cutting out a tiny cylinder with the open end of the tube and then shaking it down. If a fluffy or sticky sample plugs the tube and fails to be shaken down, try using just a tiny fragment of material at a time.

The melting point capillary is held to the thermometer by a rubber ligature[1] or a rubber band made by cutting a thin slice off the end of a piece of 3/16″ rubber tubing with scissors; the band may require replacement from time to time. Insertion of a fresh tube under the rubber band is facilitated by leaving the used tube in place. The sample should be close to and on a level with the center of the thermometer bulb, which must be fully submerged and centered in the heating flask. If the approximate melting temperature is known, the bath can be heated rapidly until the temperature is about 20° below this point, but the heating during the last 15–20° must be slow and regular (2° rise per minute). Observe both the sample and the thermometer and record the temperatures of initial and terminal melting. The Pyrex flask can be detached and cooled for a second determination as follows: with an air blast if the temperature is above 150°; if the bath is no hotter than 150° it may be submerged for 30-sec. periods in a 600-ml beaker of water from the tap.

If determinations are to be done on two or three samples that differ in melting point by as much as 10°, two or three capillaries can be secured to the thermometer together and the melting points observed in succession without removal of the thermometer from the bath. As a precaution against interchange of tubes while they are being attached, use some system of identification, such as one, two, and three dots made with a marking pencil.

---

[1] S. S. White Dental Mfg. Co., Philadelphia, Pa., small rubber ligatures, No. 241, 225 parts per box.

3. Known

Determine the melting point of either urea (mp 132.5–133°) or cinnamic acid (mp 132.5–133°). Repeat the determination and if the two determinations do not check within 1°, do a third one.

4. Mixture

Make mixtures of urea and cinnamic acid in the approximate proportions 1:4, 1:1, and 4:1 by putting side by side the correct number of equal sized small piles of the two substances and then mixing them. Note the ranges of melting of the three mixtures but use the temperatures of complete liquefaction (and the mp 132.5–133° for each pure component) to construct a rough diagram of mp *vs.* composition.

5. Unknown

Determine the melting point of one of the following unknowns selected by the instructor:

Acetanilide, mp 113.5–114°	Salicylic acid, mp 158.5–159°
Benzoic acid, mp 121.5–122°	Succinic acid, mp 184.5–185°
Cinnamic acid, mp 132.5–133°	Inositol, mp 226.5–227°

6. Rast Molecular Weight Determination[2]

A dissolved substance depresses the freezing (or melting) point of the solvent to an extent that is dependent upon the ratio of solute to solvent molecules and upon the susceptibility of the particular solvent used, as measured by the value of its cryoscopic constant. The freezing point of a fixed weight of benzene is depressed a constant number of degrees by a certain molecular proportion of any soluble, un-ionized solute, and observation of the extent of the depression (Δt) and knowledge of the cryoscopic constant C for benzene permits calculation of the molecular weight.

$$MW = \frac{\text{Solute, g} \times C}{\text{Solvent, g} \times \Delta t}$$

The constant C can be expressed in terms appropriate to the practice of using very dilute solutions for avoiding errors that otherwise decrease accuracy. One gram of a substance of molecular weight 100 produces the depressions noted when dissolved in 100 g of each of the following solvents: benzene, 0.510°; acetic acid, 0.390°; water, 0.185°. If the molecular weight of the solute is 200 or 300 the depressions are one-half or one-third those cited. With these and other common solvents the depressions are so small that molecular

[2] This technique can be demonstrated, rather than introduced as an experiment, by assigning the section for reading and then showing Part I of the film *Techniques of Organic Chemistry*, by L. F. Fieser (McGraw-Hill Book Co.).

weight determinations can be made only with use of a precision instrument such as the Beckmann thermometer. Camphor (synthetic *dl*-camphor) is a special solvent, because its cryoscopic constant is eight times that of benzene and twenty times that of water. It is a solid, mp about 173°, but in the molten state it has high solvent power for organic compounds. It also affords solid solutions whose melting points can be determined in capillary tubes with an ordinary thermometer, which suffices for observation of the depression of 4.0° produced by 1 g of a compound of molecular weight 100 dissolved in 100 g of camphor. Rast's method of molecular weight determination is to prepare such a solid solution, determine the depression, and calculate the molecular weight:

$$MW = \frac{\text{Substance, g} \times 40,000}{\text{Camphor, g} \times \Delta t}$$

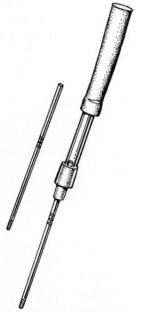

$$Molality = \frac{g \ of \ solute}{1000 \ g \ of \ solvent}$$

The melting behavior of camphor is peculiar, since even the purest material melts over a considerable temperature range: the solid softens, then it turns to a glassy semisolid, then a clear liquid appears in the upper part of the tube, and finally a skeleton of crystals remains in the lower part and slowly decreases in size until it eventually disappears. The point of disappearance of the last crystal, the melting point, can be determined with accuracy. If the bath liquid is allowed to cool very slowly, an equally accurate determination can be made of the freezing point, the temperature at which the first tiny crystals of camphor separate. That the freezing point is usually found to be about 2° below the melting point is probably due to supercooling of the unstirred solution. Dilute solid solutions (1–6%) of solid organic compounds in camphor behave similarly; the transition temperatures are lower, the melting ranges wider, and the difference between melting and freezing point somewhat greater. High accuracy in determining the extent of depression is attainable with the apparatus of Fig. 6.2 by attaching to the thermometer one tube containing pure camphor and another containing a mixture of camphor and the substance under investigation. Errors due to inaccuracy of the thermometer or to stem exposure are thus cancelled. Very slow heating and cooling and very careful observation are required. It is necessary also to seal off the capillary at a point that is below the level of the heating fluid, (Fig. 6.4) for if there is any vapor space above this level, camphor will sublime into the cooler area and leave a residual melt enriched in the solute. The capillary is evacuated prior to sealing to facilitate formation of a firm seal without increasing the fragility of the tube. The technique of sealing is of general use, since some compounds undergo air oxidation when heated in open capillary tubes, and true melting points are obtainable only when the determinations are made in evacuated tubes.

7. Problem

Suppose you have two unknowns, A and B. One unknown contains X grams of acetanilide (C_8H_9ON) per 100 g of camphor and the other contains X grams of benzophenone ($C_{13}H_{10}O$) per 100 g of camphor.[3] The problem is to determine the melting and freezing points of the two unknowns and of camphor simultaneously, establish which unknown is acetanilide and which is benzophenone, and calculate the value of X, common to both samples. Care should be taken when filling, sealing, and mounting the tubes, for once this is done properly the determinations can be repeated until you have gained sufficient experience to obtain checking results in which you can have confidence.

Push the open ends of three melting point capillary tubes into powdered unknowns A and B and camphor (contained in a short specimen tube) and tamp down the solid in each tube until you have a column about 5 mm long. Mark each tube and record what it contains. Each tube, in turn, is connected to the suction pump tubing by an adapter (Fig. 6.4) consisting of a glass tube capped with a rubber vaccine stopper[4] with a hole pierced through it with an awl or red-hot sewing needle. Lubricate the end of the hole extending toward the wide part of the stopper, grasp the capillary very close to the sealed end, and thrust it through nearly to the full length. Connect to the suction pump, restore the tube markings if necessary, and make a guiding mark 1.5 cm from the closed end where a seal is to be formed. Grasp the end of the tube with one hand and the adapter with the other and, with suction pump operating, hold the point of sealing near the small flame of a microburner, steady your hands on the bench so that neither will move much when the tube is melted, and then hold the tube in the flame until the walls collapse to form a flat seal and remove it at once. Take the tube out of the adapter and if the two straight parts of the tube are not in line, reheat to soften the seal and correct the alignment.

Mount tubes of A, B, and camphor side by side on the thermometer under a rubber band, insert the thermometer, and make sure that all the seals are completely submerged. The bath may be heated rapidly to about 130°, but from this point on the temperature rise should be as slow as possible. At the outset, a very small flame about 7 cm below the flask should produce a slow, gradual rise of 1° in 30–40 sec. Determine the point of disappearance of the last crystal for each sample in succession and then continue heating until the bath is about 2° above the melting point of camphor, but do not stop at this point to interpret the results. Instead, lower the flame by small increments until the temperature starts to fall and then adjust

---

[3] Weigh $X/10$ g of substance and 10 g of camphor in a test tube, stopper loosely, melt without undue overheating, pour the melt into a mortar, swirl until solid, then at once dislodge the solid and while it is still hot grind it lightly and not too long. Concentrations of 1–6% are satisfactory.
[4] Rubber stoppers, serum with sleeve, small size, R 7950, Scientific Glass Co., Bloomfield, N. J.

the heating so that the temperature falls 1° in 30–40 sec. Determine each freezing point in turn and then keep the bath at about 140° while calculating the depressions in melting and freezing points. If the two depressions are very divergent, or if the difference between mp and fp for either A or B is much greater than that for camphor, repeat all the determinations.

In order to assess the data, calculate the value of X for each separate determination from the equation:

$$X = \frac{\Delta t \times MW}{400}$$

Tabulate the results, put the more divergent values in parentheses, and average the others. Does any pair of determinations on A and B leave doubt as to the identity of the unknowns? When the true value of X is disclosed, calculate the molecular weights based on your best Δt values and estimate the limit of accuracy of the method. On the estimate that a 3-mm column of camphor in a capillary tube weighs 1 mg, calculate the weights of acetanilide and benzophenone used in your determinations.

Since the melting point apparatus is required for the next experiment it may be put away in assembled form. When parts are required for other purposes, the heating flask should be allowed to cool, stoppered tightly with a cork, and put away.

7 Crystallization

Mother liquor	Solvent pairs
Filtrate	Stemless funnels
Decolorizing charcoal (Norit)	Hirsch funnel
Adsorption of impurities	Neoprene adapter
Saturation and supersaturation	Filter block
Like dissolves like	Clarification

A highly effective method of purifying a solid substance consists in dissolving it in a suitable solvent at the boiling point, filtering the hot solution by gravity to remove any suspended insoluble particles, and letting crystallization proceed. Separation of crystals before the hot solution has all passed through the filter paper can be largely prevented by determining the minimum volume of solvent required to effect solution and then adding a deliberate excess (20–100%). A little more fresh hot solvent is required to wash the paper clean.

The correct quantity of solvent

The total filtrate is then evaporated to the original optimal volume and let stand undistrubed at room temperature until the solution has acquired the temperature of the surroundings and crystals have ceased to increase in number or size; the flask is then chilled in an ice bath to promote further crystallization. The crystals that have separated in this first crop are collected by suction filtration and washed free of mother liquor with a little fresh, chilled solvent. The combined mother liquor and washings can be concentrated to a small volume and let stand for separation of a second crop. The quality of each crop is ascertained by melting point determination.

Frequently a sample to be purified contains a soluble impurity that gives rise to solutions and crystals that are yellowish or brownish when they should be colorless, or of an off color rather than a pure color. Some soluble pigments can be adsorbed on finely divided carbon and removed, whereas others are not appreciably adsorbed. Colored impurities of the latter type, like colorless impurities, tend to remain in the crystallization mother liquor and are eliminated when the crystals are collected and washed. Hence if the initial solution is yellowish, brownish, or has a dull or off color it is treated with decolorizing carbon, but if the process is only partially effective it is not repeated. Norit, one commercially available de-

Norit, a decolorizing carbon

colorizing carbon (also called animal charcoal, or bone black), is often specified throughout this manual both because it is generally satisfactory and for brevity. The fine carbon particles present a

43

large, active surface for adsorption of dissolved substances, particularly the polymeric, resinous, and reactive by-products that appear in traces in most organic reaction mixtures. Norit is added to the hot solution prior to filtration and the solution is kept hot for a brief period, shaken to wet the carbon, and filtered. Adsorption occurs very rapidly and no advantage is gained by boiling the suspension for several minutes. Norit actually is less effective at a high than a low temperature, and the only point of operating at the boiling temperature is to keep the substance being crystallized in solution. It is a mistake to use more Norit than that actually needed, for an excess may adsorb some of the sample and cause losses.

The specification that the hot solution saturated with solute at the boiling point be let stand undisturbed means that crystallization is allowed to proceed without subsequently moving the flask, jarring the bench on which it is resting, or inserting a thermometer or stirring rod. For estimation of the temperature, the flask should be touched very lightly between the thumb and a finger without any movement of the flask. Crystallization does not happen at once. It usually starts after an induction period, ranging from several minutes to an hour or two, even though the temperature has dropped well below that at which the solution is saturated with solute. The

Supersaturation

phenomenon of supersaturation makes crystallization a remarkably effective means of purification and separation.

Suppose that substances A and B have exactly the same solubility in a given solvent and that a 9A:1B mixture is let crystallize undisturbed from a solution saturated at the boiling point. Molecules of the more abundant A eventually form the first crystal and this acts as seed which induces separation of further crystals of A molecules, while B remains in supersaturated solution, perhaps even after crystallization of A has ceased. The crystals, collected without any agitation of the mixture beyond that necessary, will be found enriched in A beyond the 9:1 ratio, and the mother liquor will contain most or all of the B. Supersaturation of the hot solution may be upset by rapid cooling, stirring, or even a slight motion of the flask, and then A and B may both start to crystallize and the process is spoiled. By proper, undisturbed crystallizations either one of a pair of substances of the same solubility can be isolated if it is present in sufficiently preponderant amount. Chapter 50 includes isolation, in pure form from a reaction mixture, of two compounds that differ only slightly in solubility. The more abundant and less soluble C crystallizes first, and successive crops of it are obtained by concentrating the mother liquor and adding a tiny seed crystal of C. The more soluble D initially remains in the mother liquor, but when this is so enriched that D is the dominant component, the unseeded solution slowly deposits pure D.

Solvents

The properties of solvents commonly used for crystallization are listed in Table 7.1. A general rule of solubility is that like dissolves like. Methanol (CH_3OH), ethanol (C_2H_5OH), and acetic acid (CH_3COOH) are all hydroxylic, and they are all miscible with one

Table 7.1   Properties of Some Common Crystallization Solvents

Solvent	Bp	Fp	Density	Misci-bility with Water	Flamma-bility
Water	100°	0°	1.0	+	0
Methanol	64.7°	<0°	0.79	+	+
95% Ethanol	78.1°	<0°	0.81	+	++
Acetic acid	118°	16.7°	1.05	+	+
Acetone	56.5°	<0°	0.79	+	+++
Ether	34.6°	<0°	0.71	−	++++
Petroleum ether	35–65°	<0°	0.63	−	++++
Ligroin	65–75°	<0°	0.68	−	+++
Ligroin	100–115°	<0°	0.70	−	++
Benzene	80.1°	5°	0.88	−	++++
Dichloromethane (methylene chloride)	41°	<0°	1.34	−	0
Carbon tetrachloride	76.6°	<0°	1.59	−	0
Chloroform	61.2°	<0°	1.48	−	0

another and with water; they resemble one another in that they are able to dissolve substances of specific types, particularly hydroxylic compounds, and are less effective solvents for structurally dissimilar compounds such as hydrocarbons. The solvent power usually increases with increasing boiling point of the solvent; for example, ethanol dissolves about twice as much of a given solute as does methanol. On the other hand, ethanol is more costly than methanol. Since losses entailed in dilution prior to filtration, in evaporation, and in washing are about equal, methanol is the more economical solvent of the two. The chlorinated hydrocarbons methylene chloride, chloroform, and carbon tetrachloride are particularly expensive. Ligroin is a mixture of aliphatic hydrocarbons comparable to petroleum ether but with a higher boiling point range; it has comparable but greater solvent power. A substance that dissolves easily in petroleum ether or ligroin is almost invariably insoluble in water. Acetone is similar in solvent action to ethanol, and ether is similar to benzene.

Solvent pairs   Two miscible solvents of different solvent power constitute a useful solvent pair. The hydrocarbon naphthalene is insoluble in water but so soluble in methanol that it crystallizes from this solvent only if the solution is highly concentrated and the temperature low. It can be crystallized efficiently from a suitably proportioned methanol-water mixture; the hydrocarbon is dissolved in excess methanol and water is added little by little at the boiling point until the solution is saturated.

Other useful solvent pairs are listed in Table 7.2. If solubility tests conducted with 20-mg samples of material to be crystallized show that it is readily soluble in one member of a solvent pair and sparingly soluble in the other, a solution of the sample in the first solvent can be adjusted to conditions suitable for crystallization by dilution with the second solvent. Methylene chloride (bp 41°)

Table 7.2   Solvent Pairs

Methanol–Water	Ether–Acetone
Ethanol–Water	Ether–Petroleum ether
Acetic acid–Water	Benzene–Ligroin
Acetone–Water	Methylene chloride–Methanol
Ether–Methanol	

has more solvent power for most compounds than methanol (bp 65°) and is more volatile; hence crystallization is accomplished by dissolving the solid in methylene chloride, filtering, adding methanol, and boiling down the filtrate until the product begins to crystallize.

EXPERIMENTS   **1. Apparatus**

The Erlenmeyer flask (Pyrex) is particularly well adapted to all operations of crystallization. The flask can be conveniently held at the neck with the thumb and finger and for dissolving the solid the flask can be put directly on a steam bath or hot plate or, if the liquid is water, over a free flame. The contents can be agitated by swirling, that is, imparting a circulatory motion to the contents by moving the bottom of the flask in a circle as the top remains stationary (Fig. 7.1); the conical shape of the flask prevents spilling of liquid. *Never use a beaker for crystallization.* If lumps or large crystals of solid are slow in dissolving, the process can be hastened by crushing the material against the bottom of the flask with a flattened stirring rod, (made by heating the end of a piece of 4-mm. soft glass rood to redness and pressing it onto a flat surface, such as a coin.) To pour out hot liquid cover the hand with a folded towel and grasp the flask (Fig. 7.2), or use a clamp to handle the flask.

Figure 7.1
Erlenmeyer flask being
swirled.

Figure 7.2
Pouring a hot liquid.

The funnels used in crystallization should be stemless (to avoid crystallization in the stem) and are supported merely by resting them in the mouth of the receiving flask. Proper matching sizes of flasks, funnels, and papers are indicated in Fig. 7.3.

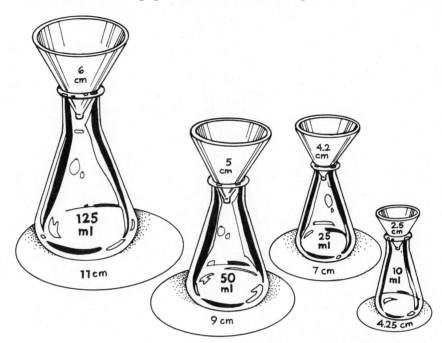

**Figure 7.3
Assemblies for gravity
filtration. Stemless
funnels have diameters
2.5, 4.2, 5.0, and 6.5 cm.
(Wilkens-Anderson,
35765.)**

After a product has crystallized at room temperature, the Erlenmeyer is usually let stand for a time in an ice-water bath to effect maximum separation of crystals. The crystals are then collected by suction filtration on a Büchner or Hirsch funnel which is placed in a neoprene adapter atop a filter flask (Fig. 7.4). The filter flask is connected through a filter trap (Figs. 2.11 and 2.12) to the water aspirator by thick-walled tubing which will not collapse under vacuum. The tubing is so heavy that the 50-ml and 125-ml filter flasks will tip over easily unless provided with a support, such as filter block (Fig. 7.5), or clamped to a ring stand. Larger filter flasks have adequate stability and require no support.

The perforated plate of the funnel is covered by a circle of filter paper of appropriate size, centered in the funnel, and moistened with the crystallization solvent. Suction is applied and the liquid or solid mixture is poured in. As soon as all the solvent has been pulled through the crystals the suction is broken, clean cold wash solvent added to the funnel, and the suction reapplied. After the crystals have been washed and while suction is maintained, the crystals are pressed on the funnel with a clean cork to squeeze out the last bit of solvent. A technique for collecting a small quantity of filtrate is illustrated in Fig. 7.6.

2. Solubility Tests

To test the solubility of a solid, transfer an amount roughly

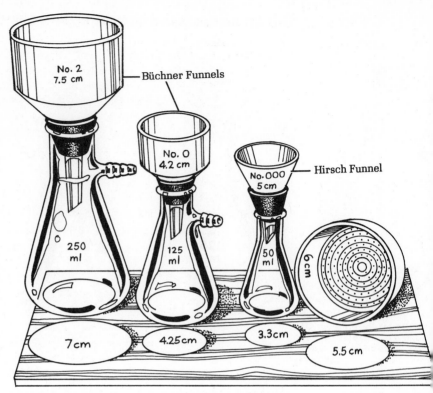

Figure 7.4
Matching filter assem-
blies. The 5.5 cm poly-
propylene Büchner
funnel (*right*) resists
breakage and can be dis-
assembled for cleaning.

Figure 7.5
Suction filter assemblies
with filter blocks for firm
support.

estimated to be about 20 mg with a small spatula into a 10 × 75-mm
test tube and add about 0.5 ml of solvent, either delivered from a
calibrated capillary dropping tube or estimated as a 1-cm column in
the small test tube. Stir with a fire-polished stirring rod (4-mm),
break up any lumps, and determine if the solid is readily soluble in
the solvent in question at room temperature. If the substance is

Hirsch funnel

Neoprene adapter

Figure 7.6
Suction filter assembly
for small quantities of
substances.

found to be readily soluble in either cold methanol, ethanol, acetone, or acetic acid, add a few drops of water from a wash bottle to see if a solid precipitates. If it does, heat the mixture, adjust the composition of the solvent pair to produce a hot solution saturated at the boiling point, let the solution stand undisturbed, and note the character of the crystals that form. If a substance fails to dissolve in a given solvent in the cold, heat the suspension and see if solution occurs; if the solvent is flammable, heat the test tube on the steam bath or in a small beaker of water kept warm on the steam bath or a hot plate. If the solid completely dissolves, it can be declared readily soluble in the hot solvent; if some but not all dissolves, it is said to be moderately soluble, and further small amounts of solvent should then be added until solution is complete. When a solution of substance in hot solvent has been obtained, cool the solution by holding the flask under the tap and, if necessary, induce crystallization by rubbing the walls of the flask with a stirring rod to make sure that the concentration is such that crystallization is possible. Then reheat to dissolve the solid, let the solution stand undisturbed, and inspect the character of the ultimate crystals.

Make solubility tests on the following compounds (see formulas) in each of the solvents listed, note the degree of solubility in the solvents, cold and hot, and suggest suitable solvents, solvent-pairs, or other expedients for crystallization of each substance. Record the crystal form, at least to the extent of distinguishing between needles (pointed crystals), plates (flat and thin), and prisms. How do your observations conform to the generalization that like dissolves like?

OH

OH

Resorcinol

Anthracene

$COOH$

Benzoic acid

$SO_3^-Na^+$

NH_2

Sodium
naphthionate

$COOH$

$COOH$

Phthalic acid

Test Compounds:

Solvents:
    Water—hydroxylic, ionic.
    Benzene—an aromatic hydrocarbon.
    Ligroin—a mixture of aliphatic hydrocarbons.

3. Clarification and Decolorization Techniques

In this section two experiments are described which demonstrate points of interest about the use of Norit, a decolorizing carbon which is so finely divided that one gram has a surface area of 800 square

OH
NO$_2$

NO$_2$

Martius Yellow

Ratio of adsorbent to adsorbate

Unsolved problem

Recovery of adsorbed dye

meters.[1] The active surface adsorbs and removes unsaturated, pigmented impurities. Both experiments use as a simulated dull-colored crude reaction product a mixture of colorless phthalic acid with 0.25% of the dye Martius Yellow.

The object of the first experiment is to determine the exact amount of Norit required to effect complete clarification of 120 mg of the crude phthalic acid. A 1-g sample of the acid is dissolved in 25 ml of water at the boiling point and 3-milliliter portions of the hot, bright yellow solution are measured with a calibrated dropping tube into four test tubes, which are kept in a bath of hot water. Each aliquot portion is for trial with a different amount of Norit. Four capillary melting point tubes contain the following amounts of Norit, weighed out accurately on a microbalance: 10, 5, 2.5, and 1.25 mg. The 10-mg portion is introduced into one test tube of yellow solution by stroking the inverted capillary with a file. The solution is boiled briefly to mix the adsorbent thoroughly with the solute and filtered into a clean test tube. The filtrate is colorless. The 5-mg and 2.5-mg samples also adsorb the dye completely. The tiny 1.25-mg sample surely reduces the color, but when the solution is viewed down the length of the tube a faint yellow tinge is detectable. Thus, 1.25 mg is not enough and the estimated saturation value is about 2 mg of Norit. Calculation from the surface area of Norit, the molecular weight of the dye ($C_{10}H_6N_2O_5$), and the assumption that the area of contact of the dye molecule is 10×20 Å, show that the saturation value found is close to that calculated for complete coverage of the surface of the adsorbent by dye molecules.

What forces are involved in the process of adsorption? How is the yellow dye held to the Norit particles? Is it by van der Waals forces; or by hydrogen bonding; or is a chemical interaction involved? The 2-mg portion of Norit must have become nearly completely covered by yellow dye molecules and yet the particles are pure black. Can the dye be recovered from the black adsorbate? Boiling with methanol removes no dye. Martius Yellow is a strong acid and forms a water-soluble yellow sodium salt, but boiling the adsorbate with aqueous alkali does not produce a yellow extract. However, exhaustive extraction of the 2-mg portion of adsorbate with hot portions of methanol (10 ml)–10% sodium hydroxide (1 ml) gave a yellow solution of Martius Yellow sodium salt. Acidification and extraction with ether gave an amount of Martius Yellow which was barely visible and too small to be weighed. Nevertheless, the dyeing properties of Martius Yellow made it possible to identify the pigment and to determine the amount recovered. Silk takes up acidic molecules of Martius Yellow until all basic sites of the silk protein are filled. One milligram of pure Martius Yellow will dye to full strength three 3×3-cm squares of silk weighing 56 mg each. The tiny bit of recovered pigment was dissolved in water, a square

[1] The experiments are recorded in Part II of the film *Techniques of Organic Chemistry* (McGraw-Hill Book Co.).

of silk was pushed into the tube, and the solution was boiled for a few minutes. Removed with a stirring rod and washed with water, the silk was found to be dyed to full strength. A second square of silk treated in the same way discharged the color of the solution but was dyed only to about half strength. That the recovery was only about 50% suggests that Martius Yellow undergoes some chemical alteration in the process of adsorption.

A second method of clarification applicable to acids and bases demonstrates the speed of adsorption and the efficiency of Norit at room temperature. One gram of the crude phthalic acid and 0.51 g of sodium bicarbonate were placed in a 50-ml Erlenmeyer and 20 ml of water was added in small portions, with swirling to break the foam (CO_2). A clear yellow solution resulted. A slurry was made by shaking 150 mg of Norit with water in a test tube, and this was filtered by suction onto a Hirsch funnel with paper, mounted with an adapter on a filter flask. A thin but even pad of decolorizing carbon resulted. The water was poured out of the filter flask, this was reconnected, and the yellow solution was filtered through the pad of Norit. Although the contact of dye with adsorbent was extremely brief, the filtrate was colorless. After acidification, scratching, and chilling, colorless phthalic acid separated.

Suction filtration through Norit pad

Activated charcoal, like Norit, is used on a large scale industrially to remove brown impurities from molasses in the refining of sugar. The resulting sucrose is one of the purest organic compounds made in large quantities.

4. Crystallization with Use of Norit

Points of technique not evident from the above demonstration can be appreciated by crystallizing a sample of the crude phthalic acid containing 0.25% of Martius Yellow.[2] First, calculate from the data of the first experiment in the preceding section the amount of Norit required to decolorize 1 g of the crude acid. To allow for exigencies, measure 1.5 times this amount in a standard melting point capillary from the approximation that a column of Norit that fills the tube to a length of 88 mm. weighs 46 mg.

Place 1 g of the acid and 5 ml of water in a 50-ml Erlenmeyer and swirl the flask over the free flame of a microburner until the boiling point is reached and then let the mixture simmer on a hot plate while adding water, 1 ml at a time, and determine the exact amount of water required to just dissolve the sample at the boiling point. Let the yellow solution cool a little (note that phthalic acid crystallizes rapidly and abundantly) and then replace the flask on the hot plate. Since the solubility of the acid in cold water is low (0.54 g per 100 g at 14°), difficulty in the filtration can be avoided by using a large excess of solvent. Add more water to make up the volume to

Note for the instructor

[2] A mixture of 100 g of phthalic anhydride, 310 mg of the ammonium salt of 2,4-dinitro-1-naphthol, 41 g of sodium hydroxide, and 500 ml of water is warmed to effect solution, acidified, cooled, and the pale yellow solid collected.

25 ml (measure this volume into a second flask for comparison), invert the tube of Norit and empty it into the flask by stroking the tube with a file, heat to the boiling point, and filter the hot solution by gravity into a second Erlenmeyer, pouring the liquid first onto the upper part of the section of paper that is reinforced by the extra folds, for this minimizes passage of carbon particles through the paper. In case the first few drops of filtrate do contain carbon, watch for the point where the filtrate is clear, change to a clean receiving flask, and reheat and refilter the first filtrate. Sometimes filtration is slow because a funnel fits so snugly into the mouth of a flask that a back pressure develops. If you note that raising the funnel by the fingers increases the flow of filtrate, make a tiny, flat spacer by folding a small strip of paper two or three times and insert it between the funnel and the flask. To rinse the emptied flask, add 1–2 ml of water, heat to boiling, and pour the hot wash liquor in a succession of drops around the upper rim of the filter paper while rotating the funnel. The emptied flask, the filter, and the filter paper containing Norit are to be put aside for processing as described below. Reheat the filtrate to dissolve any solid that has separated, and let the solu-

Let the solution stand undisturbed

tion stand undisturbed for crystallization. As crystals separate, record the crystalline form and any other distinguishing characteristics. After crystallization at room temperature has ceased, put the flask in a beaker of ice and water. In collecting the product by suction filtration use a spatula to dislodge crystals and ease them out of the flask. If crystals remain after the flask has been emptied, filtrate can be poured back into the crystallization flask for washing as often as desired, since it is saturated with solute. Fresh solvent (water from a wash bottle) should be used only in a final wash to free the crystals of mother liquor. The product should be colorless and the yield 0.83–0.85 g.

5. Purification of an Unknown[3]

With a total supply of 2.0 g of a crude unknown, you are to ascertain the properties of the major component from appropriate tests, devise a method for purification, and submit as much pure product as you can, together with evidence of its purity. The characterizing tests should include melting point determination, tests for solubility and crystallizability in organic solvents, solvent pairs, and water. If the substance is water-soluble, determine if it is acidic or basic. Conserve your material by using very small samples for the tests. After a test in a volatile solvent has been com-

Cool a hot solution before adding charcoal

pleted, the solvent can be evaporated by heating the tube on the steam bath and the residue used for another test. In case you use Norit for clarification of a solution of substance in an organic solvent take the precaution to let the boiling solution cool slightly and

Note for the instructor

[3] Many accumulated student preparations can be rendered suitable for use as unknowns; example: 500 g p-nitroaniline, 50 g p-bromoacetanilide, 20 g sand, 10 g ashes.

swirl it well to relieve superheating before addition of the decolorizing carbon; otherwise the fine particles are liable to initiate vigorous boiling with loss of liquid out of the mouth of the flask.

Large crystals will result as crystallization occurs if you allow the hot solution to cool spontaneously to room temperature undisturbed. Shaking or moving the flask as crystallization takes place produces more nuclei to start crystallization and thus large numbers of small crystals; rapid cooling does the same. With a large surface-to-volume ratio, fine crystals are more difficult to wash free of contaminating mother liquor.

Sometimes obtaining crystals from a supersaturated solution is quite difficult. In later experiments crystallization is initiated by scratching the inside surface of the flask with a glass rod or by the addition of seed crystals.

Crystallization is part of the art of organic chemistry

8 Extraction

KEYWORDS Use of ether
Equilibration
Drying ether using saturated salt (NaCl) solution
Distribution ratio
Soxhlet extractor
Separatory funnel
Separation of acidic, basic, and neutral substances
Use of drying agents

Extinguish all flames when working with ether!

**Figure 8.1
Correct positions for holding a separatory funnel when shaking.**

A frequently used method of working up a reaction mixture is to dilute the mixture with water and extract with ether in a separatory funnel (Fig. 8.1). When the stoppered funnel is shaken to distribute the components between the immiscible solvents ether and water, pressure always develops through volatilization of ether from the heat of the hands, and liberation of a gas (CO_2) will increase the pressure. Consequently, the funnel is so grasped that the stopper is held in place by one hand and the stopcock by the other, as illustrated. After a brief shake or two the funnel is held in the inverted position shown and the stopcock opened cautiously (with the funnel stem pointed away from nearby people) to test the extent of pressure and to release it. After the test the mixture can be shaken more vigorously and tested further. When equilibration is judged to be complete, the slight, constant terminal pressure due to ether is released, the stopper is rinsed with a few drops of ether delivered by a Calcutta wash bottle or a capillary dropping tube, and the layers are let settle. The organic reaction product is distributed wholly or largely into the upper ether layer, whereas inorganic salts, acids, and bases pass into the water layer, which can be drawn off and discarded. If the reaction was run in alcohol or acetone solution the bulk of the solvent is removed in the water layer and the rest can be eliminated in two or three washings with 1–2 volumes of water conducted with the techniques used in the first equilibration. Ways of supporting a separatory funnel are shown in Figures 1.5 and 8.2.

Acids

Acetic acid is also distributed largely into the aqueous phase, but if the reaction product is a neutral substance the residual acetic acid in the ether can be removed by one washing with excess 5% sodium bicarbonate solution. If the reaction product is a higher MW acid, for example benzoic acid (C_6H_5COOH), it will stay in the ether layer, while acetic acid is being removed by repeated washing with

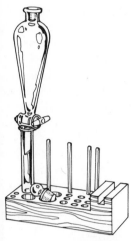

Figure 8.2
Filter block used to support a separatory funnel.

Ether-water solubility

Drying an extract

Distribution ratio

water; the benzoic acid can then be separated from neutral by-products by extraction with bicarbonate solution and acidification of the extract. Acids of high molecular weight are extracted only slowly by sodium bicarbonate and sodium carbonate is used in its place; however, carbonate is more prone than bicarbonate to produce emulsions. Sometimes an emulsion in the lower layer can be settled by twirling the stem of the funnel. An emulsion in the upper layer can be broken by grasping the funnel by the neck and swirling it. Since the tendency to emulsify increases with removal of electrolytes and solvents, a little sodium chloride or hydrochloric acid solution is added with each portion of wash water. If the layers are largely clear but an emulsion persists at the interface, the clear part of the water layer can be drawn off and the emulsion run into a second funnel and shaken with fresh ether.

At room temperature water dissolves 7.5% of ether (by weight) and ether dissolves 1.5% of water. Equilibration of 100 ml each of ether and water gives an upper layer of 90 ml and a lower layer of 105 ml; if the upper layer is separated and washed with 100-ml portions of water the volume is reduced about 10 ml each time. The reduction in volume can be prevented by using wash water saturated with ether, or fresh ether can be added to make up for that lost. In one of the experiments to follow, an ethereal solution is washed extensively with water with the express purpose of reducing the volume to a desired point. Whereas water dissolves in ether to the extent of 1.5%, water saturated with sodium chloride (36.7 g/100 g) has no appreciable solubility. Hence the bulk of the water present in a wet ethereal extract can be removed by shaking the extract with saturated sodium chloride solution and drawing off the lower layer. Drying is completed by filtering the ethereal solution through granular anhydrous sodium sulfate contained on a folded filter paper in a funnel. This salt abstracts water by formation of the pentahydrate.

When a given substance is partitioned between ether and water the ratio of concentrations in the ether and water layers is a constant called the distribution ratio, K, and is specific to the solute, solvent pair, and temperature in question. If the compound has finite solubility in both solvents, K can be calculated from the ratio of solubilities, since

$$K = \frac{[\text{g}/100 \text{ ml}]^{\text{Ether layer}}}{[\text{g}/100 \text{ ml}]^{\text{Water layer}}} = \frac{\text{Solubility, g}/100 \text{ ml Ether}}{\text{Solubility, g}/100 \text{ ml Water}}$$

For a substance, such as acetic acid, which is miscible in all proportions with both ether and water, K can be determined by distributing a known weight of acid between measured volumes of the two solvents and determining the concentration in the aqueous layer by titration with standard alkali. Calculation will show that in consequence of the constancy of the ratio of concentrations in the two phases, regardless of the relative volumes, extraction of a substance from an aqueous solution is more efficient if several small portions of ether are used than if the same total volume of ether is employed in one portion. The relation is illustrated by the problem presented in Section 4 of this chapter.

Extraction solvents

Extraction with a water-immiscible solvent is useful for isolation of natural products that occur in animal and plant tissues having high water content. Diethyl ether, often simply called ether $CH_3CH_2OCH_2CH_3$, has high solvent power for hydrocarbons and for oxygen-containing compounds and is so highly volatile (bp 34.6°) that it is easily removed from an extract at a temperature so low that even highly sensitive compounds are not likely to decompose. Although often preferred for research work because of these properties, ether is avoided in industrial processes because of its fire hazard, high solubility in water, losses in solvent recovery incident to its volatility, and oxidation of ether on long exposure to air to a peroxide, which in a dry state may explode. Alternative water-immiscible solvents sometimes preferred, even though they do not match all the favorable properties of ether, are: petroleum ether, ligroin, benzene, carbon tetrachloride, chloroform, methylene chloride, ethylene dichloride (1,2-dichloroethylene), 1-butanol. The chlorinated hydrocarbon solvents are heavier than water rather than lighter and hence, after equilibration of the aqueous and non-aqueous phases, the heavier lower layer is drawn off into a second separatory funnel for washing and the upper aqueous layer is extracted further and discarded. Chlorinated hydrocarbon solvents have the advantage of freedom from fire hazard, but the higher cost militates against their general use.

For exhaustive extraction of solid mixtures, and even of dried leaves or seeds, the solid is packed into a filter paper thimble placed in a Soxhlet extractor (Fig. 8.3). Solvent vapor rises in the large diameter tube on the right, and condensed solvent drops onto the solid contained in a filter paper thimble, leaches out soluble material and, after initiating an automatic siphon, carries it to the boiling flask where nonvolatile extracted material accumulates. The same solvent is used over and over, and even substances of very slight solubility can be extracted by prolonging the operation.

EXPERIMENTS

1. Equipment

(a) A 125-ml separatory funnel with a Teflon stopcock, which requires no grease, or same size funnel with glass stopcock carefully greased. Avoid excess grease, which dissolves in ether and will contaminate product. Use a ground-glass stopper if available but do not grease it; otherwise use a rubber stopper. The funnel is conveniently supported as shown in Fig. 8.4.

(b) Saturated sodium chloride solution. Put 130 g of common salt and 400 ml of distilled water in a 16-oz tincture bottle, grease the neck to prevent creeping, and shake occasionally until the solid is dissolved.

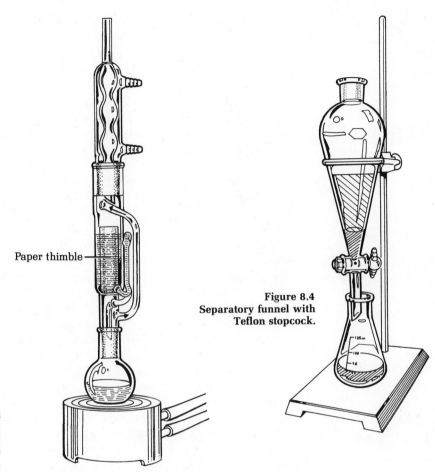

Paper thimble

**Figure 8.3
Soxhlet extractor. See
Figure 50.2 caption for a
description of how the
extractor works.**

**Figure 8.4
Separatory funnel with
Teflon stopcock.**

(c) Calibrated and uncalibrated capillary dropping tubes.
(d) Calcutta wash bottle of ether.

2. Basic and Neutral Substances

$$pH = -\log [H^+]$$
$pK_a = acidity\ constant$
$pK_b = basicity\ constant$

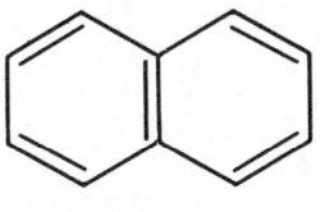

Naphthalene
Mp 80°, neutral

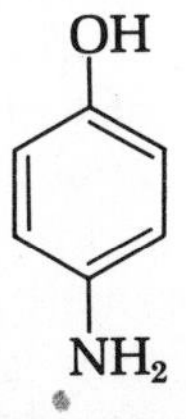

***p*-Aminophenol**
Mp 186°, pK$_a$ 8.2, pK$_b$ 8.8
(Hydrochloride, mp 302°, dec)

A mixture of naphthalene, a neutral hydrocarbon, and *p*-ami-
nophenol, an amphoteric compound, is to be separated by extraction from ether with dilute hydrochloric acid. *p*-Aminophenol
darkens on storage due to air oxidation, but this colored impurity
is removed in the extraction process. Note the detailed directions

for extraction, carefully. In the next experiment you are to work out your own extraction procedure.

To a 125-ml separatory funnel add 30 ml of water, 30 ml of ether, 1 ml of concd hydrochloric acid, and 2 g of a mixture of equal parts of naphthalene and practical grade p-aminophenol (dark brown). Stopper the funnel and shake thoroughly to distribute the hydrocarbon into the ether phase and p-aminophenol into the aqueous phase as the hydrochloride. Each time you remove the stopper, rinse it and the mouth of the funnel with ether from a Calcutta wash bottle. When you think equilibrium has been reached, add a small spatulaful of stannous chloride crystals to reduce impurities in the practical grade p-aminophenol and shake vigorously for a few minutes. If the brown color is not largely discharged from the aqueous layer, add another small portion of stannous chloride (avoid excess) and shake again. Draw off the nearly colorless aqueous solution of the hydrochloride into a 125-ml Erlenmeyer. Extract again with 10 ml of water and 1 ml of concd hydrochloric acid. To recover more of the hydrochloride, rinse down the walls of the funnel by directing a stream of water from a wash bottle around the mouth of the funnel (a few ml is enough). Drain off the water layer into the Erlenmeyer and put the flask on the steam bath to drive off the dissolved ether. The ethereal layer is usually dark and contains suspended scum, but it can be clarified easily. To do so, add about 3 small spatulafuls of decolorizing charcoal and shake vigorously for a minute or two. Rest a 40-mm stemless funnel fitted with a filter paper in a tared 50-ml Erlenmeyer flask, fill a dry 10 × 75-mm test tube with granular anhydrous sodium sulfate (a convenient measure of approximately 5 g), and pour the drying agent

Technique of drying

into the cone of paper. Then run the ethereal solution out of the stem onto the drying agent in the filter as rapidly as filtration permits. Note that residual drops of aqueous solution adhere nicely to the walls and stem of the funnel. If filtration is at all delayed, raise the filter to see if there is an air lock. Use a Calcutta wash bottle for rinsing the funnel after it has been emptied and for rinsing the sodium sulfate and paper. Add a carborundum boiling stone and evaporate on the steam bath under an aspirator tube for entrainment of vapor. After removal of the bulk of the ether, evacuate briefly to remove a trace of solvent. The molten naphthalene should solidify readily on cooling; yield 0.8–0.9 g. Dissolve the crude product in a small volume of methanol with heating (steam bath) and, if the solution is at all colored, add excess solvent, clarify with decolorizing charcoal, and either concentrate the solution to the point of saturation or add water by drops to produce a saturated solution. Naphthalene crystallizes in large white plates, which should be washed with chilled methanol.

If the hot aqueous solution containing p-aminophenol hydrochloride and stannous chloride is not clear and colorless, add one small spatulaful of decolorizing charcoal and filter the hot solution into an Erlenmeyer having a side arm; rinse and wash as usual.

Then add 10 ml of concd hydrochloric acid, close the flask with a rubber stopper, clamp it within the rings of a steam bath, and wrap a towel around it for efficient heating. Connect the side arm to an efficient suction pump and turn it on full force. The solution is to be concentrated until crystals of p-aminophenol hydrochloride begin to separate (10–15 min) and then cooled, or else taken to dryness. In either case, rub up the crystals with acetone and use this solvent for transferring the solid to a suction filter; yield 0.7–0.8 g. The quality of this salt can be judged better from the color than from the mp. Dissolve a small sample in water and add sodium bicarbonate (solid or aqueous solution) and compare the color of the free aminophenol with that of the starting material.

3. Acidic and Neutral Substances

A mixture of equal parts of the following substances is to be separated by extraction[3]:

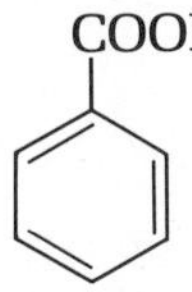

Benzoic acid
Mp 123°, pK$_a$ 4.17

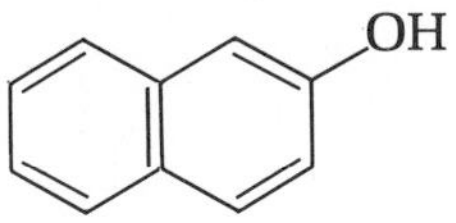

2-Naphthol
Mp 123°, pK$_a$ 8.0

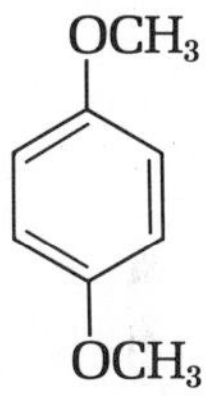

Hydroquinone dimethyl ether
Mp 57°

As can be inferred from the pK$_a$ values, benzoic acid reacts with sodium bicarbonate to form a water-soluble sodium salt, whereas 2-naphthol does not; 2-naphthol, in turn, is soluble in aqueous sodium hydroxide solution and hydroquinone dimethyl ether is insoluble in aqueous alkali. The solubility of benzoic acid in 100 g of cold water is 0.2 g; of 2-naphthol, 0.1 g.

Each student is to plan a procedure for separating 3.0 g of the mixture into the components and to have the plan checked by the instructor before proceeding. A flow sheet is a convenient way to present the plan. In each case select a solvent for crystallization on the basis of solubility tests.

4. Problem

Suppose a reaction mixture, when diluted with water, afforded 300 ml of an aqueous solution of 30 g of the reaction product malononitrile, CH$_2$(CN)$_2$, which is to be isolated by extraction with ether. The solubility of malononitrile in ether at room temperature is 20.0 g per 100 ml and that in water is 13.3 g per 100 ml. What

Note for the Instructor

[3] Heat a mixture of the components on the steam bath and stir until a homogeneous melt results; then cool with stirring and break up the warm solid with a spatula.

pK$_a$ = acidity constant

weight of malononitrile would be recovered by extraction with
(a) three 100-ml portions of ether; (b) one 300-ml portion of ether.
Suggestion: For each extraction let x equal the weight extracted into
the ether layer. In case (a) the concentration in the ether layer is
x/100 and that in the water layer (30 − x)/300; the ratio of these
quantities is equal to $K = 20/13.3$.

9 Steam Distillation

KEYWORDS   Immiscible liquids   Dalton's law
                Partial pressure   Water trap
                Vapor pressure   Isolation of citral
                Boiling point

The experience of distilling a mixture of ethanol and water (Chapter 4, section 3) will have shown that the boiling point of the mixture is below that of the more volatile component. Since the two liquids are essentially insoluble in each other, the benzene molecules in a droplet of benzene are not diluted by water molecules from nearby water droplets, and hence the vapor pressure exerted by the benzene is the same as that of benzene alone at the existing temperature. The same is true of the water present. Because they are immiscible, the two liquids independently exert pressures against the common external pressure, and when the sum of the two partial pressures equals the external pressure boiling occurs. Benzene has a vapor pressure of 760 torr at 80.1°, and if it is mixed with water the combined vapor pressure must equal 760 torr at some temperature below 80.1°. This temperature, the boiling point of the mixture, can be calculated from known values of the vapor pressures of the separate liquids at that temperature. Vapor pressures found for water and benzene in the range 50–80° are plotted in Fig. 9.1. The dotted line cuts the two curves at points where the sum of the vapor pressures is 760 torr; hence this is the boiling point of the mixture (69.3°).

Practical use can sometimes be made of the fact that many water-insoluble liquids and solids behave like benzene when mixed with water, volatilizing at temperatures below their boiling points. Thus, naphthalene, a solid, boils at 218° but distils with water at a temperature below 100°. Since naphthalene is not very volatile, considerable water is required to entrain it and the conventional way of conducting the operation of distillation is to pass steam into a boiling flask containing naphthalene and water. The process is called *steam distillation*. With more volatile compounds, or with a small amount of material, the substance can be heated with water in a simple distillation flask and the steam generated *in situ*.

Some high-boiling substances decompose before the boiling point is reached and, if impure, cannot be purified by ordinary distillation. However, they can be freed from contaminating substances by steam distillation at a lower temperature at which they

Figure 9.1
Vapor pressure *vs.* tem-
perature curves for water
and benzene.

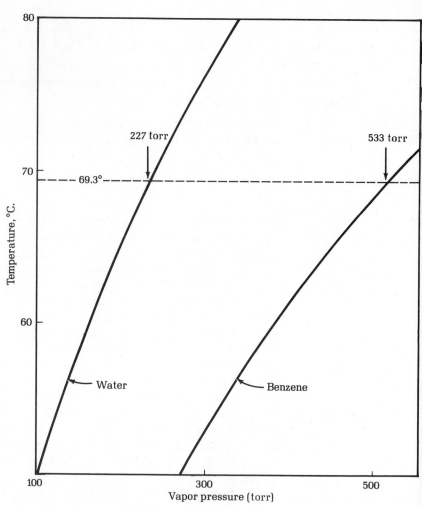

are stable. Steam distillation also offers the advantage of selectivity, since some water-insoluble substances are volatile with steam and others are not, and some volatilize so very slowly that sharp separa-tion is possible. The technique is useful in processing natural oils and resins, which can be separated into steam-volatile and nonsteam-volatile fractions. It is useful for recovery of a nonsteam-volatile solid from its solution in a high-boiling solvent such as nitrobenzene, bp 210°; all traces of the solvent can be eliminated and the temperature can be kept low.

Used for isolation of
perfume and flavor oils

The boiling point remains constant during a steam distillation so long as adequate amounts of both water and the organic com-ponent are present to saturate the vapor space. Determination of the boiling point and correction for any deviation from normal atmo-spheric pressure permits calculation of the amount of water required for distillation of a given amount of organic substance. According to Dalton's law the molecular proportion of the two components in the distillate is equal to the ratio of their vapor pressures (*p*) in the boiling mixture; the more volatile component contributes the greater number of molecules to the vapor phase. Thus,

$$\frac{\text{Moles of water}}{\text{Moles of substance}} = \frac{p^{\text{Water}}}{p^{\text{Substance}}}$$

The vapor pressure of water (p^{Water}) at the boiling temperature in question can be found by interpolation of the data of Table 9.1 and

Table 9.1   Vapor Pressure of Water in mm. of Mercury (torr)

$t°$	p	$t°$	p	$t°$	p	$t°$	p
60	149.3	70	233.7	80	355.1	90	525.8
61	156.4	71	243.9	81	369.7	91	546.0
62	163.8	72	254.6	82	384.9	92	567.0
63	171.4	73	265.7	83	400.6	93	588.6
64	179.3	74	277.2	84	416.8	94	610.9
65	187.5	75	289.1	85	433.6	95	633.9
66	196.1	76	301.4	86	450.9	96	657.6
67	205.0	77	314.1	87	468.7	97	682.1
68	214.2	78	327.3	88	487.1	98	707.3
69	223.7	79	341.0	89	506.1	99	733.2

that of the organic substance is, of course, equal to $760 - p^{\text{Water}}$. Hence, the weight of water required per gram of substance is given by the expression

$$\frac{\text{Wt. of water per}}{\text{g of substance}} = \frac{18 \times p^{\text{Water}}}{\text{MW of substance} \times (760 - p^{\text{Water}})}$$

From the data given in Fig. 9.1 for benzene–water, and the molecular weight 78.11 for benzene, the water required for steam distillation of 1 g of benzene is only $227 \times {}^{18}\!/_{533} \times {}^{1}\!/_{78} = 0.10$ g. Nitrobenzene (bp 210°, MW 123.11) steam distils at 99° and requires 4.0 g of water per gram. The low molecular weight of water makes water a favorable liquid for two-phase distillation of organic compounds.

EXPERIMENTS   1. Apparatus

In the assembly shown in Fig. 9.2, steam is passed into a 250-ml, round-bottomed flask through a section of 6-mm glass tubing fitted into a stillhead with a piece of $^{3}/_{16}$-in rubber tubing connected to a trap, which in turn is connected to the steam line. The trap serves two purposes: it allows water, which is in the steam line, to be removed before it has a chance to reach the round-bottomed flask and adjustment of the clamp on the hose at the bottom of the trap allows precise control of the steam flow. The stopper in the trap should be wired on, as shown, as a precaution. A bent adapter attached to a *long* condenser delivers the condensate into a 250-ml Erlenmeyer flask.

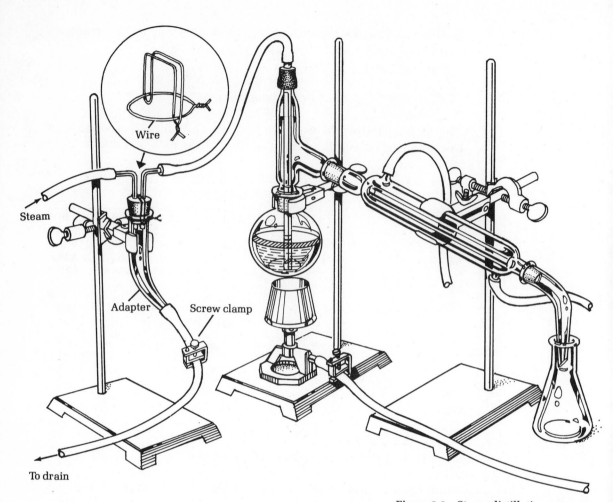

Figure 9.2   Steam distillation apparatus.

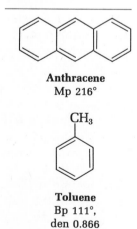

Anthracene
Mp 216°

CH₃

Toluene
Bp 111°,
den 0.866

Note for the instructor

2. Recovery of a Dissolved Substance

Measure 50 ml of a 0.2% solution of anthracene in toluene[1] into the 250-ml round-bottomed flask and add 100 ml of water. For an initial distillation to determine the boiling point and composition of the toluene–water azeotrope, fit the stillhead with a thermometer instead of the steam-inlet tube. Heat the mixture with a microburner, distil about 50 ml of the azeotrope and record a best value for the boiling point. After removing the flame, pour the distillate into a graduate and measure the volumes of toluene and water. Calculate the weight of water per gram of toluene and compare the result with the theoretical value calculated from the vapor pressure of water at the observed boiling point (see Table 9.1).

Replace the thermometer with the steam-inlet tube. To start the steam distillation, heat the flask containing the mixture with a small

[1] Pure, fluorescent anthracene should be used. Preparation: Chapter 40.

flame to prevent water from condensing in the flask to the point where water and product splash over into the receiver. Then turn on the steam valve, making sure the screw clamp on the bottom of the trap is open. Slowly close the clamp, and allow steam to pass into the flask. Unlike ordinary distillations, steam distillations are usually run as fast as possible, with proper care to avoid having material splash into the receiver and to avoid having steam escape uncondensed.

Continue distillation by passing in steam until the distillate is clear and then until fluorescence appearing in the stillhead indicates that a trace of anthracene is beginning to distil. Stop the steam distillation by opening the clamp at the bottom of the trap and then turning off the steam valve. Grasp the round-bottomed flask with a towel when disconnecting it and, using the clamp to support it, cool it under the tap. The bulk of the anthracene can be dislodged from the flask walls and collected on a small suction filter. To recover any remaining anthracene, add a little acetone to the flask, warm on the steam bath to dissolve the material, add water to precipitate it, and collect the precipitate on the same suction filter. About 80% of the hydrocarbon in the original toluene solution should be recoverable. When dry, crystallize the material from about 1 ml of benzene and observe that the crystals are more intensely fluorescent than the solution or the amorphous solid. The characteristic fluorescence is quenched by mere traces of impurities.

3. Isolation of a Natural Product

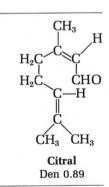

Citral
Den 0.89

Citral, a fragrant terpene aldehyde made up of two isoprene units, is the main component of the steam-volatile fraction of lemon grass oil and is used in a commercial synthesis of vitamin A. Lemon grass tea is a popular drink in Mexico.

Using a graduate, measure out 10 ml of lemon grass oil (**not** lemon oil) into the 250-ml boiling flask. Rinse the remaining contents of the graduate into the flask with a little ether. Add 100 ml of water, make connections as in Fig. 9.2, heat the flask with a small flame, and pass in steam. Distil as rapidly as the cooling facilities allow and continue until droplets of oil no longer appear at the tip of the condenser (about 250 ml of distillate).

Pour 50 ml of ether into a 125-ml separatory funnel, cool the distillate if necessary, and pour a part of it into the funnel. Shake, let the layers separate, discard the lower layer, add another portion of distillate, and repeat. When the last portion of distillate has been added, rinse the flask with a little ether to recover adhering citral. Use the techniques described in Section 2, of Chapter 8, for drying, filtering, and evaporating the ether. Take the tare of a 1-g tincture bottle, transfer the citral to it with a capillary dropping tube, and determine the weight and the yield from the lemon grass oil. Label the bottle and store (in the dark) for later testing for the presence of functional groups.

QUESTIONS   1. A mixture of ethyl iodide (C_2H_5I, bp 72.3°) and water boils at 63.7°. What weight of ethyl iodide would be carried over by 1 g of steam during steam distillation?

2. Iodobenzene (C_6H_5I, bp 188°) steam distils at a temperature of 98.2°. How many molecules of water are required to carry over one molecule of iodobenzene? How many grams per gram of iodobenzene?

10 Nuclear Magnetic Resonance Spectroscopy

KEYWORDS

Integral

Downfield, upfield

TMS (tetramethylsilane)

Coupling constant

Chemical shift

Shift reagents [Eu(dpm)$_3$]

Nmr solvents (CS$_2$, CDCl$_3$)

Phase

Spinning side bands

Saturation

Magnetic impurities, removal

Homogeneity

Nmr—Determination of the number, kind, and relative locations of hydrogen atoms (protons) in a molecule

Nuclear magnetic resonance (nmr) spectroscopy is a means of determining the number, kind, and relative locations of certain atoms, principally hydrogen, in molecules. Experimentally, the sample, 0.3 ml of a 20% solution in a 5-mm O.D. dia glass tube, is placed in the probe of the spectrometer between the faces of a powerful (14,000 Gauss) permanent or electromagnet and irradiated with radiofrequency energy (60 MHz; 60,000,000 cps). The absorption of radiofrequency energy *vs.* magnetic field strength is plotted by the spectrometer to give a spectrum.

In a typical spectrum (Fig. 10.1, ethyl iodide) the relative numbers

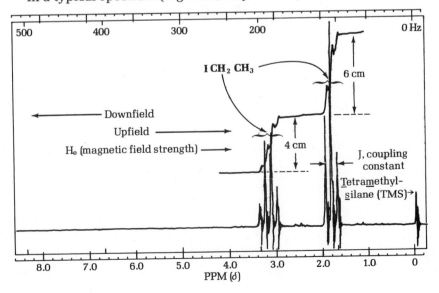

Figure 10.1 Nmr spectrum of ethyl iodide. The staircase-like line is the integral. In the integral mode of operation the recorder pen moves from left to right and moves vertically a distance proportional to the areas of the peaks over which it passes. Hence, the relative area of the quartet of peaks at 3.20 ppm to the triplet of peaks at 1.83 ppm is given by the relative heights of the integral (4 cm is to 6 m as 2 is to 3). The relative numbers of hydrogen atoms are proportional to the peak areas (2H and 3H).

of hydrogen atoms (protons) in the molecule are determined from the *integral*, the stair-step line over the peaks. The height of the step is proportional to the area under the nmr peak, and in nmr spectroscopy (contrasted with infrared, for instance) the area of each group of peaks is directly proportional to the number of hydrogen atoms causing the peaks. Integrators are part of all nmr spectrometers, and running the integral takes no more time than running the spectrum. The different kinds of protons are indicated by their *chemical shifts*.[1] For ethyl iodide the two protons adjacent to the electronegative iodine atom are *downfield* (at lower magnetic field strength) ($\delta = 3.20$ ppm) from the three methyl protons at $\delta = 1.83$ ppm. Tables of chemical shifts for protons in various environments can be found in reference books.[2]

The relative locations of the five protons in ethyl iodide are indicated by the pattern of peaks on the spectrum. The three peaks indicate methyl protons adjacent to two protons; four peaks indicate methylene protons adjacent to three methyl protons. In general, in molecules of this type, a given set of protons will appear as $n + 1$ peaks if they are adjacent to n protons. The distance between adjacent peaks in the quartet and triplet is the *coupling constant*, *J*.

Not all nmr spectra are so easily analyzed as the spectrum for ethyl iodide. Consider the one for 3-hexanol (Fig. 10.2). Twelve protons give rise to an unintelligible group of peaks between 1.0 and 2.2 ppm. It is not clear from a 3-hexanol spectrum which of the

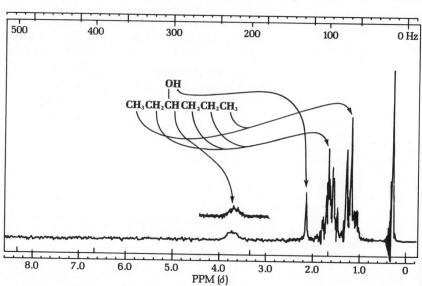

**Figure 10.2
Nmr spectrum of
3-hexanol.**

[1] Chemical shift is a measure of a peak's position relative to the peak of a standard substance (e.g., TMS, tetramethylsilane), which is assigned a chemical shift value of 0.0.
[2] For a complete and detailed discussion of all aspects of nmr see L. M. Jackman and S. Sternhell, *Applications of Nuclear Magnetic Resonance Spectroscopy in Organic Chemistry*, 2nd ed., Pergamon Press, N.Y., 1969. See also K. L. Williamson, *Basic NMR Spectroscopy*, Communication Skills Corp., 1220 Post Road, Fairfield Conn. 06430, 1973. This is an audio-visual program in six parts (the nmr phenomenon, chemical shifts, coupling constants, preparing the sample, running the spectrum, and applications).

two low-field peaks (on the left-hand side of the spectrum) should be assigned to the hydroxyl proton and which should be assigned to the proton on C-3.

Deuterium atoms give no peaks in the nmr spectrum; thus the peak in the 3-hexanol spectrum for the proton on C-3 will be evident if the spectrum of the alcohol is one in which the hydroxyl proton of the alcohol has been replaced by deuterium. The hydroxyl proton of 3-hexanol is acidic and will exchange rapidly with the deuterium of D_2O.

Addition of a few milligrams of a hexacoordinate complex of europium (tris(dipivaloylmethanato)europiumIII, [Eu(dpm)$_3$]) to an nmr sample, which contains a Lewis base center (an amine or basic oxygen, such as a hydroxyl group), has a dramatic effect on the spectrum. This rare earth compound (soluble in CS_2 and $CDCl_3$) causes large shifts in the positions of peaks arising from the protons near the metal atom in this molecule and is therefore referred to as a shift reagent. It produces the shifts by complexing with the un-shared electrons of the hydroxyl oxygen, the amine nitrogen, or other Lewis base centers. As part of the complex the europium atom exerts a large dipolar effect on nearby hydrogens, with resulting changes in the nmr spectrum.

With no shift reagent present, the nmr spectrum of 2-methyl-3-pentanol (Fig. 10.3A) is not readily analyzed. Addition of about 10 mg of Eu(dpm)$_3$ (the shift reagent) to the sample (0.5 ml of a 0.4 M solution) causes very large downfield shifts of peaks due to protons near the coordination site (Fig. 10.3B).[3] Further additions of 10-mg portions of the shift reagent cause further downfield shifts (Figs. 10.3D, 10.3E, and 10.3F), so that in Fig. 10.3G only the methyl peaks appear within 500 Hz of TMS. Finally, when the mole ratio of Eu(dpm)$_3$ to alcohol is 1:1 we find the spectrum is as shown in Fig. 10.4. The two protons on C-4 and the two methyls on C-2 are magnetically nonequivalent because they are adjacent to a chiral center (an asymmetric carbon atom, C-3), and therefore each gives a separate set of peaks. With shift reagent added, this spectrum can be analyzed by inspection.

Quantitative information about molecular geometry can be obtained from shifted spectra. The shift induced by the shift reagent, $\Delta H/H$, is related to the distance (r) and the angle (θ) which the proton bears to the europium atom:

$$\frac{\Delta H}{H} = \frac{3\cos^2\theta - 1}{r^3}$$

[3] See also K. L. Williamson, D. R. Clutter, R. Emch, M. Alexander, A. E. Burroughs, C. Chua, and M. E. Bogel, *J. Amer. Chem. Soc.*, **96**, 1471 (1974).

Figure 10.3
The nmr spectrum of 2-methyl-3-pentanol (0.4M in CS_2) with various amounts of shift reagent present. (A) No shift reagent present. All methyl peaks are superimposed. The peak for the proton adjacent to the hydroxyl group is downfield from the others because it is adjacent to the electronegative oxygen atom. (B) 2-Methyl-3-pentanol + Eu(dpm)$_3$. Mole ratio of Eu(dpm)$_3$ to alcohol = 0.05. The hydroxyl proton peak at 1.6 ppm in spectrum (A) appears at 6.2 ppm in spectrum (B) because it is closest to the Eu in the complex formed between Eu(dpm)$_3$ and the alcohol. The next closest proton, the one on the hydroxyl bearing carbon atom, gives a peak at 4.4 ppm. Peaks due to the three different methyl groups at 1.1–1.5 ppm begin to differentiate. (C) Mole ratio of Eu(dpm)$_3$ to alcohol = 0.1. The hydroxyl proton does not appear in this spectrum because its chemical shift is greater than 8.6 ppm with this much shift reagent present. (D) Mole ratio of Eu(dpm)$_3$ to alcohol = 0.25. Further differentiation of methyl peaks (2.3–3.0 ppm) is evident. (E) Mole ratio of Eu(dpm)$_3$ to alcohol = 0.5. Separate groups of peaks begin to appear in the region 4.2–6.2 ppm. (F) Mole ratio of Eu(dpm)$_3$ to alcohol = 0.7. Three groups of peaks (at 6.8, 7.7, and 8.1 ppm) due to the protons on C-2 and C-4 are evident, and three different methyls are now apparent. (G) Mole ratio of Eu(dpm)$_3$ to alcohol = 0.9. Only the methyl peaks appear on the spectrum. The two doublets come from the methyls attached to C-2 and the triplet comes from the terminal methyl at C-5.

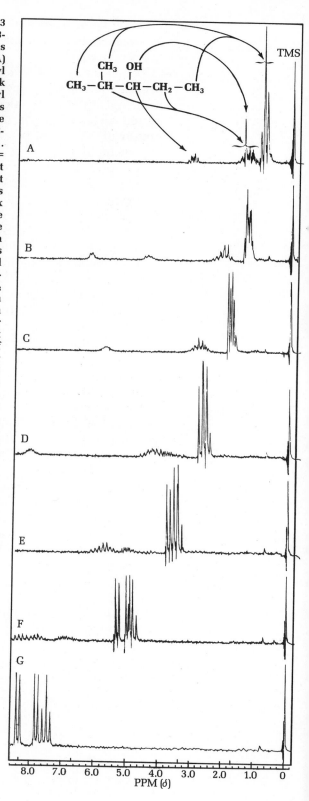

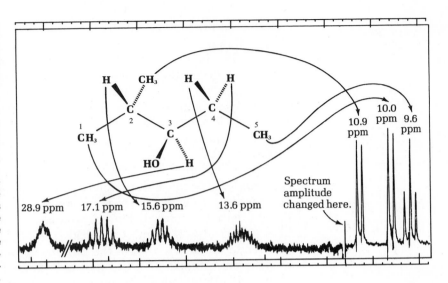

Figure 10.4 Nmr spectrum of 2-methyl-3-pentanol containing Eu(dpm)₃ Mole ratio of Eu(dpm)₃ to alcohol = 1.0. Compare this spectrum to those shown in Fig. 10.3. Protons nearest the hydroxyl group are shifted most. Methyl groups are recorded at reduced spectrum amplitude. Note the large chemical shift difference between the two protons on C-4. The average conformation of the molecule is the one shown and was calculated using the equation on p. 69.

Eu(dpm)₃ is a metal chelate (*chele*, Greek, claw). The β-diketone dipivaloylmethane (2,2,6,6-tetramethylheptane-3,5-dione) is a ligand, in this case a bidentate ligand, attached to the europium at two places. Since Eu^{3+} is hexacoordinate, three dipivaloylmethane molecules cluster about this metal atom. However, when a molecule with a basic group like an amine or an alcohol is in solution with Eu(dpm)₃ the europium will expand its coordination sphere to complex with this additional molecule. Such a complex is weak and so its nmr spectrum is an average of the complexed and uncomplexed molecule.

Tris(dipivaloylmethanato)europiumIII, Eu(dpm)₃
MW 701.78, Mp 188–189°

EXPERIMENTS   Procedure

A typical nmr sample is 0.3 to 0.5 ml of a nonviscous liquid of a

Sample size: 0.3 ml of 10–20% solution plus TMS

10–20% solution or a solid in a proton-free solvent contained in a 5-mm dia glass tube. The sample tube must be of uniform outside and inside diameter with uniform wall thickness. Test a sample tube by rolling it down a very slightly inclined piece of plate glass. Reject all tubes that roll unevenly.

The ideal solvent, from the nmr standpoint, is carbon tetrachloride. It is proton-free and nonpolar but unfortunately a poor solvent. Carbon disulfide is an excellent compromise. It will, however, react with amines.

CS_2, the solvent of choice

Deuteriochloroform ($CDCl_3$) is one of the most widely used nmr solvents. Although more expensive than nondeuteriated solvents, it will dissolve a wider range of samples than carbon disulfide or carbon tetrachloride. Residual protons in the $CDCl_3$ will always give a peak at 7.27 ppm. Chemical shifts of protons are measured relative to the sharp peak of the protons in tetramethylsilane (taken as 0.0 ppm). Stock solutions of 3 to 5 percent tetramethylsilane in carbon disulfide and in deuteriochloroform are useful for preparing routine samples.

A wide variety of completely deuteriated solvents are commercially available, e.g., deuteroacetone (CD_3COCD_3), deuterodimethylsulfoxide (CD_3SOCD_3), deuterobenzene (C_6D_6), although they are expensive. For highly polar samples a mixture of the expensive deuterodimethylsulfoxide with the less expensive deuteriochloroform will often be satisfactory. Water soluble samples are dissolved in deuteriated water containing a water soluble salt [DSS, $(CH_3)_3SiCH_2CH_2CH_2SO_3^-Na^+$] as a reference substance. The protons on the three methyl groups bound to the silicon in this salt absorb at 0.0 ppm.

Erratic spectra from ferromagnetic impurities; remove by filtration

Solid impurities in nmr samples will cause very erratic spectra. If two successive spectra taken within minutes of each other are not identical, suspect solid impurities, especially ferromagnetic ones. These can be removed by filtration of the sample through a tightly packed wad of glass wool in a capillary pipet (Fig. 10.5). If very high resolution spectra (all lines very sharp) are desired, oxygen, a paramagnetic impurity, must be removed by bubbling a fine stream of pure nitrogen through the sample for 60 sec. Routine samples do not require this treatment.

The usual nmr sample has a volume of 0.3 ml to 0.5 ml, even though the volume sensed by the spectrometer receiver coils (referred to as the active volume) is much smaller (Fig. 10.6). To average the magnetic fields produced by the spectrometer within the sample, the tube is spun by an air turbine at thirty to forty revolutions per second while taking the spectrum. Too rapid spinning or an insufficient amount of sample will cause the vortex produced by the spinning to penetrate the active volume, giving erratic nonreproducible spectra. A variety of microcells are available for holding and proper positioning of small samples with respect to the receiver coils of the spectrometer (Fig. 10.7). The vertical positioning of these cells in the spectrometer is critical.

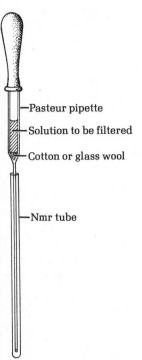

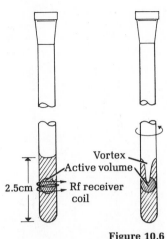

Figure 10.6
Effect of too rapid spinning or insufficient sample. The active volume is the only part of the sample detected by the spectrometer.

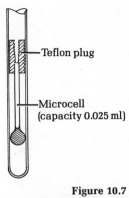

Figure 10.7
Nmr microcell positioned in an nmr tube. The Teflon plug supports the microcell and allows the sample to be centered in the active volume of the spectrometer.

Figure 10.5
Micro filter for nmr samples. Solution to be filtered is placed in the top of the Pasteur pipette, rubber bulb put in place, and pressure applied to force sample through the cotton or glass wool into an nmr sample tube.

Sweep zero

Resolution, sharp, narrow peaks with good ringing = high resolution

Saturation

If microcells are used, only one or two mg of the sample are needed to give satisfactory spectra, in contrast to the 20–30 mg usually needed.

Adjusting the Spectrometer

To be certain the spectrometer is correctly adjusted and working properly, record the spectrum of the standard sample of chloroform and tetramethylsilane (TMS) usually found with the spectrometer. While recording the spectrum from left to right, the $CHCl_3$ peak should be brought to 7.27 ppm and the TMS peak to 0.0 ppm with the sweep zero control. The most important adjustment, the resolution control (also called homogeneity, or Y-control), should be adjusted for each sample so the TMS peak is as high and narrow as possible, with good ringing (Fig. 10.8). The signal (the peak traced by the spectrometer) should also be properly phased; it will then have the same appearance in both forward and backward scans (Fig. 10.8).

Small peaks symmetrically placed on each side of a principal peak are artifacts called spinning side bands (Fig. 10.9). They are recognized as such by changing the spin speed (see again Fig. 10.9), which causes the spinning side bands to change positions. Two controls on the spectrometer determine the height of a signal as it is recorded on the paper. One, the spectrum amplitude control, increases the size of the signal as well as the baseline noise (the jitter of the pen when no signal is present). The other, the radio frequency (rf) power control, increases the size of the signal alone, but only to a point, after which saturation occurs (Fig. 10.10).

Figure 10.8
Effect of phasing on signal shape. The TMS peaks on both forward and backward scans are quite high and narrow, with good ringing and perfect symmetry.

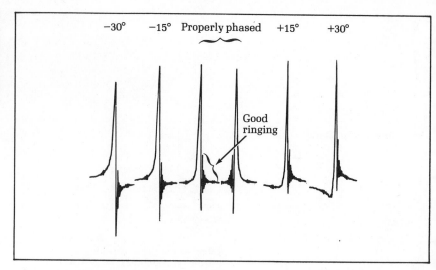

−30°   −15°   Properly phased   +15°   +30°

Good ringing

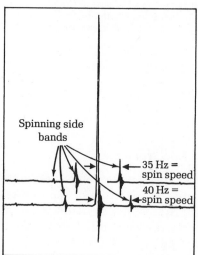

Spinning side bands

35 Hz = spin speed
40 Hz = spin speed

Figure 10.9
Spinning side bands.

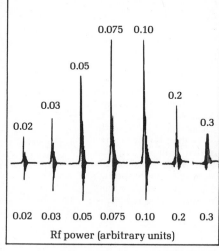

0.075   0.10

0.05

0.03

0.02

0.2

0.3

0.02   0.03   0.05   0.075   0.10   0.2   0.3
Rf power (arbitrary units)

Figure 10.10
Saturation of the nmr signal for chloroform. At the optimum *rf* power level (0.075–0.10) the signal reaches maximum intensity. Further increase of power causes the signal to become smaller and broader, with poor ringing —saturation.

Applying more than the optimum rf power will cause the peak to become distorted and of low intensity.

Unknowns

Using a stock solution of 4% TMS in carbon disulfide, prepare 0.5 ml of a 0.4 M (or 10%) solution of an unknown alcohol or amine. Filter the solution, if necessary, into a clean, *dry* nmr tube. Set the

TMS peak at 0.0 ppm, check the phasing, maximize the resolution, and run a spectrum over a 500-Hz range using a 250-sec sweep time. To the unknown solution add about 5 mg of Eu(dpm)$_3$ (the shift reagent), shake thoroughly to dissolve, and run another spectrum. Use the technique described in Chapter 3 and estimate the 5 mg. Continue adding Eu(dpm)$_3$ in 10-mg portions until the spectrum is shifted enough for easy analysis. Integrate peaks and groups of peaks if in doubt about their relative areas. To protect the Eu(dpm)$_3$ from moisture store in a desiccator.

11 *Infrared Spectroscopy*

KEYWORDS

Wavenumber
Reciprocal cm, microns
Band spectra
Transmittance, absorbance
Double beam spectrometer
Molecular vibrations
Stretching, bending
Group frequencies

Fingerprint region
Correlation tables
Neat solutions, mulls, KBr discs
Demountable and solution cells
Nujol, paraffin oil
Polystyrene calibration film
Carbonyl frequencies

The presence and also the environment of functional groups in organic molecules can be identified by infrared spectroscopy. Like nuclear magnetic resonance and ultraviolet spectroscopy, infrared spectroscopy is nondestructive. Moreover, the small quantity of sample needed, the speed with which a spectrum can be obtained, the relatively small cost of the spectrometer, and the wide applicability of the method combine to make infrared spectroscopy one of the most useful tools available to the organic chemist.

An infrared spectrum is a record of the absorption of electromagnetic radiation by the sample in the region 4000 to 650 wavenumbers (reciprocal cm, cm^{-1}), which is equal to the wavelength region of 2 to 15 microns (μ) ($1\mu = 1 \times 10^{-4}$ cm).

$$cm^{-1} = \frac{10,000}{\mu}$$

Spectra are often presented with an abscissa linear in cm^{-1} and an ordinate linear in percentage transmittance with 100% at the top. The ordinate is also often presented in absorbance, where absorbance, A, is the logarithm of the reciprocal of the transmittance

$$A = \log\left(\frac{1}{T}\right)$$

(T). Solution spectra consist of very closely spaced unresolved lines resulting in a *band* spectrum. By comparison with nmr and ultraviolet spectra, infrared absorption bands are upside down.

Figure 11.1 is a schematic representation of a typical double beam,

The infrared spectrometer

optical null, infrared spectrometer. An electrically heated metal rod serves as the radiation source; the radiation passes through both the sample cell and reference cell, through combs and a beam

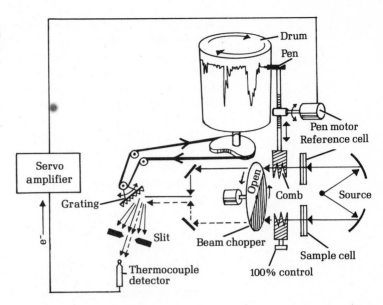

Figure 11.1
Schematic diagram of a double
beam, optical null, infrared
spectrometer.

chopper to the dispersion grating. In some instruments the radiation is dispersed by a prism made of sodium chloride, which is transparent to infrared radiation. Of the electromagnetic radiation frequencies spread out by the grating (or prism) only a small range of frequencies is allowed to pass through the slit to the detector, which is a thermocouple with a very rapid response time.

The spectrometer works on the optical null principle. The detector senses infrared light coming alternately through the substances in the sample and reference cells. If the amount of light is the same from both beams the detector produces a direct current and nothing happens, but if less light comes through the sample beam than the reference beam (because of absorption of radiation by the sample molecules), then the detector senses an alternating current which alternates at the rate the chopper is turning. This current is amplified in the servo amplifier which activates the pen motor. The pen motor moves the pen down the paper drawing an absorption band and at the same time drives a comb into the reference beam just far enough so that the detector will again sense a null, i.e., no alternating current. The motion of the drum holding the paper is linked to the grating so that as the drum moves the grating moves with it to scan the entire range of frequencies.

Absorption of infrared radiation results in changes in the amplitude of vibration of various parts of an organic molecule. Each band in the spectrum results from the absorption of radiation of a specific frequency which excites a particular bond or group of bonds in the molecule causing the atoms to vibrate with respect to each other. Most of the important peaks in infrared spectra result from changes in the frequencies of bond stretching or bending.

Although it is possible to calculate the exact frequencies of vibrations of all the atoms in extremely simple molecules, the principal usefulness of infrared spectroscopy lies in empirical correlations of thousands of spectra of complex molecules. Certain group frequencies are well characterized and are known to appear in very narrow frequency ranges in the spectrum. The C—H bond stretching vibrations, for example, occur in the 3600 to 2700 cm^{-1} range, with different types of C—H bonds at specific frequencies within this range. For instance, saturated tertiary carbon-proton bonds absorb at 2890 cm^{-1}, isolated carbon-vinyl proton bonds at 3020 cm^{-1}, aromatic carbon-proton bonds at 3070 cm^{-1}, and acetylenic bonds at 3300 cm^{-1}.

Absorption of infrared radiation requires that a bond undergo a change in dipole moment upon excitation. The C=C stretching mode of ethylene shows up as an exceedingly weak peak near 1650 cm^{-1} because stretching the C=C bond does not change the dipole moment of the molecule. Some of the most intense (and most useful) bands in an infrared spectrum come from the carbonyl group. This is because the carbon and the oxygen atoms are of different electronegativity, bound by a double bond, and the change in dipole moment is large when the bond stretches; consequently, the infrared absorption is intense. In general, the intensities of infrared bands are proportional to the change in dipole moment causing the absorption band and not to the number of atoms causing the absorption, as in nmr.

The amount of energy (and hence the frequency of the radiation) required to excite the stretching and bending modes depends upon the strength of the bond and the masses of the atoms attached to the bond; the amount of energy required decreases with decreasing bond order (C≡C > C=C > C—C) and decreases with increasing mass (e.g., C—F > C—Cl > C—Br). An infrared spectrum is conveniently divided into two regions; the region from 4000 to *ca.* 1550 cm^{-1}, which includes vibrational frequencies characteristic of particular functional groups (*e.g.*, O—H, N—H, C—H at 3600 to 2700 cm^{-1}; O—H····X hydrogen bonds at 3300 to 2500 cm^{-1}; C=C, C=N at 2400 to 2000 cm^{-1}; and C=C, C=O at 1850 to 1550 cm^{-1}). The region from 1550 to 650 cm^{-1} is characteristic of the complex vibrations for the whole molecule. Bands in this region are difficult to interpret, but nevertheless are highly characteristic of the molecule; hence this region is designated the "fingerprint region." If two samples give identical infrared spectra, including the entire fingerprint region, they can be presumed to be identical substances. Infrared spectroscopy is much more useful than melting points, for example, in proving that two substances are identical.

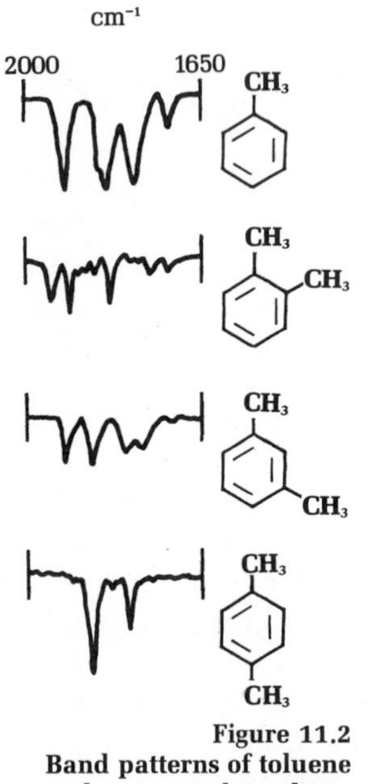

Extensive correlation tables and discussions of characteristic group frequencies can be found in specialized references. As one example, consider the band patterns of toluene, and of *o*-, *m*-, and *p*-xylene which appear in the frequency range 2000 to 1650 cm^{-1} (Fig. 11.2). These band patterns are due to changes in the dipole moment accompanying changes in vibrational modes of the aromatic

Characteristic group frequencies

Figure 11.2
Band patterns of toluene and *o*-, *m*-, and *p*-xylene.

ring and are surprisingly similar to those for mono substituted and other o-, m-, and p-disubstituted benzenes.

Infrared spectra can be run on neat (undiluted) liquids, on solutions with an appropriate solvent, and on solids as mulls and KBr pellets. Glass is opaque to infrared radiation; therefore, the sample and reference cells used in infrared spectroscopy are constructed so as to use sodium chloride plates. The sodium chloride plates are fragile and can be attacked by moisture. *Handle only by the edges.*

*Sample, solvents, and equipment **must** be dry*

Spectra of Neat Liquids

To run a spectrum of a neat liquid (free of water!) remove a demountable cell (Fig. 11.3) from the desiccator and place a drop of the

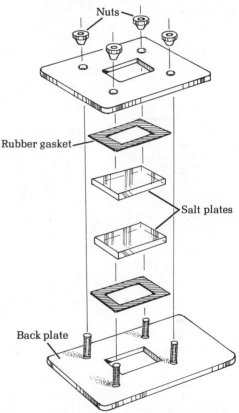

Figure 11.3
Exploded view of a demountable infrared cell for analyzing neat liquids.

Nuts

Rubber gasket

Salt plates

Back plate

liquid between the salt plates, press the plates together to remove any air bubbles and add the top rubber gasket and metal top plate. Next, put on all four of the nuts and *gently* tighten them to apply an even pressure to the top plate. Place the cell in the sample compartment (nearest the front of the spectrometer) and run the spectrum.

Although running a spectrum on a neat liquid is convenient and results in no extraneous bands to interpret, it is not possible to control the path length of the light through the liquid in a demountable cell. A low viscosity liquid when squeezed between the salt plates may be so thin that the short path length gives peaks that are too weak. A viscous liquid, on the other hand, may give peaks that are too intense. A properly run spectrum will have the most intense peak with an absorbance of about 1.0. Unlike the nmr spectrometer the infrared spectrometer does not have a control to adjust the peak intensities; this is done entirely by adjusting the sample concentration.

Spectra of Solutions

The most widely applicable method of running spectra of solutions involves dissolving an amount of the liquid or solid sample in an appropriate solvent to give a 10% solution. Just as in nmr spectroscopy, the best solvents to use are carbon disulfide and carbon tetrachloride, but since these compounds are not polar enough to dissolve many substances, chloroform is used as a compromise. Unlike nmr solvents, no solvent suitable in infrared spectroscopy is entirely free of absorption bands in the frequency range of interest (Figs. 11.4a and b). In chloroform, for instance, no

Solvents: $CHCl_3$, CS_2, CCl_4

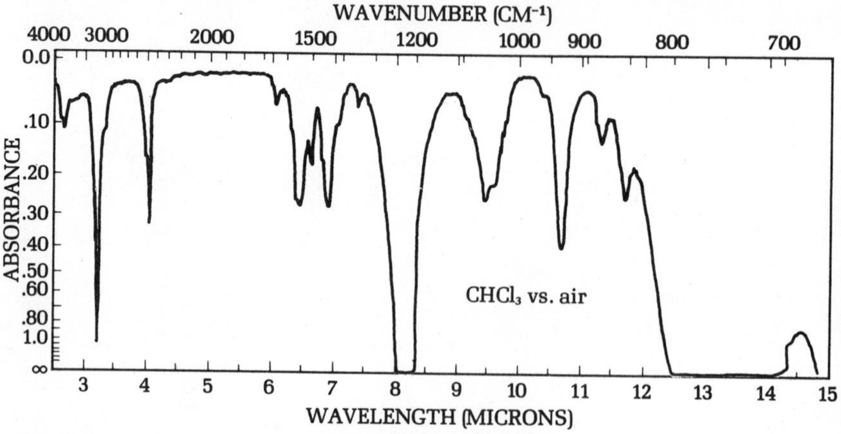

Figure 11.4a
Spectrum of chloroform in sample cell, air in reference cell. No infrared light passes through chloroform between 1200 and 1250 cm^{-1}, and between 650 and 800 cm^{-1}, therefore no information about sample absorption in those regions can be obtained.

Figure 11.4b
Spectrum of carbon disulfide in sample cell, air in reference cell. No infrared light passes through carbon disulfide between 1430 and 1550 cm^{-1}.

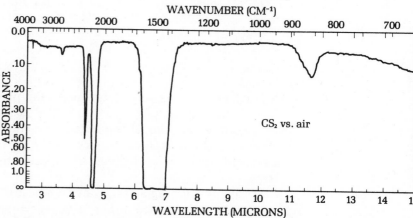

light passes through the cell between 650 and 800 cm^{-1}. As can be seen from the figures, spectra obtained using carbon disulfide and chloroform cover the entire infrared frequency range. In practice, a base line is run with the same solvent in both cells to ascertain if the cells are clean and matched (Fig. 11.4c). Often it is necessary to

Figure 11.4c
Spectrum of chloroform in both sample and reference cells. A typical baseline.

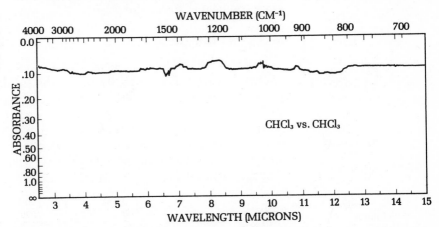

obtain only one spectrum employing one solvent, depending on which region of the spectrum you need to use.

Three large drops of solution will fill the usual sealed infrared cell (Fig. 11.5). A 10 percent solution of a liquid sample can be

Figure 11.5
Sealed infrared sample cell.

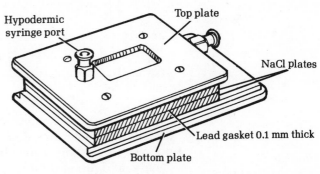

approximated by dilution of one drop of the liquid sample with nine of the solvent. Since weights are more difficult to estimate, solid samples should be weighed to obtain a 10 percent solution.

The infrared cell is filled by inclining it slightly and placing about three drops of the solution in the lower hypodermic port, using a capillary dropper. The liquid can be seen rising between the salt plates through the window. In the most common sealed cell, the salt plates are spaced 0.1 mm apart. Make sure that the cell is filled past the window and that no air bubbles are present. Then place the Teflon stopper lightly but firmly in the hypodermic port. Be particularly careful not to spill any of the sample on the outside of the cell windows.

Fill the reference cell from a clean hypodermic syringe in the same manner as the sample cell and place both cells in the spectrometer, with the sample cell toward the front of the instrument. After running the spectrum, force clean solvent through the sample

Filling the cell

cell, using a syringe attached to the top port of the cell (Fig. 11.6) Finally, with the syringe pull the last bit of solvent from both cells, blow clean dry compressed air through the cells to dry them, and store them in a desiccator.

Mulls and KBr Discs

Solids insoluble in the usual solvents can be run as either mulls or KBr discs. In preparing the mull, the solid sample is ground to a particle size less than the wavelength of light going through the sample (2.5 microns), in order to avoid scattering the light. About 15 to 20 mg of the sample is ground for 3 to 10 minutes in an agate mortar until it is spread over the entire inner surface of the mortar and has a caked and glassy appearance. Then, to make a mull, 1 or 2 drops of paraffin oil (Nujol) (Fig. 11.7) is added, and the sample

Figure 11.6
Flushing the infrared sample cell. The solvent used to dissolve the sample is used in this process.

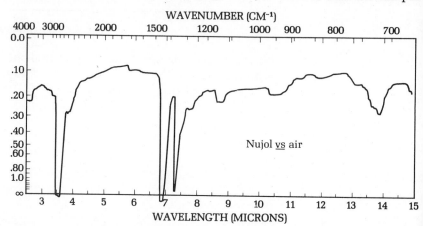

Figure 11.7
Infrared spectrum of Nujol (paraffin oil).

Sample must be finely ground

Handle salt plates with care

ground two to five more minutes. The mull is transferred to the bottom salt plate of a demountable cell (Fig. 11.3) using a rubber policeman, the top plate added and twisted to distribute the sample evenly and to eliminate all air pockets, and the spectrum run. Since the bands from Nujol obscure certain frequency regions, running another mull using Fluorolube as the mulling agent will allow the entire infrared spectral region to be covered. If the sample has not been ground sufficiently fine, in either case there will be marked loss of transmittance at the short-wavelength end of the spectrum. After running the spectrum, the salt plates are wiped clean with a cloth saturated with an appropriate solvent.

The spectrum of a solid sample can also be run by incorporating the sample in a KBr disc. This procedure has the advantage of needing only one disc to cover the entire spectral range, since KBr is completely transparent to infrared radiation. Although very little sample is required, making the disc calls for special equipment and time to prepare it. Since KBr is hygroscopic, water is a problem. The sample is first ground as for a mull and 1.5 mg of this is added to 300 mg of spectroscopic grade KBr (previously dried in an oven and stored in a desiccator). The two are gently mixed (not

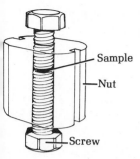

Figure 11.8
KBr disc die. Pressure is
applied by tightening
the machine screws with
a wrench.

ground) and quickly placed in a 13-mm die and subjected to 14,000–16,000 lb/sq in. pressure for 3–6 min while under vacuum in a specially constructed hydraulic press. A transparent disc is produced which is removed from the die with tweezers and placed in a special holder, prior to running the spectrum.

A simple low cost small press is illustrated by Fig. 11.8. The press consists of a large nut and two machine screws. The sample is placed between the two machine screws (which have polished faces), and the machine screws are tightened with a wrench with the nut held in a vise. The screws are then loosened and removed. The KBr disc is left in the nut, which is then mounted in the spectrometer to run the spectrum.

Running the Spectrum

Satisfactory spectra are easily obtained with the lower cost spectrometers, even those which have only a few controls and require few adjustments. To run a spectrum the paper must be positioned accurately, the pen set between 90% and 100% transmittance with the 100% control (0.0 and 0.05 absorbance), and the speed control set for a fast scan (usually one of about three minutes). The calibration of a given spectrum can be checked by backing up the drum and superimposing a spectrum of a thin polystyrene film.

Calibration with
polystyrene film

This film, mounted in a cardboard holder which has the frequencies of important peaks printed on it, will be found near most spectrometers. The film is held in the sample beam and parts of the spectrum to be calibrated are rerun. The spectrometer gain (amplification) should be checked frequently and adjusted when necessary. To check the gain, put the pen on the 90% transmittance line with the 100% control. Place your finger in the sample beam so that the pen goes down to 70% T. Then, quickly remove your finger. The pen should overshoot the 90% T line by 2%.

Throughout the remainder of this book representative infrared spectra of starting materials and products will be presented and the important bands in each spectrum identified.

EXPERIMENT   **Unknown Carbonyl Compound**

Run the infrared spectrum of an unknown carbonyl compound obtained from the laboratory instructor. Be particularly careful that all apparatus and solvents are completely free of water, which will damage the sodium chloride cell plates. Immediately after running the spectrum, position the spectrometer pen at a wavelength of about 6.2 μ without disturbing the paper, and rerun the spectrum in the region from 6.2 to 6.4 μ while holding the polystyrene calibration film in the sample beam. This will superimpose a sharp calibration peak at 6.246 μ (1601 cm^{-1}) and a less intense peak at 6.317 μ (1583 cm^{-1}) on the spectrum. Determine the frequency of the carbonyl peak and list the possible types of compounds which could correspond to this frequency (Table 11.1).

Table 11.1   Characteristic Infrared Carbonyl Stretching Frequencies (chloroform solutions)

	Wavenumber cm^{-1}	Wavelength μ
Aliphatic ketones	1725–1705	5.80–5.87
Acid chlorides	1815–1785	5.51–5.60
α,β-Unsaturated ketones	1685–1666	5.93–6.00
Aryl ketones	1700–1680	5.88–5.95
Cyclobutanones	1775	5.64
Cyclopentanones	1750–1740	5.72–5.75
Cyclohexanones	1725–1705	5.80–5.87
β-Diketones	1640–1540	6.10–6.50
Aliphatic aldehydes	1740–1720	5.75–5.82
α,β-Unsaturated aldehydes	1705–1685	5.80–5.88
Aryl aldehydes	1715–1695	5.83–5.90
Aliphatic acids	1725–1700	5.80–5.88
α,β-Unsaturated acids	1700–1680	5.88–5.95
Aryl acids	1700–1680	5.88–5.95
Aliphatic esters	1740	5.75
α,β-Unsaturated esters	1730–1715	5.78–5.83
Aryl esters	1730–1715	5.78–5.83
Formate esters	1730–1715	5.78–5.83
Vinyl and phenyl acetate	1776	5.63
δ-Lactones	1740	5.75
γ-Lactones	1770	5.65
Acyclic anhydrides (two peaks)	1840–1800 1780–1740	5.44–5.56 5.62–5.75
Primary amides	1694–1650	5.90–6.06
Secondary amides	1700–1670	5.88–6.01
Tertiary amides	1670–1630	5.99–6.14

12 Gas Chromatography

KEYWORDS   Gas, vapor phase, and gas-liquid chromatography
Injector
Carrier gas, helium
Thermal detector
Retention time
Stationary phase, silicone oil, Carbowax
Column

Separation of mixtures of volatile samples

Gas chromatography (gc), also called vapor phase chromatography (vpc) and gas-liquid chromatography (glc), is a means of separating volatile mixtures, the components of which may differ in boiling points by only a few tenths of a degree. The gc process is similar to fractional distillation, but instead of a glass column 10 in. long packed with a stainless-steel sponge the gc column used is a 10–50 foot long coiled metal tube (dia ¼ in.), packed with ground firebrick. The firebrick serves as an inert support for a very high-boiling liquid (essentially nonvolatile), such as silicone oil and low molecular weight polymers like Carbowax. These are the *liquids* of gas-*liquid* chromatography and are referred to as the stationary

High-boiling liquid phase

phase. The sample (1–25 *microliters*) is injected through a silicone rubber septum into the column, which is being swept with a current of helium (ca. 200 ml/min). The sample first dissolves in the high-boiling liquid phase and then the more volatile components of the sample evaporate from the liquid and pass into the gas phase. Helium, the carrier gas, carries these components along the column a short distance where they again dissolve in the liquid phase before reevaporation (Fig. 12.1).

The thermal conductivity detector

Eventually the carrier gas, which is a very good thermal conductor, and the sample reach the detector, an electrically heated tungsten wire. As long as pure helium is flowing over the detector the temperature of the wire is rather low and the wire has a high resistance to the flow of electric current. Organic molecules have lower heat capacities than helium. Hence, when a mixture of helium and an organic sample flows over the detector wire it is cooled less efficiently and heats up. When hot the electrical resistance of the wire becomes lower and offers less resistance to the flow of current. The detector wire actually is one leg of a Wheatstone bridge, con-

**Figure 12.1
Gas chromatography:
Diagrammatic partition
between gas and liquid
phases.**

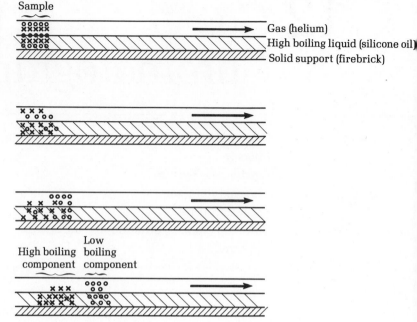

nected to a chart recorder which records, as a peak, the amount of current necessary to again balance the bridge. The record produced by the recorder is called a **chromatogram.**

A gas chromatogram is simply a recording of current *vs.* time (Fig. 12.2). In the illustration the smaller peak from component A

**Figure 12.2
Gas chromatogram.**

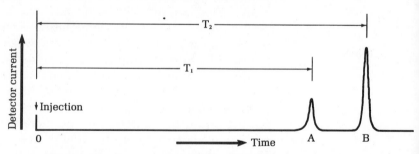

*The number and
relative amounts of
components in a
mixture*

has the shorter retention time, T_1, and so is a more volatile substance than component B (if the stationary phase is an inert liquid such as silicone oil). The areas under the two peaks are directly proportional to the amounts of A and B in the mixture. The retention time of a given component is a function of the column temperature, the helium flow rate, and the nature of the stationary phase. Hundreds of stationary phases are available; picking the correct one to carry out a given analysis is somewhat of an art, but widely-used stationary phases are silicone oil and silicone rubber, both of which can be used at temperatures up to 300° and separate mixtures on the basis of boiling point differences of the mixture's components. More specialized stationary phases will, for instance, allow alkanes to pass through readily (short retention time) while holding back (long retention time) alcohols by hydrogen bonding to the liquid phase.

A diagram of a typical gas chromatograph is shown in Fig. 12.3.

**Figure 12.3
Diagram of a gas
chromatograph.**

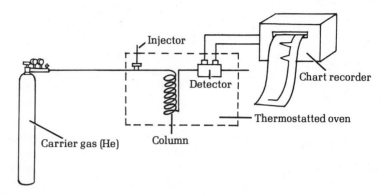

---

*The gas
chromatograph*

The carrier gas, usually helium, enters the chromatograph at ca. 60 lb/sq in. The sample (1 to 25 microliters) is injected through a rubber septum using a small hypodermic syringe (Fig. 12.4). The sample

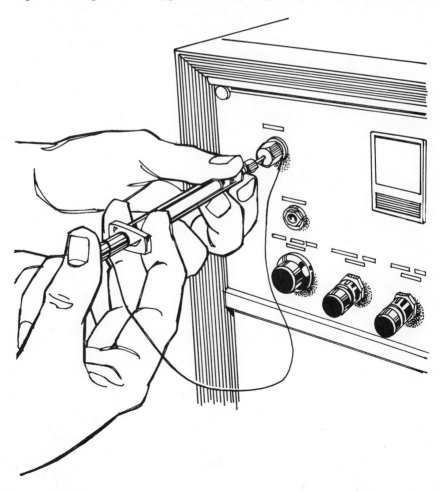

**Figure 12.4
One of the commercially
available chromato-
graphs.**

immediately passes through the column and then the detector. Injector, column, and detector are all enclosed in a thermostatted oven, which can be maintained at any temperature up to 300°. In this way samples which would not volatilize enough at room temperature can be analyzed.

Gas chromatography is a means of determining the number of components and their relative amounts in a very small sample. The small sample size is an advantage in many cases, but it precludes isolating the separated components. Some specialized chromatographs can separate samples as large as 0.5 ml per injection and automatically collect each fraction in a separate container. At the other extreme gas chromatographs equipped with flame ionization detectors can detect micrograms of sample and are used to analyze for traces of pesticides in food or traces of drugs in blood and urine. Clearly a gas chromatogram gives little information about the chemical nature of the sample being detected. However, it is sometimes possible to collect enough sample at the exit port of the chromatograph to obtain an infrared spectrum. As the peak for the compound of interest appears on the chart paper, a 2-mm dia glass tube, 3 in. long and packed with glass wool, is inserted into the rubber septum at the exit port. The sample, if it is not too volatile, will condense in the cold glass tube. Subsequently, the sample is washed out with a drop or two of solvent and an infrared spectrum obtained.

Collecting a sample for an infrared spectrum

13 Olefins from Alcohols; Analysis of a Mixture by Gas Chromatography

KEYWORDS
Dehydration

Pyrolysis

Alumina (silica gel)

Phosphoric acid

Cyclohexanol, cyclohexene

Chaser solvent

Xylene

2-Methyl-2-butanol (*t*-amyl alcohol)

2-Methyl-1-butene

2-Methyl-2-butene

Fractional distillation

Anhydrous sodium sulfate

Integration

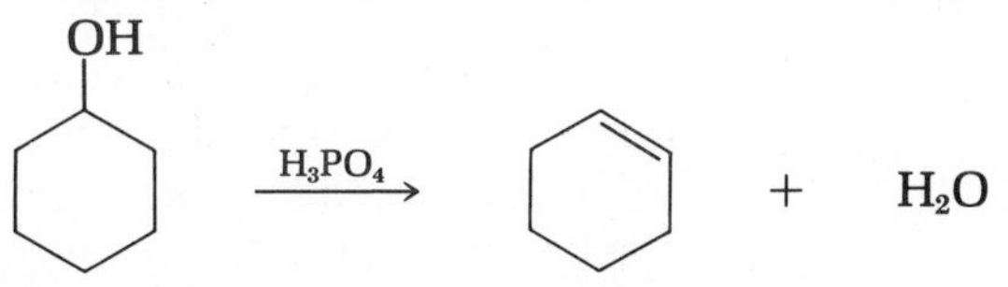

Cyclohexanol
Mp 25°, bp 161°,
den 0.96, MW 100.16

Cyclohexene
Bp 83°,
den 0.81, MW 82.14

Dehydration of cyclohexanol to cyclohexene can be accomplished by pyrolysis of the cyclic secondary alcohol with an acid catalyst at a moderate temperature or by distillation over alumina or silica gel. The procedure selected for this experiment involves catalysis by phosphoric acid; sulfuric acid is no more efficient, causes charring, and gives rise to sulfur dioxide. When a mixture of cyclohexanol and phosphoric acid is heated in a flask equipped with a fractionating column, the formation of water is soon evident. On further heating, the water and the cyclohexene formed distil together by the principle of steam distillation, and any high-boiling cyclohexanol that may volatilize is returned to the flask. However, after dehydration is complete and the bulk of the product has distilled, the column remains saturated with water–cyclohexene that merely refluxes and does not distil. Hence, for recovery of otherwise lost reaction product, a chaser solvent is added and distillation is continued. A suitable chaser solvent is water-immiscible technical xylene, bp about 140°; as it steam distils it carries over the more volatile cyclohexene. When the total water-insoluble layer is separated, dried, and redistilled through the dried column the chaser again drives the cyclohexene out of the column; the difference in boiling points is such that a sharp separation is possible. The holdup in the metal sponge-packed column is so great that if a chaser solvent is not used in the procedure the yield will be only about one-third that reported in the literature.

An alternative experiment to the dehydration of cyclohexanol is the dehydration of 2-methyl-2-butanol (*t*-amyl alcohol) with dilute sulfuric acid, to give a mixture of 2-methyl-1-butene and 2-methyl-2-butene. This mixture can be analyzed by gas chromatography.

$$
\underset{\substack{\text{2-Methyl-2-butanol} \\ \text{Bp } 102°, \text{ den } 0.805, \\ \text{MW } 88.15}}{\underset{\substack{|\\ \text{OH}}}{\overset{\overset{\displaystyle CH_3}{\overset{|}{}}}{CH_3CH_2CCH_3}}} \xrightarrow{H_2SO_4} \underset{\substack{\text{2-Methyl-2-butene} \\ \text{Bp } 38.57°, \text{ den } 0.662, \\ \text{MW } 70.14}}{\underset{H_3C}{\overset{H}{}}C=\underset{CH_3}{\overset{CH_3}{}}} \quad + \quad \underset{\substack{\text{2-Methyl-1-butene} \\ \text{Bp } 31.16°, \text{ den } 0.662, \\ \text{MW } 70.14}}{\underset{CH_3CH_2}{\overset{H_3C}{}}C=\underset{H}{\overset{H}{}}}
$$

EXPERIMENTS

1. Preparation of Cyclohexene

Introduce 20.0 g of cyclohexanol (technical grade), 5 ml of 85% phosphoric acid, and a boiling stone into a 125-ml round-bottomed boiling flask and shake to mix the layers. Use the arrangement for fractional distillation shown in Fig. 5.2 but modified by use of a bent adapter delivering into an ice-cooled test tube in a 125-ml Erlenmeyer receiver, as shown in Fig. 13.1.

Use of a chaser

Note the initial effect of heating the mixture, and then distil until the residue in the flask has a volume of 5–10 ml and very little distillate is being formed; note the temperature range. Then let the assembly cool a little, remove the thermometer briefly, and pour 20 ml of technical xylene (the chaser solvent) into the top of the column through a long-stemmed funnel. Note the amount of the upper layer in the boiling flask and distil again until the volume of the layer has been reduced by about half. Pour the contents of the test tube into a small separatory funnel and rinse with a little chaser solvent; use this solvent for rinsing in subsequent operations. Wash the mixture with an equal volume of saturated sodium chloride solution, separate the water layer, run the upper layer into a clean flask, and add 5 g of anhydrous sodium sulfate (10 × 75-mm test tube-full) to dry it. Before the final distillation note the barometric pressure, apply any thermometer corrections necessary, and

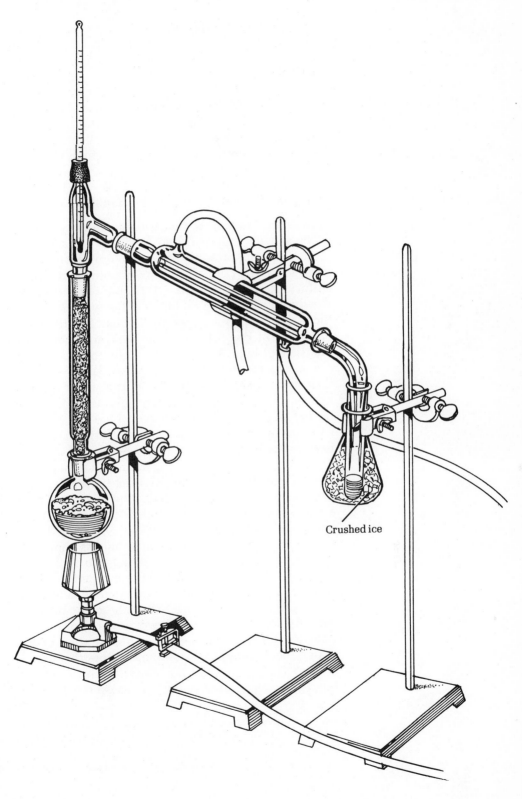

Crushed ice

Figure 13.1
Fractionation
into an ice-
cooled receiver.

determine the reading expected for a boiling point of 83°. Dry the boiling flask, column, and condenser, decant the dried liquid into the flask through a stemless funnel plugged with a bit of cotton, and fractionally distil, with all precautions against evaporation losses. The rim of condensate should rise very slowly as it approaches the top of the fractionating column in order that the thermometer may record the true boiling point soon after distillation starts. Record both the corrected boiling point of the bulk of the cyclohexene fraction and the temperature range, which should not be more than 2°. Yield, 13.2 g.

2. 2-Methyl-1-butene and 2-Methyl-2-butene

Look up the boiling points[1] and calculate the molecular weights[2] of the starting materials and products.

Pour 36 ml of water into a 250-ml round-bottomed flask, cool in an ice-water bath while **slowly** pouring in 18 ml of concentrated sulfuric acid. Cool this "1:2" acid further with swirling while slowly pouring in 36 ml (30 g) of t-amyl alcohol. Shake the mixture thoroughly and then mount the flask for distillation over a micro-burner as in Fig. 5.1, but with the arrangement for ice cooling of the distillate of Fig. 13.1, since the olefin to be produced is very volatile. Use a long condenser and a rapid stream of cooling water. Heat the flask with a small flame of a microburner until distillation of the hydrocarbon is complete. Transfer the distillate to a separatory funnel and shake with about 10 ml of 10% sodium hydroxide solution to remove any traces of sulfurous acid. The aqueous solution sinks to the bottom and is drawn off and discarded. Dry the hydrocarbon layer by filtering it through a cone of anhydrous sodium sulfate and distil the dried product through a fractionating column from a clean, dry flask, taking the same precautions as before to avoid evaporation losses. Collect in a tared (previously weighed) bottle the portion boiling at 37–43°. The yield reported in the literature is 84%; the average student yield is about 50%.

Inject a few microliters of product into a gas chromatograph maintained at room temperature and equipped with a ¼-in. dia × 10-ft column packed with 10% SE-30 silicone rubber on Chromosorb-W or a similar inert packing. Mark the chart paper at the time of injection. In a few minutes two peaks should appear. From your knowledge of the mechanism of dehydration of secondary alcohols which olefin should predominate? Does this agree with the boiling points? (In general, the compound with the shorter retention time has the lower boiling point.) Measure the relative areas under the

---

[1] First try the *Handbook of Chemistry and Physics*, Chemical Rubber Publishing Co., Cleveland, Ohio, or Lange's *Handbook of Chemistry*, Handbook Publishers, Inc., Sandusky, Ohio. If you cannot find the compounds in either of these, try Heilbron's *Dictionary of Organic Compounds*, Oxford University Press, New York, 4th edition, 1965–1973, or Rodd's *Chemistry of Carbon Compounds*, 2nd edition, edited by S. Coffey, Elsevier, Amsterdam, 1964–1970.
[2] See the back end-papers of this book for a convenient table to facilitate calculation of molecular weights.

two peaks. One way to perform this integration is to cut out the peaks with scissors and weigh the two pieces of paper separately on an analytical balance. Although time-consuming, this method gives very precise results. If the peaks are symmetrical, their areas can be approximated by simply multiplying the height of the peak by its width at half-height.

14 Alkanes and Alkenes

KEYWORDS Ligroin (saturated alkane) Pyridinium hydrobromide perbromide
Cyclohexene N-Bromosuccinimide
Bromine water Bromohydrin
Bromine in CCL$_4$ Pinene
Acid permanganate Rubber

The following tests demonstrate properties characteristic of saturated and unsaturated hydrocarbons, provide means of distinguishing between compounds of the two types, and distinguish between pure and impure alkanes. Use your own preparation of cyclohexene as a typical alkene, purified 66–75° ligroin (Eastman Organic Chemicals No. 513) as a typical alkane mixture (the bp of hexane is 69°), and unpurified ligroin (Eastman No. P513) as an impure alkane. Write equations for all tests that are positive.

EXPERIMENTS ## 1. Bromine Water

Measure 3 ml of a 3% aqueous solution of bromine into each of three 13 × 100-mm test tubes, add 1-ml portions of purified ligroin to two of the tubes and 1 ml of cyclohexene to the third. Shake each tube and record the initial results. Put one of the ligroin-containing tubes in the desk out of the light and expose the other to bright sunlight or hold it close to a light bulb. When a change is noted compare the appearance with that of the mixture kept in the dark.

2. Bromine in Nonaqueous Solution

Treat 1-ml samples of purified ligroin, unpurified ligroin, and cyclohexene with 5–6 drops of a 3% solution of bromine in carbon tetrachloride. In case decolorization occurs, breathe across the mouth of the tube to see if hydrogen bromide can be detected. If the bromine color persists illuminate the solution and, if a reaction occurs, test as before for hydrogen bromide.

3. Acid Permanganate Test

To 1-ml portions of purified ligroin, unpurified ligroin, and cyclohexene add a drop of an aqueous solution containing 1%

potassium permanganate and 10% sulfuric acid and shake. If the initial portion of reagent is decolorized, add further portions.

4. Sulfuric Acid

Cool 1-ml portions of purified ligroin and cyclohexene in ice, treat each with 3 ml of concd sulfuric acid, and shake. Observe and interpret the results. Is any reaction apparent? Any warming? If the mixture separates into two layers, identify them.

5. Bromination with Pyridinium Hydrobromide Perbromide $(C_5H_5\overset{+}{N}HBr_3{}^-)$[1]

**Pyridinium hydrobromide
perbromide,
MW 319.84**

This substance is a crystalline, nonvolatile, odorless complex of high molecular weight (319.86), which, in the presence of a bromine acceptor such as an alkene, dissociates to liberate one mole of bromine. For small-scale experiments it is much more convenient and agreeable to measure and use than free bromine.

Weigh one millimole (320 mg) of the reagent as accurately as you can on a triple beam balance, put it into a 10-ml Erlenmeyer flask and add 2 ml of acetic acid. Swirl the mixture and note that the solid is sparingly soluble. Determine the mg per micro drop of cyclohexene delivered by a capillary dropping tube and add one millimole of the alkene to the suspension of reagent. Swirl, crush any remaining crystals with a flattened stirring rod, and if after a time the amount of cyclohexene appears insufficient to exhaust the reagent, add a little more. When the solid is all dissolved, dilute with water and note the character of the product. By what property can you be sure that it is a reaction product and not starting material?

6. Formation of a Bromohydrin

N-Bromosuccinimide in an aqueous solution is in equilibrium with hypobromous acid and may be used to effect addition of this reagent:

$$
\begin{array}{ccc}
\text{CH}_2\text{CO} & & \text{CH}_2\text{CO} \\
\diagdown & & \diagdown \\
\quad\quad \text{NBr} + \text{HOH} \rightleftharpoons & \quad\quad \text{NH} + \text{HOBr} \\
\diagup & & \diagup \\
\text{CH}_2\text{CO} & & \text{CH}_2\text{CO}
\end{array}
$$

N-Bromosuccinimide   **Succinimide**
MW 178.00

[1] Crystalline material suitable for small-scale experiments is supplied by Arapahoe Chemicals, Inc. Massive crystals commercially available should be recrystallized from acetic acid (4 ml per g). Preparation: Mix 15 ml of pyridine with 30 ml of 48% hydrobromic acid and cool; add 25 g of bromine gradually with swirling, cool, collect the product with use of acetic acid for rinsing and washing. Without drying the solid, crystallize it from 100 ml. of acetic acid. Yield of orange needles, 33 g (69%).

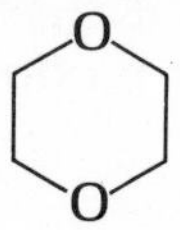

Dioxane
Bp 101.5°, den 1.04
(Dissolves organic
compounds, miscible
with water.)

Weigh 178 mg of N-bromosuccinimide approximately (triple beam balance), put it into a 13 × 100-mm test tube and add 0.5 ml of dioxane and 1 millimole of cyclohexene. In another tube chill 0.2 ml of water and add to it 1 millimole of concd sulfuric acid. Transfer the cold dilute solution to the first tube with the capillary dropper. Note the result and the nature of the product that separates on dilution with water.

7. Tests for Unsaturation

Determine which of the following hydrocarbons are saturated and which are unsaturated or contain unsaturated material. Use any of the above tests that seem appropriate.

Pinene, the principal constituent of turpentine oil
Paraffin oil, a purified petroleum product
Gasoline produced by cracking
Cyclohexane
Rubber (The adhesive Grippit and other rubber cements are solutions of unvulcanized rubber in benzene. Squeeze a drop of it onto a stirring rod and dissolve it in more benzene. For tests with permanganate or with bromine in carbon tetrachloride use only a drop of the former and just enough of the latter to produce coloration.)

15 *Ultraviolet Spectroscopy*

KEYWORDS

Electronic transitions
π-electrons
Nanometers, nm
Molecular orbitals, bonding
and antibonding
Band spectrum
Energy levels
Quantitative analysis
Molar extinction coefficient, ϵ
Beer-Lambert law
Absorbance

Wavelength of maximum
absorption, λ_{max}
Spectro grade solvents
Woodward rules—dienes and
polyenes
Fieser rules—enones and
dienones
Exocyclic double bond
Homoannular diene
Solvent correction

Ultraviolet spectroscopy gives information about electronic transitions within molecules. Whereas absorption of low energy infrared radiation causes bonds in a molecule to stretch and bend, the absorption of short wavelength, high energy ultraviolet radiation causes electrons to move from one energy level to another with energies that are often capable of breaking chemical bonds.

We shall be most concerned with transitions of π-electrons in conjugated and aromatic ring systems. These transitions occur in the wavelength region 200 to 800 nm (nanometers, 10^{-9} meters, formerly known as $m\mu$, millimicrons). Most common ultraviolet spectrometers cover the region 200 to 400 nm as well as the visible spectral region 400 to 800 nm. Below 200 nm air (oxygen) absorbs UV radiation; spectra in that region must therefore be obtained in a vacuum or in an atmosphere of pure nitrogen.

Consider ethylene, even though it absorbs UV radiation in the normally inaccessible region at 163 nm. The double bond in ethylene has two s electrons in a σ-molecular orbital and two, less tightly held, p electrons in a π-molecular orbital. Two unoccupied, high energy level, antibonding orbitals are associated with these orbitals. When ethylene absorbs UV radiation, one electron moves up from the bonding π-molecular orbital to the antibonding π^*-molecular orbital (Fig. 15.1). As the diagram indicates, this change requires less energy than the excitation of an electron from the σ to the σ^* orbital:

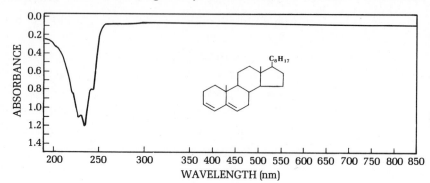

Figure 15.1
Electronic energy levels
of ethylene.

By comparison with infrared spectra and nmr spectra, UV spectra are fairly featureless (Fig. 15.2). This condition results as molecules

Figure 15.2
The ultraviolet spectrum
of cholesta-3,5-diene
in ethanol.

in a number of different vibrational states undergo the same electronic transition, to produce a band spectrum instead of a line spectrum.

Unlike IR spectroscopy, ultraviolet spectroscopy lends itself to precise quantitative analysis of substances. The intensity of an absorption band is usually given by the molar extinction coefficient, ϵ, which according to the Beer-Lambert Law is equal to the absorbance, A, divided by the product of the molar concentration, c, and the path length, l, in centimeters.

$$\epsilon = \frac{A}{cl}$$

The wavelength of maximum absorption (the tip of the peak) is given by λ_{max}. Since UV spectra are so featureless it is common practice to describe a spectrum like that of cholesta-3,5-diene (Fig. 15.2) as λ_{max} 234 nm ($\epsilon = 20{,}000$), and not bother to reproduce the actual spectrum.

The extinction coefficients of conjugated dienes and enones are in the range 10,000–20,000, so only very dilute solutions are needed for spectra. In the example of Fig. 15.2, the absorbance at the tip of the peak, A, is 1.2, and the path length is the usual 1 cm; so the molar concentration needed for this spectrum is 6×10^{-5} mole per liter,

$$c = \frac{A}{l\epsilon} = \frac{1.2}{20{,}000} = 6 \times 10^{-5} \text{ mole per liter}$$

which is 0.221 mg per 10 ml of solvent.

The usual solvents used for running UV spectra are 95% ethanol, methanol, water, and saturated hydrocarbons such as hexane, trimethylpentane, and isooctane; the three hydrocarbons are often specially purified to remove impurities that absorb in the UV region. Any transparent solvent can be used when running spectra in the visible region.

Spectro grade solvents

Sample cells for spectra run in the visible region are made of glass, but UV cells must be of the more expensive fused quartz, since glass absorbs UV radiation. The cells and solvents must be clean and pure, since it takes very little of a substance to produce a UV spectrum. A single fingerprint will give a spectrum!

Ethylene has λ_{max} 163 nm ($\epsilon = 15,000$) and butadiene has λ_{max} 217 nm ($\epsilon = 20,900$). As the conjugated system is extended, the wavelength of maximum absorption moves to longer wavelengths (toward the visible region): for example, lycopene with 11 conjugated double bonds in a row has λ_{max} 470 nm ($\epsilon = 185,000$), Fig. 15.3. Since lycopene absorbs blue visible light at 470 nm the

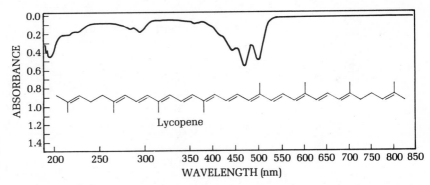

Figure 15.3
The ultraviolet-visible spectrum of lycopene in isooctane.

substance appears bright red. It is responsible for the color of tomatoes; its isolation is described in Chapter 58.

The wavelengths of maximum absorption of conjugated dienes and polyenes and conjugated enones and dienones are given by the Woodward and Fieser Rules, Tables 15.1 and 15.2.

Table 15.1   Rules for the Prediction of λ_{max} for Conjugated Dienes and Polyenes

	Increment (nm)
Parent heteroannular diene	214
Parent homoannular diene	253
Double bond extending the conjugation	30
Alkyl substituent or ring residue	5
Exocyclic location of double bond to any ring	5
Groups: OAc, OR	0
Solvent correction	0
λ_{max} = Total	

Table 15.2   Rules for the Prediction of λ_{max} for Conjugated Enones and Dienones

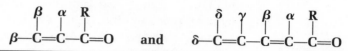

	Increment (nm)
Parent α,β-unsaturated system	215
Double bond extending the conjugation	30
R (alkyl or ring residue), OR, OCOCH$_3$ α	10
β	12
γ, and higher	18
α-Hydroxyl, enolic	35
α-Cl	15
α-Br	23
Exo location of double bond to any ring	5
Homoannular diene component	39
Solvent correction, see Table 15.3	...
λ_{max}^{EtOH} = Total	

Table 15.3   Solvent Correction

Solvent	Factor for Correction to ethanol
Hexane	+11
Ether	+7
Dioxane	+5
Chloroform	+1
Methanol	0
Ethanol	0
Water	−8

The application of the rules in the above tables is demonstrated by the spectra of pulegone (**1**) and carvone (**2**), Fig. 15.4, with the calculations given in Tables 15.4 and 15.5.

Table 15.4   Calculation of λ_{max} for Pulegone

Parent α,β-unsaturated system	215 nm
α-Ring residue, R	10
β-Alkyl group (two methyls)	24
Exocyclic double bond	5
Solvent correction (hexane)	−11

Calcd λ_{max} 243 nm; found 244 nm

Table 15.5   Calculation of λ_{max} for Carvone

Parent α,β-unsaturated system	215 nm
α-Alkyl group (methyl)	10
β-Ring residue	12
Solvent correction (hexane)	−11

Calcd λ_{max} 226 nm; found 229 nm

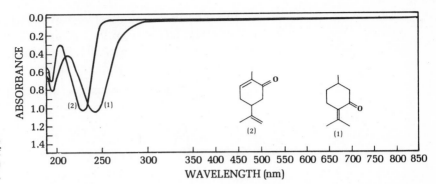

Figure 15.4
Ultraviolet spectra of
(1) pulegone and (2)
carvone in hexane.

These rules will be applied in the next experiment where cholesterol is converted into an α,β-unsaturated ketone.

No simple rules exist for calculation of aromatic ring spectra, but several generalizations can be made. From Fig. 15.5 it is obvious that as polynuclear aromatic rings are extended linearly, λ_{max} shifts to longer wavelengths.

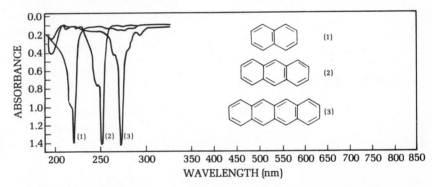

Figure 15.5
The ultraviolet spectra
of (1) naphthalene, (2)
anthracene, and (3)
tetracene.

As alkyl groups are added to benzene, λ_{max} shifts from 255 nm for benzene to 261 nm for toluene to 272 nm for hexamethylbenzene. Substituents bearing nonbonding electrons also cause shifts of λ_{max} to longer wavelengths, e.g., from 255 nm for benzene to 257 nm for chlorobenzene, 270 nm for phenol, and 280 nm for aniline ($\epsilon =$ 6,200–8,600). That these effects are the result of interaction of the π-electron system with the nonbonded electrons is seen dramatically in the spectra of vanillin and the derived anion (Fig. 15.6). Addition

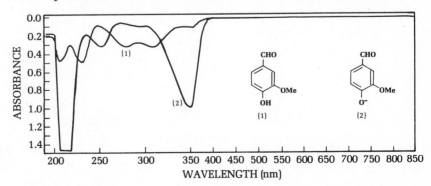

Figure 15.6
Ultraviolet spectrum of
(1) neutral vanillin and
(2) the anion of vanillin.

of two more nonbonding electrons in the anion causes λ_{max} to shift from 279 nm to 351 nm and ϵ to increase in value. Removing the electrons from the nitrogen of aniline by making the anilinium cation causes λ_{max} to decrease from 280 nm to 254 nm (Fig. 15.7).

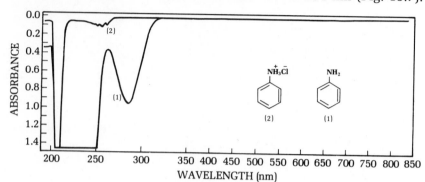

Figure 15.7
Ultraviolet spectrum of
(1) aniline and (2) aniline
hydrochloride.

These changes of λ_{max} as a function of pH have obvious analytical applications.

Intense bands result from π-π conjugation of double bonds and carbonyl groups with the aromatic ring. Styrene, for example, has λ_{max} 244 nm ($\epsilon = 12,000$) and benzaldehyde λ_{max} 244 nm ($\epsilon = 15,000$).

EXPERIMENT   **Unknown**

Determine whether an unknown compound obtained from the instructor is acidic, basic, or neutral by running the ultraviolet spectra in the presence of acid and base as well as in neutral media.

16 Cholesterol

KEYWORDS   Gallstones
                      Bilirubin
                      Zinc dust debromination
                      Jones, Collins reagents
                      4- and 5-Cholesten-3-one
                      Dioxane
                      Bromine

Cholesterol dibromide
Isomerization
Liebermann-Burchard test
Acetic anhydride + sulfuric acid
Selenium dioxide test
Tetranitromethane test, unsaturation

Human gallstones are composed primarily of the crystalline unsaturated alcohol, cholesterol, along with the bile pigment bilirubin, a metabolite of hemoglobin. Cholesterol is also the substance which thickens arterial walls leading to decreased blood flow in the disease atherosclerosis. The average adult human male contains about 200 g of cholesterol located principally in the brain and nerve tissue; the commercial substance is isolated from the spinal cords of cattle.

In this experiment cholesterol is isolated from human gallstones and freed from various steroidal impurities by conversion to the dibromide, followed by zinc dust debromination. The pure cholesterol, a secondary alcohol, can be oxidized to the corresponding ketone by very versatile procedures: oxidation with Jones reagent or Collins reagent. This ketone, 5-cholesten-3-one, is quite labile (unstable) and can be isomerized, via the enol, to the conjugated ketone 4-cholesten-3-one. These conversions can be followed spectroscopically. (See the following page.)

Cholesterol isolated from natural sources contains small amounts (0.1–3%) of 3β-cholestanol, 7-cholesten-3β-ol, and 5,7-cholestadien-3β-ol.[1] These are so very similar to cholesterol in solubility that their removal by crystallization is not feasible. However, complete purification can be accomplished through the sparingly soluble dibromo derivative 5α,6β-dibromocholestan-3β-ol. 3β-Cholestanol being saturated does not react with bromine and remains in the mother liquor. 7-Cholesten-3β-ol and 5,7-cholestadien-3β-ol are dehydrogenated by bromine to dienes and trienes that likewise remain in the mother liquor and are eliminated along with colored by-products.

The cholesterol dibromide that crystallizes from the reaction solution is collected, washed free of the impurities or their dehydrogena-

---

[1] A fourth companion, cerebrosterol, or 24-hydroxycholesterol, is easily eliminated by crystallization from alcohol.

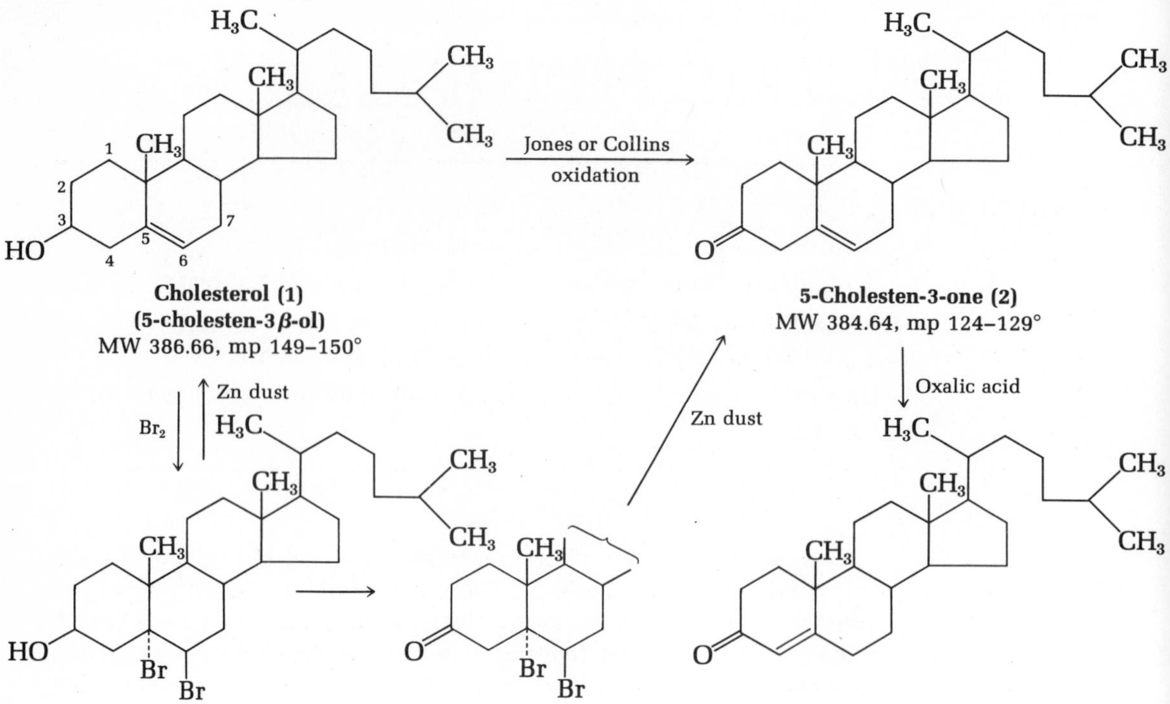

Cholesterol (1)
(5-cholesten-3β-ol)
MW 386.66, mp 149–150°

Jones or Collins oxidation

5-Cholesten-3-one (2)
MW 384.64, mp 124–129°

Br₂ / Zn dust

Zn dust

Oxalic acid

5α,6β-Dibromocholestan-3β-ol (3)

4-Cholesten-3-one (4)
Mp 81–82°, λ_{max}^{EtOH}

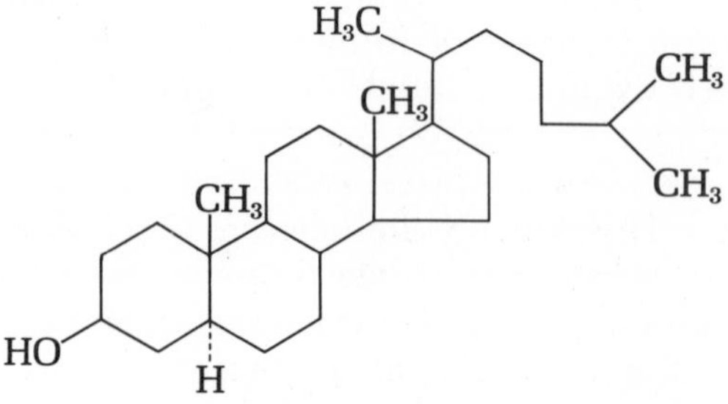

3β-Cholestanol

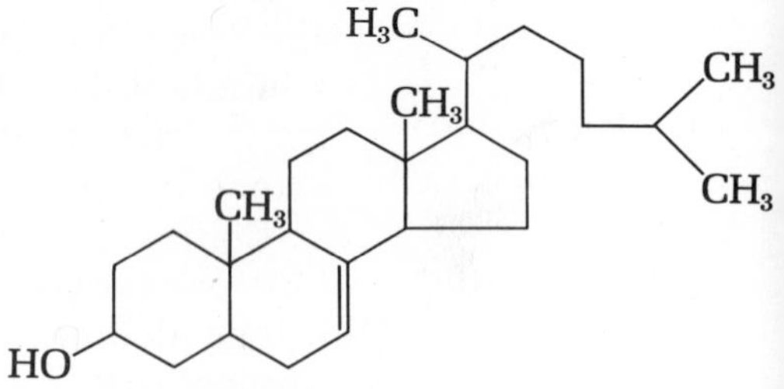

7-Cholesten-3β-ol

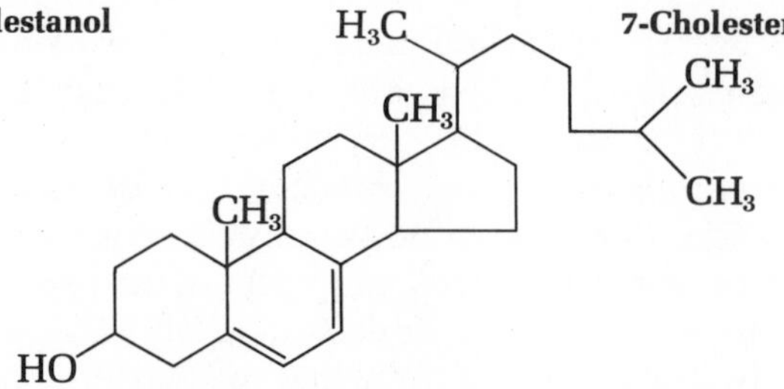

5,7-Cholestadien-3β-ol

tion products, and debrominated with zinc dust, with regeneration of cholesterol in pure form. Specific color tests can differentiate between pure cholesterol and tissue cholesterol purified by ordinary methods.

EXPERIMENTS[2]   1. Cholesterol from Gallstones[3]

Swirl 2 g of crushed gallstones in a 25-ml Erlenmeyer flask with 10 ml of dioxane on a hot plate for a few minutes until the solid has disintegrated and the cholesterol has dissolved. Filter the solution while hot. Its dirty-yellow appearance is from a brown residue of the bile pigment bilirubin, a metabolite of hemoglobin. Dilute the

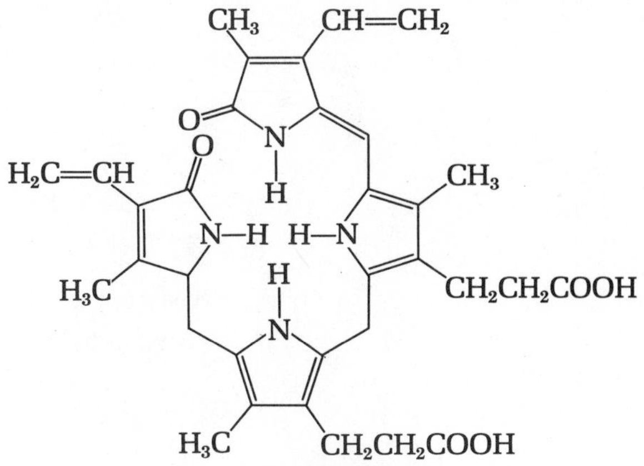

Bilirubin,
$\lambda_{max}^{CHCl_3}$ 450 nm

filtrate with 10 ml of methanol, clarify by warming the solution with a little decolorizing charcoal for a few minutes; filter, using a funnel that has been warmed on the steam bath, reheat the pale greenish-yellow filtrate to the boiling point, add a little water gradually until the solution is saturated at the boiling point, and let the solution stand for crystallization. Collect the crystals, wash, dry, and take the melting point. Typical result: 1.5 g of large colorless plates, mp 146–147.5°. Use 1 g for purification and save the rest for color tests.

In a 25-ml Erlenmeyer flask dissolve 1 g of gallstone cholesterol or of commercial cholesterol (content of 7-cholesten-3β-ol about 0.6%) in 7 ml of ether by gentle warming and, with a pipette fitted with a pipetter (Fig. 16.1), or with a plastic syringe (Fig. 16.2), or from a burette in the hood, add 5 ml of a solution of bromine and sodium acetate in acetic acid.[4] Cholesterol dibromide begins to crystallize in a minute or two. Cool in an ice bath and stir the crystalline paste with a stirring rod for about 10 minutes to ensure complete crystallization, and at the same time cool a mixture of 3 ml of ether and 7 ml of acetic acid in ice. Then collect the crystals on a small suction

[2] The isolation can be demonstrated with Part II of the film *Techniques of Organic Chemistry,* by L. F. Fieser (McGraw-Hill Book Co.), and the color tests done by the students with samples supplied.
[3] Obtainable from the department of surgery of a hospital. Wrap the stones in a towel and crush by light pounding with a hammer. (See also *Instructor's Manual for Organic Experiments,* 3rd. ed.)
[4] Weigh a 125-ml Erlenmeyer flask on a balance placed in the hood, add 4.5 g of bromine by a capillary dropping tube (avoid breathing the vapor), and add 50 ml of acetic acid and 0.4 g of sodium acetate (anhydrous).

Notes for the instructor

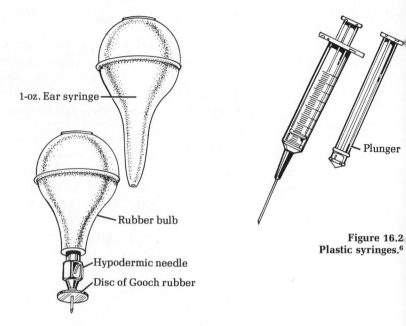

1-oz. Ear syringe

Rubber bulb

Hypodermic needle

Disc of Gooch rubber

Figure 16.1
Pipetters.[5]

Plunger

Figure 16.2
Plastic syringes.[6]

funnel and wash with the iced ether-acetic acid solution to remove the yellow mother liquor. Finally wash with a little methanol, continuing to apply suction, and transfer the white solid without drying it (dry weight 1.2 g) to a 50-ml Erlenmeyer flask. Add 20 ml of ether, 5 ml of acetic acid, and 0.2 g of zinc dust[7] and swirl. In about 3 min the dibromide dissolves; after 5–10 min swirling, zinc acetate usually separates to form a white paste (the dilution sometimes is such that no separation occurs). Stir for 5 min more and then add water by drops (about 0.5 ml) until any solid present (zinc acetate) dissolves to make a clear solution. Decant the solution from the zinc into a separatory funnel, wash the ethereal solution twice with water and then with 10% sodium hydroxide (to remove traces of acetic acid). Then shake the solution with an equal volume of saturated sodium chloride solution to reduce the water content, filter it by gravity through a filter paper cone containing anhydrous sodium sulfate, add 10 ml of methanol (and a boiling stone), and evaporate the solution on the steam bath to the point where most of the ether is removed and the purified cholesterol begins to crystallize. Remove the solution, let crystallization proceed at room temperature and then in an ice bath, and collect the crystals and wash with methanol; yield 0.6–0.7 g, mp 149–150°. Save about 0.1 g of the pure cholesterol for tests.

Aspirator tube

Notes for the instructor

[5] The pipetter at the top is a 1-oz ear syringe (#526, Davol, Inc., Providence, R.I.) available at most drug stores. The pipetter below is made from a rubber bulb, a hypodermic needle, and a disc of Gooch rubber. A commercial version of the latter pipetter is available from Wilkens-Anderson Co.
[6] Becton, Dickinson Co. (Rutherford, N. J.) disposable 2.5 ml plastic syringe "Plastipak."
[7] If the reaction is slow add more zinc dust. The amount specified is adequate if material is taken from a freshly opened bottle, but zinc dust deteriorates on exposure to air.

2. Oxidation of Cholesterol with Jones Reagent

Oxidation of the secondary hydroxyl group of cholesterol to form a ketone (e.g., Oppenauer oxidation) leads to the α,β-unsaturated ketone 4-cholesten-3-one (**4**); the intermediate 5-cholesten-3-one (**2**) is labile and easily isomerized under the reaction conditions to the thermodynamically more stable α,β-unsaturated isomer. One way to avoid this isomerization is to brominate the double bond, oxidize the alcohol, and then debrominate with zinc dust.[8] In this experiment cholesterol is oxidized with either Jones reagent or Collins reagent. The initially formed 5-isomer does not isomerize under the reaction conditions, even though Jones reagent contains sulfuric acid and Collins reagent contains pyridine, both of which will catalyze the isomerization. In a separate experiment the 5-compound is isomerized to the 4-isomer using oxalic acid as the catalyst.

Jones Reagent:
CrO_3—H_2SO_4—H_2O,
acetone solvent

Procedure

5-Cholesten-3-one. Dissolve 0.5 g of purified cholesterol in 65 ml of acetone by warming on the steam bath. Cool to 15° and add Jones reagent[9] dropwise with swirling until the orange color persists (about 0.56 ml of the reagent is required). After 2 to 5 min the reaction mixture is poured into 300 ml of 5% sodium bicarbonate solution. The aqueous solution is extracted three times with 30-ml portions of ether, the combined ether extracts are washed with an equal volume of water, and then shaken with 50 ml of saturated sodium chloride solution. The ether solution is dried for a few minutes over anhydrous sodium sulfate and evaporated to dryness. The residual oil is often difficult to crystallize; reserve about 60 mg for spectroscopic examination. Dissolve the oil in 5 ml of hot methanol, cool slowly to room temperature and then to 0–4°. The large colorless prisms are collected by suction filtration; yield 0.3 g, mp 124–129° (camphorlike odor).

3. Oxidation of Cholesterol with Collins Reagent

Collins reagent, a chromium trioxide-pyridine complex, is another very versatile oxidizing agent, which usually gives high yields of pure products. It can be used to oxidize primary allylic and benzylic alcohols to aldehydes in 50–80% yield, and it is one of the best reagents available for the oxidation of secondary alcohols to ketones. In this experiment it is employed for the oxidation of cholesterol to the corresponding ketone without isomerization of the double bond.

The reagent is prepared *in situ*[10] in methylene chloride, in which

[8] L. F. Fieser, *Organic Syntheses*, Collect. Vol. 4, John Wiley & Sons, Inc., New York, N.Y., 1963, p. 195.
[9] Jones reagent is prepared by dissolving 26.72 g of chromium trioxide in 23 ml of concd sulfuric acid and diluting carefully with water to a total volume of 100 ml.
[10] J. C. Collins, W. W. Hess, and F. J. Frank, *Tetrahedron Lett.*, 3363 (1968); and R. Ratcliffe and R. Rhodehorst, *J. Org. Chem.*, **35** 4000 (1970). See also Louis F. Fieser and Mary Fieser, *Reagents for Organic Synthesis*, John Wiley and Sons, Inc., New York, N.Y., 1967, pp. 142–147.

it dissolves, to give a dark red solution. A one-sixth molar equivalent of the alcohol to be oxidized (in methylene chloride) is added at or near room temperature. After a few minutes a dark sticky precipitate of chromium salts separates; the product is isolated from the methylene chloride solution by the usual process of extracting, washing, drying, and evaporating the solvent.

Procedure

5-Cholesten-3-one. Add 1.2 g (12 mmoles) of anhydrous chromium trioxide with swirling to a solution of 1.9 g (24 mmoles) of pyridine in 30 ml of dry methylene chloride in a 125-ml Erlenmeyer flask. Continue swirling until all the chromium trioxide has dissolved (10 min), then cool the dark red solution to 10° and add 0.77g (2 mmoles) of cholesterol dissolved in 10 ml of methylene chloride. Maintain the temperature of the mixture at 10° for 20 min, then decant the methylene chloride solution from the tarry black residue into a separatory funnel. Rinse the reaction flask with 20 ml of ether and add these washings to the separatory funnel. Wash the ether-methylene chloride solution with three 20-ml portions of 5% aqueous sodium hydroxide solution, two 20-ml portions of 3% hydrochloric acid, one 20-ml portion of 5% sodium bicarbonate solution, and 20 ml of saturated sodium chloride solution. Dry the organic layer over anhydrous sodium sulfate and evaporate the solvent (preferably under reduced pressure). Dissolve the residual oil in 8 ml of hot methanol and cool the solution slowly to room temperature and finally to 0–4°. The yield of colorless prisms, which are collected by suction filtration, is 0.5 g, mp 124–129°.

4. Isomerization of 5-Cholesten-3-one

The ketone 5-cholesten-3-one, resulting from the oxidation of cholesterol, is easily isomerized to the conjugated ketone, 4-cholesten-3-one, with oxalic acid.

Procedure

4-Cholesten-3-one. A mixture of 0.25 g of 5-cholesten-3-one (either crystalline material or the oil from the preceding experiment), 25 mg of anhydrous oxalic acid, and 2 ml of 95% ethanol is heated on the steam bath until all the solid has dissolved and for 10 min longer, and then allowed to stand at room temperature. If crystallization does not start after 30 min the solution is seeded or scratched. After crystallization has proceeded at room temperature and then at 0–4°, the large colorless prismatic needles that separate are collected by suction filtration; yield about 0.2 g, mp 81–82°. This compound is also difficult to crystallize; noncrystalline material can be used for spectroscopic examination.

From an examination of the IR (Chapter 11), UV (Chapter 15) and nmr spectra (Chapter 10) of your cholesterol and the two isomeric ketones, assess your success in carrying out these transformations.

5. Liebermann-Burchard Color Test

Cholesterol reacts with acetic anhydride and sulfuric acid to give a transient purple color changing to blue and then green. This color reaction is the basis for the most common clinical determination of cholesterol in blood; serum is extracted with ether, the extract mixed with acetic anhydride and sulfuric acid, and the absorbance of the mixture at 660 nm read on a visible spectrometer after a predetermined time. The results are compared with a set of standards; normal human blood contains 160–200 mg of cholesterol per 100 ml. Of this about 25% is free, the remainder esterified.

Procedure

Introduce a very small quantity of your purified cholesterol, commercial cholesterol, and unpurified gallstone cholesterol into separate melting point capillaries (to give a layer no more than 0.5 mm thick) and add chloroform (from a micro pipette made by drawing out another melting point capillary) to give a column 10 mm high. Shake the chloroform down into the capillaries the same way you would "whip" down the mercury column in a clinical thermometer. With a second micro pipette introduce a 3-mm column of the test reagent into the melting point tubes without allowing it to mix with the cholesterol solution. The test reagent is made by chilling 1 ml of acetic anhydride in an ice bath and adding one drop of concd sulfuric acid (the solution is stable for 3–4 hr if kept on ice). Grasp all three tubes at the open end and mix the two solutions by a quick whip. Note the time to the nearest second and place the tubes on a white background.

Gallstone cholesterol contains 2–3% of 7-cholesten-3β-ol, which reacts very rapidly with the test reagent to give an initial blue color changing to green; pure cholesterol reacts more slowly and the initial transient color is purplish.

The test is still more sensitive when conducted at 0° because at this temperature 7-cholesten-3β-ol reacts to the point of maximum color density in 5 min, whereas cholesterol is negative indefinitely. See if, by this modified test, you can differentiate between pure cholesterol and commercial cholesterol, which contains about 0.6% 7-cholesten-3β-ol. The only change necessary is to cool the chloroform solution in an ice bath and keep the tube in ice while introducing the acid reagent into the open end; whip the tube and return it to the ice bath.

A 3-mm column of cholesterol in a melting point capillary weighs approximately 1 mg. Estimate the weight of 7-cholesten-3β-ol that you are just able to detect.

6. Selenium Dioxide Test

Selenium dioxide in acetic acid oxidizes 7-cholesten-3β-ol at 25° (or even at 0°) to an acetoxy derivative with liberation of selenium detectable first as a yellow colloidal solution and then as a red precipitate. Cholesterol is oxidized only at 50–60°. Cholesterol containing as little as 3% of 7-cholesten-3β-ol gives a faint yellow color in about 5 min and red selenium in 15–20 min.

Introduce a 3-mm column (tamped down; about 1 mg) of gall-stone cholesterol in one capillary tube and the same amount of pure cholesterol in another. Add enough toluene to each tube to extend the column to 6 mm, and heat on the steam bath until all the solid is dissolved. Now introduce 0.1 M selenium dioxide solution[11] to give a total column of 10 mm, place the tubes on a piece of white paper, and note the time and the ultimate result.

7. Tetranitromethane Test

Whereas the two tests above are specific to sterols, the test with tetranitromethane [$C(NO_2)_4$, bp 126°] is a generally applicable method of detecting unsaturation. Compounds containing one or more ethylenic double bonds, providing these are not conjugated with carbonyl groups, give a yellow color with the reagent as the result of π-complex formation. The microtechnique is particularly appropriate because the reagent is costly and must be used sparingly. It is conveniently mixed with an equal volume of chloroform and kept in a specimen vial with a capillary pipette sealed (liquid solder) into a hole drilled through the Bakelite screw cap.

To a 0.5-mm layer of cholesterol in a melting point capillary, introduce a 5-mm column of tetranitromethane solution and view the tube against white paper. Test your samples of cyclohexene and citral for unsaturation, and for comparison, test purified ligroin.

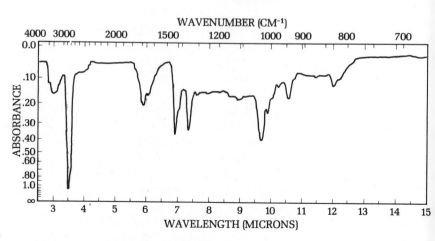

Figure 16.3 Infrared spectrum of cholesterol.

Note for the instructor   [11] Dissolve 1.29 g of selenious acid or 1.11 g of selenium dioxide in 2 ml of water by heating and dilute with acetic acid to a volume of 100 ml.

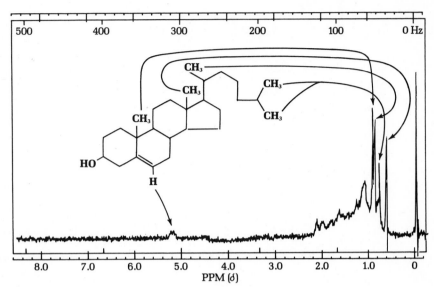

Figure 16.4
Nmr spectrum of
cholesterol.

17 n-Butyl Bromide

KEYWORDS   Preparative experiment
Stoichiometry
Laboratory notebook
Hydrobromic acid
n-Butyl alcohol

1-Butene
Dibutyl ether
Anhydrous calcium chloride
Washing in separatory funnel
Decant (pour off)

$$CH_3CH_2CH_2CH_2OH \xrightarrow{\text{NaBr, H}_2\text{SO}_4} CH_3CH_2CH_2CH_2Br + NaHSO_4 + H_2O$$

n-Butyl alcohol
Bp 118°,
den 0.810,
MW 74.12

n-Butyl bromide
Bp 101.6°,
den 1.275,
MW 137.03

Choice of reagents

A primary alkyl bromide can be prepared by heating the corresponding alcohol with (a) constant-boiling hydrobromic acid (47% HBr); (b) an aqueous solution of sodium bromide and excess sulfuric acid, which is an equilibrium mixture containing hydrobromic acid; or (c) with a solution of hydrobromic acid produced by bubbling sulfur dioxide into a suspension of bromine in water. Reagents (b) and (c) contain sulfuric acid at a concentration high enough to dehydrate secondary and tertiary alcohols to undesirable by-products (alkenes and ethers) and hence the HBr method (a) is preferred for preparation of halides of the types R_2CHBr and R_3CBr. Primary alcohols are more resistant to dehydration and can be converted efficiently to the bromides by the more economical methods (b) and (c), unless they are of such high molecular weight as to lack adequate solubility in the aqueous mixtures. The $NaBr-H_2SO_4$ method is preferred to the Br_2-SO_2 method because of the unpleasant, choking property of sulfur dioxide. The overall equation is given above, along with key properties of the starting material and principal product.

The procedure that follows specifies a certain proportion of n-butyl alcohol, sodium bromide, sulfuric acid, and water; defines the reaction temperature and time; and describes operations to be performed in working up the reaction mixture. The prescription of quantities is based upon considerations of stoichiometry as modified by the results of experimentation. Before undertaking a preparative experiment you should analyze the directions and calculate the molecular properties of the reagents. Construction of tables (see next page) of properties of starting material, reagents, products, and by-products provides guidance in regulation of temperature and in separation and purification of the product and should be entered in the laboratory notebook.

The laboratory notebook

Reagents

Reagent	MW	Den	Bp	Wt used (g)	Moles	
					Theory	Used
n-C$_4$H$_9$OH	74.12	0.810	118°	16.2	0.22	0.22
NaBr	102.91	—	—	27.0	.22	.26
H$_2$SO$_4$	98.08	1.84	—	42.3	.22	.44

Product and By-Products

Compound	MW	Den	Bp		Yield	
			Given	Found	Theory	%
n-C$_4$H$_9$Br	137.03	1.275	101.6°		30.1	
CH$_3$CH$_2$CH=CH$_2$			−6.3°			
C$_4$H$_9$OC$_4$H$_9$			141°			

One mole of *n*-butyl alcohol theoretically requires one mole each of sodium bromide and sulfuric acid, but the prescription calls for use of a slight excess of bromide and twice the theoretical amount of acid. Excess acid is used to shift the equilibrium in favor of a high concentration of hydrobromic acid. The amount of sodium bromide taken, arbitrarily set at 1.2 times the theory as an insurance measure, is calculated as follows:

$$\frac{16.2 \text{ (g of C}_4\text{H}_9\text{OH)}}{} \left| \frac{102.91 \text{ (MW of NaBr)}}{74.12 \text{ (MW of C}_4\text{H}_9\text{OH)}} \right| \frac{1.2}{} = 27.0 \text{ g of NaBr}$$

The theoretical yield is 0.22 mole of product, corresponding to the 0.22 mole of butyl alcohol taken; the maximal weight of product is calculated thus:

0.22 (mole of alcohol) × 137.03 (MW of product) =

30.1 g butyl bromide

The probable by-products are 1-butene, dibutyl ether, and the starting alcohol. The alkene is easily separable by distillation, but the other substances are in the same boiling point range as the product. However, all three possible by-products can be eliminated by extraction with concd sulfuric acid.

EXPERIMENT   Put 27.0 g of sodium bromide, 30 ml of water, and 20 ml of *n*-butyl alcohol in a 250-ml round-bottomed flask, cool the mixture in an ice-water bath, and slowly add 23 ml of concentrated sulfuric acid with swirling and cooling. Mount the flask over a microburner and fit it with a short condenser for reflux condensation (Fig. 17.1). Heat to the boiling point, note the time, and adjust the flame for brisk, steady refluxing. The upper layer that soon separates is the alkyl bromide, since the aqueous solution of inorganic salts has a greater density. Reflux for 30 minutes, remove the flame, and let the

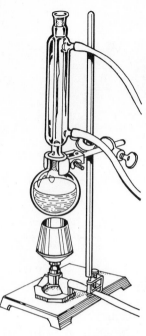

Figure 17.1
Refluxing a reaction
mixture.

condenser drain for a few minutes (extension of the reaction period to 1 hr increases the yield by only 1–2%). Remove the condenser mount, a stillhead in the flask, and set the condenser for downward distillation through a bent adapter into a 125-ml Erlenmeyer. Distil the mixture, make frequent readings of the temperature, and distil until no more water-insoluble droplets come over, by which time the temperature should have reached 115° (collect a few drops of distillate in a test tube and see if it is water soluble). The increasing boiling point is due to azeotropic distillation of n-butyl bromide with water containing increasing amounts of sulfuric acid, which raises the boiling point.

Pour the distillate into a separatory funnel, shake with about 20 ml of water, and note that n-butyl bromide now forms the lower layer. A pink coloration in this layer due to a trace of bromine can be discharged by adding a pinch of sodium bisulfite and shaking again. Drain the lower layer of n-butyl bromide into a clean flask, clean and dry the separatory funnel, and return the n-butyl bromide to it. Then cool 20 ml of concd sulfuric acid thoroughly in an ice bath and add the acid to the funnel, shake well, and allow 5 min for separation the layers. The relative densities given in the tables presented in the introduction of this experiment identify the two layers; an empirical method of telling the layers apart is to draw off a few drops of the lower layer into a test tube and see whether the material is soluble in water (H_2SO_4) or insoluble in water (butyl bromide). Separate the layers, allow 5 min for further drainage, and separate again. Then wash the n-butyl bromide with 20 ml of 10% sodium hydroxide solution to remove traces of acid, separate, and be careful to save the proper layer.

Calcium chloride: removes water and alcohol from a solution

Dry the cloudy n-butyl bromide by adding 2 g of anhydrous calcium chloride and warming the mixture gently on the steam bath with swirling until the liquid clears. Decant the dried liquid into a 50-ml flask through a funnel fitted with a small loose plug of cotton, add a boiling stone, distil, and collect material boiling in the range 99–103°. Yield 21–25 g. Note the approximate volumes of forerun and residue.

A proper label is important

Put the sample in a narrow-mouth bottle of appropriate size; make a neatly printed label giving the name and formula of the product and your name. Press the label onto the bottle under a piece of filter paper for a full minute and make sure that it is secure. After all the time spent on the preparation, the final product should be worthy of a carefully executed and secured label.

QUESTIONS   1. *What experimental method would you recommend for the preparation of* n-octyl bromide? t-Butyl bromide?

2. *Explain why the crude product is likely to contain certain definite organic impurities.*

3. *How does each of these impurities react with sulfuric acid when the n-butyl bromide is shaken with this reagent?*

18 2,7-Dimethyl-3,5-octadiyn-2,7-diol

KEYWORDS   2-Methyl-3-butyn-2-ol
Oxidative coupling, Glaser reaction
Cuprous chloride-pyridine complex (catalyst)
Pressure balloon

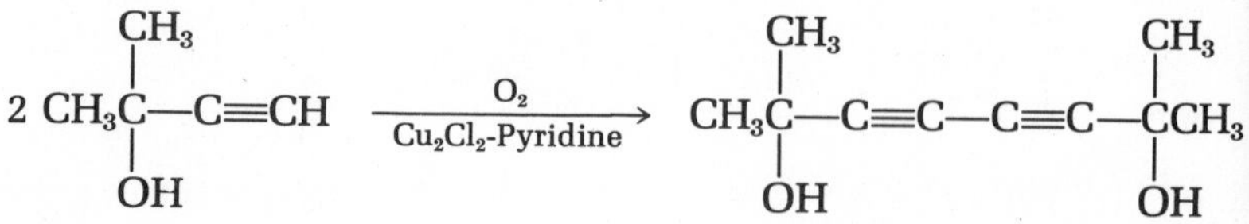

$$2 \; CH_3\underset{\underset{OH}{|}}{\overset{\overset{CH_3}{|}}{C}}\!-\!C\!\equiv\!CH \quad \xrightarrow[Cu_2Cl_2\text{-Pyridine}]{O_2} \quad CH_3\underset{\underset{OH}{|}}{\overset{\overset{CH_3}{|}}{C}}\!-\!C\!\equiv\!C\!-\!C\!\equiv\!C\!-\!\underset{\underset{OH}{|}}{\overset{\overset{CH_3}{|}}{C}}CH_3$$

2-Methyl-3-butyn-2-ol
MW 84.11, den 0.868, bp 103°

2,7-Dimethyl-3,5-octadiyn-2,7-diol
MW 166.21, mp 130°

The starting material, 2-methyl-3-butyn-2-ol, is made commercially from acetone and acetylene and is convertible into isoprene. This experiment illustrates the oxidative coupling of a terminal acetylene to produce a diacetylene, the Glaser reaction. The catalyst, a cuprous chloride–pyridine complex, probably functions by converting the acetylene into a copper derivative, which is oxidized by oxygen to the diacetylene and cuprous oxide. The procedure is based on one reported by H. A. Stansbury, Jr., and W. R. Proops (1962), but the original reaction time is shortened by use of more catalyst and by supplying oxygen under the pressure of a balloon; the state of the balloon provides an index of the course of the reaction.

EXPERIMENT   The reaction vessel is a 125-ml filter flask or Erlenmeyer flask with side tube with a rubber bulb[1] secured to the side arm with copper wire, evident from Figs. 18.1 or 18.2. Add 10 ml of 2-methyl-3-butyn-2-ol, 10 ml of methanol, 3 ml of pyridine, and 0.5 g of cuprous chloride. Before going to the oxygen station,[2] practice capping the flask with a large serum stopper[3] until you can do this quickly. You are to flush out the flask with oxygen and cap it before air can

---

[1] White rubber pipette bulb for medicine dropper, Will Corp. No. 23266.
[2] The valves of a cylinder of oxygen should be set to deliver gas at a pressure of 10 lbs/sq in when the terminal valve is opened. The barrel of a 2.5 ml plastic syringe is cut off and thrust into the end of a $\frac{1}{4} \times \frac{3}{16}$-in. rubber delivery tube.
[3] Red rubber serum stopper (large), Scientific Glass Co., R7950.

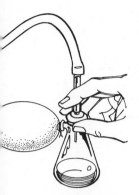

Figure 18.1
Balloon technique of
oxygenation. About 10
lb oxygen pressure are
needed to inflate the
pipette bulbs.

Figure 18.2
Appearance of pipette
bulb after oxygenation.

diffuse in; the reaction is about twice as fast in an atmosphere of oxygen as in air. Insert the oxygen delivery syringe in the flask with the needle under the surface of the liquid, open the valve and let oxygen bubble through the solution in a brisk stream for 2 min. Close the valve and quickly cap the flask and wire the cap. Next thrust the needle of the delivery syringe through the center of the serum stopper, open the valve, and run in oxygen until you have produced a sizeable inflated bulb (see Fig. 18.1). Close the valve and withdraw the needle (the hole is self-sealing), note the time and start swirling the reaction mixture. The rate of oxygen uptake depends on the efficiency of mixing of the liquid and gas phases. By vigorous and continuous swirling it is possible to effect deflation of the balloon to a 5-cm sphere (see Fig. 18.2) in 20–25 min. The reaction mixture warms up and becomes deep green. Introduce a second charge of oxygen of the same size as the first, note the time, and swirl. In 25–30 min the balloon reaches a constant volume (e.g., a 5-cm sphere), and the reaction is complete. A pair of calipers is helpful in recognizing the constant size of the balloon and thus the end point of the reaction.[4]

Open the reaction flask, cool if warm, add 5 ml of concentrated hydrochloric acid to neutralize the pyridine and keep copper compounds in solution, and cool again; the color changes from green to yellow. Use a spatula to dislodge copper salts adhering to the walls, leaving it in the flask. Then add 25 ml of saturated sodium chloride solution, to precipitate the diol, and stir and cool the thick paste that results. Scrape out the paste onto a small Büchner funnel with the spatula, press down the material to an even cake. Rinse out the flask with water and wash the filter cake with enough more water to remove all the color from the cake and leave a colorless product. Since the moist product dries slowly, drying is accomplished in ether solution, and the operation is combined with recovery of diol retained in the mother liquor. Transfer the moist product to a 125-ml Erlenmeyer flask, extract the mother liquor in the filter flask with one portion of ether, wash the extract once with water, and run the ethereal solution into the flask containing the solid product. Add enough more ether to dissolve the product, transfer the solution to a separatory funnel, drain off the water layer, wash with saturated sodium chloride solution, filter through sodium sulfate into a tared flask. The solvent is evaporated on the steam bath and the solid residue is heated and evacuated until the weight is constant. Crystallization from benzene then gives colorless needles of the diol, mp 129–130°. The yield of crude product is 7–8 g. In the crystallization from benzene, the recovery is practically quantitative.

---

[4] A magnetic stirrer does not materially shorten the reaction time.

19 cis-Norbornene-5,6-endo-dicarboxylic Anhydride

KEYWORDS

Diels-Alder reaction
Dicyclopentadiene, dimer
Reverse Diels-Alder reaction
Cyclopentadiene, monomer

Maleic anhydride
Endo-, exo-
Compound X

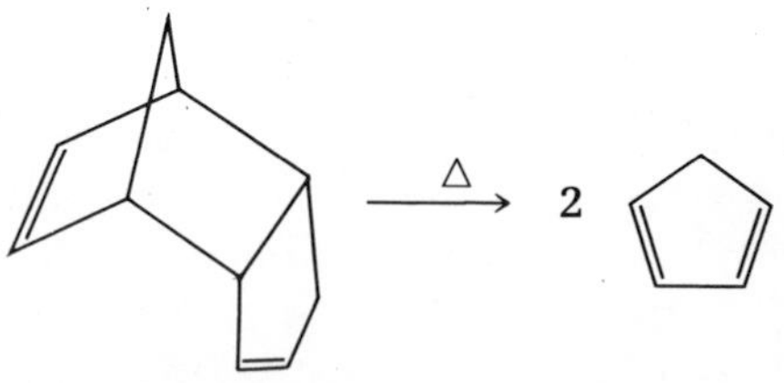

Dicyclopentadiene
Den 0.98,
MW 132.20

Cyclopentadiene
Bp 41°, Den 0.80,
MW 66.10

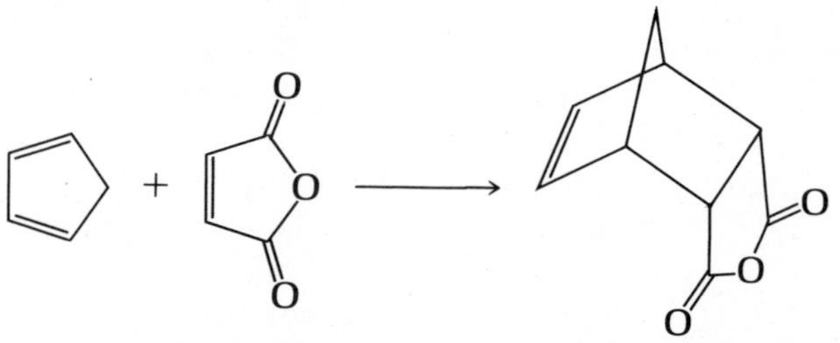

Maleic anhydride
Mp 53°, MW 98.06

cis-**Norbornene-5,6-***endo***
-dicarboxylic anhydride**
Mp 165°, MW 164.16

    Cyclopentadiene is obtained from the light oil from coal tar and stored as the stable dimer, dicyclopentadiene, which is the Diels-Alder adduct from two molecules of the diene. Thus, generation of cyclopentadiene by pyrolysis of the dimer represents a reverse

Diels-Alder reaction. In the Diels-Alder addition of cyclopentadiene and maleic anhydride the two molecules approach each other in the orientation shown in the drawing, as this orientation provides maximal overlap of π-bonds of the two reactants and favors formation of an initial π-complex and then the final *endo*-product. Dicyclopentadiene also has the *endo*-configuration.

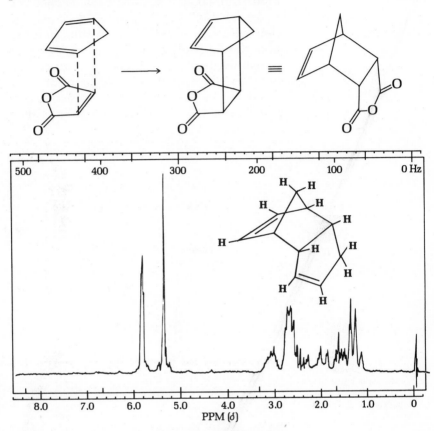

Figure 19.1
Nmr spectrum of dicyclopentadiene.

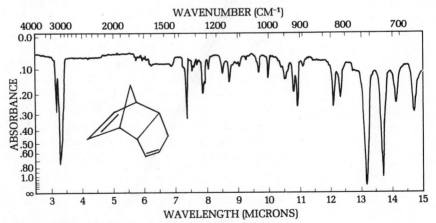

Figure 19.2
Infrared spectrum of dicyclopentadiene.

EXPERIMENTS   **1. *cis*-Norbornene-5,6-*endo*-dicarboxylic Anhydride[1]**

Measure 30 ml of technical dicyclopentadiene (85% pure) into a 125-ml flask and arrange for fractional distillation into an ice-cooled receiver as in Fig. 13.1. Heat the dimer with a microburner until it refluxes briskly and at such a rate that the monomeric diene begins to distil in about 5 min and soon reaches a steady boiling point in the range 40–42°. Apply heat continuously to promote rapid distillation without exceeding the boiling point of 42°. Distillation for 45 min should provide the 12 ml of cyclopentadiene required for two preparations of the adduct; continued distillation for another half hour gives a total of about 20 ml of monomer.

Reverse Diels-Alder

While the distillation is in progress, place 6 g of maleic anhydride[2] in a 125-ml Erlenmeyer flask and dissolve the anhydride in

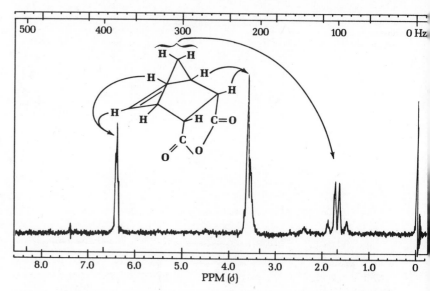

Figure 19.3
Nmr spectrum of *cis*-norbornene-5,6-*endo*-dicarboxylic anhydride.

20 ml of ethyl acetate by heating on a hot plate or steam bath. Add 20 ml of ligroin, bp 60–90°, cool the solution thoroughly in an ice-water bath, and leave it in the bath (some anhydride may crystallize).

The distilled cyclopentadiene may be slightly cloudy, because of the condensation of moisture in the cooled receiver. Add about 1 g of calcium chloride to remove the moisture. Measure 6 ml of dry cyclopentadiene, and add it to the ice-cold solution of maleic anhydride. Swirl the solution in the ice bath for a few minutes until the exothermic reaction is over and the adduct separates as a white solid. Then heat the mixture on a hot plate or steam bath until the solid is all dissolved.[3] If you let the solution stand undisturbed, you

Rapid addition at 0°

---

[1] Based on an experiment reported by William J. Sheppard, *J. Chem. Education*, **40**, 40 (1963).
[2] Material available in the form of cast 7/8-in rods should be crushed in a mortar and preferably crystallized from benzene-ligroin.
[3] In case moisture has gotten into the system, a little of the corresponding diacid may remain undissolved at this point and should be removed by filtration of the hot solution.

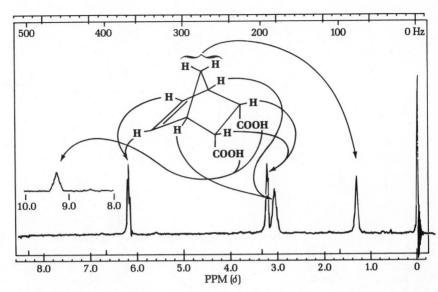

Figure 19.4
Nmr spectrum of *cis*-norbornene-5,6-*endo*-dicarboxylic acid.

will be rewarded with a beautiful display of crystal formation. The anhydride crystallizes in long spars, mp 164–165°; yield, 8.2 g.[4]

endo-cis-Diacid

For preparation of the *endo-cis*-diacid, place 4.0 g of bicyclic anhydride and 50 ml of distilled water in a 125-ml Erlenmeyer flask, grasp this with a clamp, swirl the flask over a free flame, and bring the contents to the boiling point, at which point the solid partly dissolves and partly melts. Continue to heat until the oil is all dissolved and let the solution stand undisturbed. Since the diacid has a strong tendency to remain in supersaturated solution, allow half an hour or more for the solution to cool to room temperature and then drop in a carborundum boiling stone. Observe the stone and its surroundings carefully, waiting several minutes before applying the more effective method of making one scratch with a stirring rod on the inner wall of the flask at the air-liquid interface. Let crystallization proceed spontaneously to give large needles, which will dry quickly, then cool in ice and collect. Yield, 4.0 g; mp 180–185°, dec (anhydride formation).[5]

CO₂H
CO₂H

endo-cis-Diacid

The temperature of decomposition is variable

---

[4] The student need not work up the mother liquor but may be interested in learning the result. Concentration of the solution to a small volume is not satisfactory because of the presence of dicyclopentadiene, formed by dimerization of excess monomer; the dimer has high solvent power. Hence the bulk of the solvent is evaporated on the steam bath, the flask is connected to the water pump with a rubber stopper and glass tube and heated under vacuum on the steam bath until dicyclopentadiene is removed and the residue solidifies. Crystallization from 1:1 ethyl acetate–ligroin affords 1.3 g of adduct, mp 156–158°; total yield, 95%.

[5] The *endo-cis*-diacid is stable to alkali but can be isomerized to the *trans*-diacid (mp 192°) by conversion to the dimethyl ester (3 g of acid, 10 ml methanol, 0.5 ml concd H_2SO_4; reflux 1 hr). This ester is equilibrated with sodium ethoxide in refluxing ethanol for 3 days and saponified. For an account of a related epimerization and discussion of the mechanism, see J. Meinwald and P. G. Gassman, *J. Amer. Chem. Soc.*, **82**, 5445 (1960).

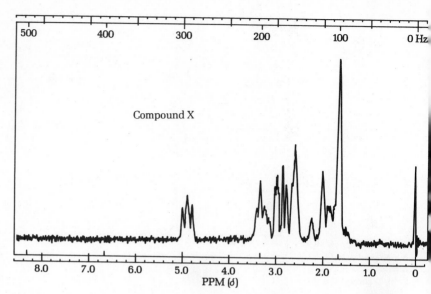

**Figure 19.5
Nmr spectrum of
Compound X, prepared
from *endo-cis*-diacid.**

Problem:[6] Compound X

For preparation of X, place 1 g of the *endo-cis*-diacid and 5 ml of concentrated sulfuric acid in a 50-ml Erlenmeyer and heat gently on the hot plate for a minute or two until the crystals are all dissolved. Then cool in an ice bath, add a small piece of ice, swirl to dissolve, and add further ice until the volume is about 20 ml. Heat to the boiling point and let the solution simmer on the hot plate for 5 minutes. Cool well in ice, scratch to induce crystallization, and allow for some delay in complete separation. Collect, wash with water, and crystallize from water. Compound X (0.7 g) forms large prisms, mp 203°.

What is X?

Write formulas for all possible products of the reaction and devise tests to distinguish between them.

[6] Introduced by James A. Deyrup.

20 Catalytic Hydrogenation

KEYWORDS   Platinum catalyst   Brown[2] procedure

Sodium borohydride   Pressure balloon

$$\xrightarrow[\text{PtCl}_4]{\text{NaBH}_4\text{—HCl}}$$

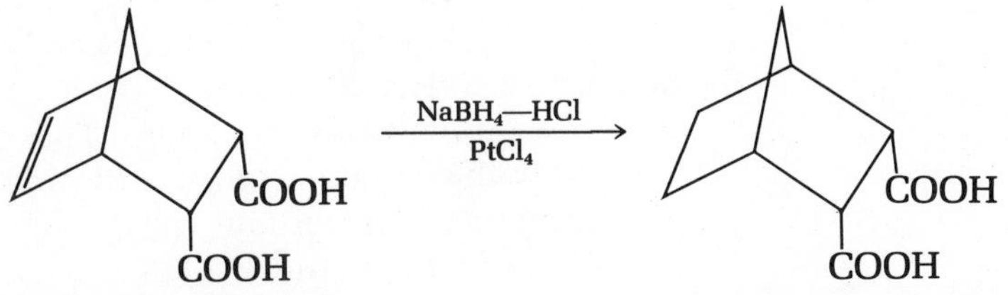

cis-**Norbornene-5,6-***endo*
-dicarboxylic acid
Mp 180–190°, dec;
MW 182.17

cis-**Norbornane-5,6-***endo*
-dicarboxylic acid
Mp 170–175°, dec;
MW 184.19

Hydrogen generated in situ

Conventional procedures for hydrogenation in the presence of a platinum catalyst employ hydrogen drawn from a cylinder of compressed gas and require elaborate equipment. H. C. Brown and C. A. Brown[1] introduced the simple procedure of generating hydrogen *in situ* from sodium borohydride and hydrochloric acid and prepared a highly active supported catalyst by reduction of platinum chloride with sodium borohydride in the presence of decolorizing carbon. The special apparatus described by these authors is here dispensed with in favor of the balloon technique employed for catalytic oxygenation (Chapter 18).

EXPERIMENT

The reaction vessel is a 125-ml filter flask with a white rubber pipette bulb wired onto the side arm. Introduce 10 ml of water, 1 ml of platinum chloride solution,[2] and 0.5 g of decolorizing charcoal and swirl during addition of 3 ml of stabilized 1 *M* sodium borohydride solution.[3] While allowing 5 minutes for formation of the catalyst, dissolve 1 g of *cis*-norbornene-5,6-*endo*-dicarboxylic acid in

---

[1] H. C. Brown and C. A. Brown, *J. Amer. Chem. Soc.*, **84**, 1495 (1962).

[2] A solution of 1 g of PtCl$_4$ in 20 ml of water.

[3] Dissolve 1.6 g of sodium borohydride and 0.3 g of sodium hydroxide (stabilizer) in 40 ml of water. When not in use, the solution should be stored in a refrigerator. If left for some time at room temperature in a tightly stoppered container, gas pressure may develop sufficient to break the vessel.

10 ml of hot water. Pour 4 ml of concd hydrochloric acid into the reaction flask, followed by the hot solution of the unsaturated acid. Cap the flask with a large serum stopper and wire it on. Draw 1.5 ml of the stabilized sodium borohydride solution into the barrel of a plastic syringe, thrust the needle through the center of the stopper and add the solution dropwise with swirling. The initial uptake of hydrogen is so rapid that the balloon may not inflate until you start injecting a second 1.5 ml of borohydride solution through the stopper. When the addition is complete and the reaction appears to be reaching an end point (about 5 min), heat the flask on the steam bath with swirling and try to estimate the time at which the balloon is deflated to a constant size (about 5 min). When balloon size is constant, heat and swirl for 5 min more and then release the pressure by injecting the needle of an open syringe through the stopper.

Reaction time about 15 min

Filter the hot solution by suction and place the catalyst in a jar marked "Catalyst Recovery."[4] Cool the filtrate and extract it with three 15-ml portions of ether. The combined extracts are to be washed with saturated sodium chloride solution and filtered through anhydrous sodium sulfate into a 125-ml flask. Evaporation gives 0.8–0.9 g of white solid.

Common ion effect

The only solvent of promise for crystallization of the saturated *cis*-diacid is water, and the diacid is very soluble in water and crystallizes extremely slowly and with poor recovery. However, the situation is materially improved by addition of a little hydrochloric acid to decrease the solubility of the diacid.

Scrape out the bulk of the solid product and transfer it to a 25-ml Erlenmeyer flask. Add 1–2 ml of water to the 125-ml flask, heat to boiling to dissolve residual solid, and pour the solution into the 25-ml flask. Bring the material into solution at the boiling point with a total of not more than 3 ml of water (as a guide, measure 3 ml of water into a second 25-ml Erlenmeyer). With a capillary dropping tube add 3 microdrops of concentrated hydrochloric acid and let the solution stand for crystallization. Clusters of heavy prismatic needles soon separate; the recovery is about 90%. The product should give a negative test for unsaturation with acidified permanganate solution.

Observe what happens when a sample of the product is heated in a melting point capillary to about 170°. Account for the result. You may be able to confirm your inference by letting the oil bath cool until the sample solidifies and then noting the mp temperature and behavior on remelting.

[4] Used catalyst can be sent for recovery of the $PtCl_4$ to Engelhard Industries, 865 Ramsey St., Hillside, N. J., 07205.

21 Triphenylcarbinol

KEYWORDS   Grignard synthesis   Calcium chloride drying tube
              Fischer esterification   Absolute ether
              Lewis acid catalyst, $BF_3 \cdot Et_2O$ Starting the Grignard reaction
              Methyl benzoate   Triphenylmethyl
              Phenylmagnesium bromide   Beilstein test, copper wire
              Biphenyl, by-product

Benzoic acid
Mp 122°,
MW 122.12

Methyl benzoate
Bp 199°,
MW 136.14

$[(C_6H_5)_3COH]$

Triphenylcarbinol
Mp 162°,
MW 260.32

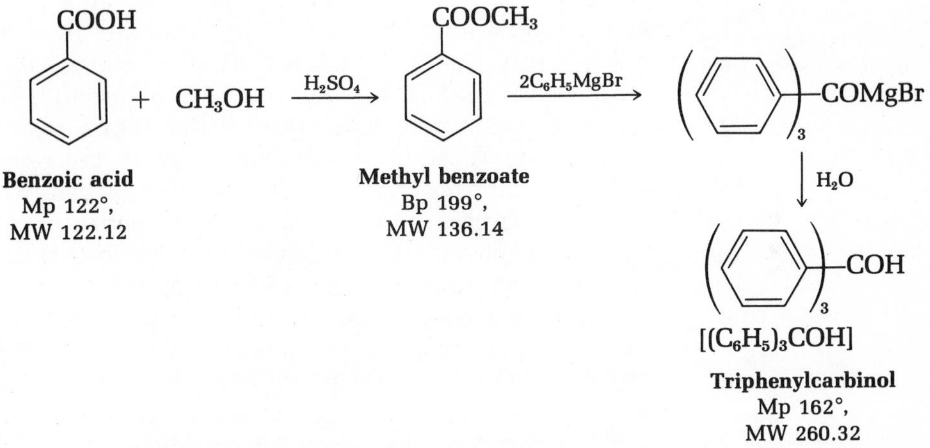

Benzoic acid is esterified by the Fischer method, and the methyl benzoate produced is used for the Grignard synthesis of triphenylcarbinol by interaction with two equivalents of phenylmagnesium bromide.

In Fischer esterification, methanol is used in excess both to shift the equilibrium in favor of the product and to serve as solvent for the solid acid component. Of the acid and Lewis acid catalysts commonly employed, sulfuric acid is preferable to hydrogen chloride because of greater convenience and to boron fluoride etherate because it is a more effective catalyst. In case either the alcohol or the acid involved in an esterification contains reactive double bonds or is sensitive to dehydration, boron trifluoride etherate becomes the catalyst of choice.

In this two-step synthesis, directions are given for the Grignard reaction of 5.0 g (0.037 mole) of pure methyl benzoate and the student is to adjust the quantities of reagents according to the amount of intermediate available. Satisfactory yield and quality of the final product require care in the preparation and purification of the intermediate ester, as well as in the second step. The chief impurity in the Grignard reaction mixture is the hydrocarbon biphenyl, formed by the coupling reaction:

$$C_6H_5MgBr + C_6H_5Br \rightarrow C_6H_5{-}C_6H_5 + MgBr_2$$

This hydrocarbon by-product is easily eliminated, however, since it is much more soluble in hydrocarbon solvents than the hydroxylic major product.

EXPERIMENTS

1. Methyl Benzoate

1-Hr reflux period; this can be done in advance and the mixture let stand

Place 10.0 g of benzoic acid and 25 ml of methanol in a 125-ml round-bottomed flask, and pour 3 ml of concd sulfuric acid *slowly and carefully down the walls of the flask,* and then swirl to mix the components. Attach a reflux condenser and reflux the mixture gently with a microburner for 1 hr. Cool the solution, decant it into a separatory funnel containing 50 ml of water, and rinse the flask with a part of 35 ml of ether that is to be used for extraction of the product. Shake and drain off the water, which removes the sulfuric acid and the bulk of the methanol. Wash the remaining solution with a second portion of water (25 ml) and then with 25 ml of 5% sodium bicarbonate to remove unreacted benzoic acid. Again shake, with frequent release of pressure, until no further reaction is apparent, and then drain off and acidify the extract. Any benzoic acid that precipitates is collected and the amount recovered should be allowed for in calculation of the percentage yield. Repeat the washing with bicarbonate until no precipitate forms on acidification of the aqueous layer. Wash with saturated sodium chloride solution and filter the solution through anhydrous sodium sulfate into an Erlenmeyer flask.

Remove the ether by evaporation on the steam bath under an aspirator tube. When evaporation ceases, add 2–3 g of anhydrous sodium sulfate to the residual oil and heat for about 5 min longer. Then decant the methyl benzoate into a 50-ml round-bottomed flask, attach a stillhead, and distil downward through an empty fractionation column which functions as an air condenser; the boiling point of the ester is so high (199°) that a water-cooled condenser is liable to crack. Use a tared 25-ml Erlenmeyer as the receiver and collect material boiling above 190°. Yield 8 g.

Add a drop of the ester to a few drops of concd sulfuric acid and see if the ester dissolves, as an oxygen-containing compound should, and if the solution is colorless; a yellow coloration indicates lack of purity. Keep the flask stoppered until the ester is to be used.

2. Phenylmagnesium Bromide (Phenyl Grignard Reagent)

The Grignard reagent is prepared in a dry 125-ml round-bottomed flask fitted with a long reflux condenser. A calcium chloride drying tube inserted in a cork that will fit either the flask or the top of the

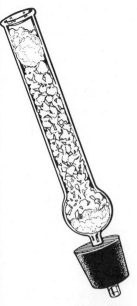

Figure 21.1
Calcium chloride drying
tube fitted with a rubber
stopper.

condenser is also made ready.[1] The flask and condenser should be as dry as possible to being with, and then, as a further precaution to eliminate a possible film of moisture, the magnesium to be used (2 g = 0.082 mole of magnesium turnings) is placed in the flask, the calcium chloride tube is attached directly, and the flask is heated gently but thoroughly on the steam bath while being rotated. The flask on cooling pulls dry air through the calcium chloride. Cool to room temperature before proceeding! **Extinguish all flames!**

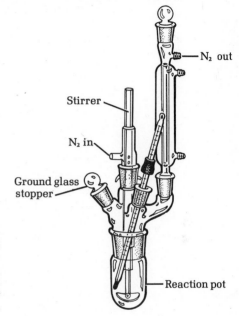

N₂ out

Stirrer

N₂ in

Ground glass
stopper

Reaction pot

Figure 21.2
Semimicro-scale,
research-type apparatus
for Grignard reaction,
with provision for a
motor-driven stirrer and
an inlet and outlet for
dry nitrogen.

*Specially dried ether
is required*

Make an ice bath ready in case control of the reaction becomes necessary, remove the drying tube and fit it to the top of the condenser. Then pour into the flask 15 ml of *absolute* ether (absolutely dry, anhydrous) and 9 ml (13.5 g = 0.086 mole) of bromobenzene. (More ether is to be added as soon as the reaction starts, but at the outset the concentration of bromobenzene is kept high to promote easy starting.) If there is no immediate sign of reaction, insert a *dry* stirring rod with a flattened end and crush a piece of magnesium firmly against the bottom of the flask under the surface of the liquid, giving a twisting motion to the rod. When this is done properly the liquid becomes slightly cloudy, and ebullition commences at the surface of the turning that has been compressed. Attach the condenser at once, swirl the flask to provide fresh surfaces for contact, and, as soon as you are sure that the reaction has started, add 25 ml more absolute ether through the top of the condenser before spontaneous boiling becomes too vigorous (replace the drying tube). Cool in ice if necessary to slow the reaction. If difficulty is ex-

[1] Insert a small piece of cotton or glass wool into the bulb, add 8-mesh anhydrous calcium chloride, and then insert another wad of cotton in the top. Store for future use with cork in top, pipette bulb on bottom.

perienced in initiating the reaction, try in succession the expedients described in the footnote.[2] If spontaneous boiling in the diluted mixture is slow or becomes slow, mount the flask and condenser on the steam bath (one clamp supporting the condenser suffices) and reflux gently until the magnesium has disintegrated and the solution has acquired a cloudy or brownish appearance. The reaction is complete when only a few small remnants of metal (or metal contaminants) remain. Since the solution of Grignard reagent deteriorates on standing, the next step should be started at once.

Use minimum steam to avoid condensation on condenser

3. Triphenylcarbinol

Put 5 g (0.037 mole) of methyl benzoate and 15 ml of absolute ether into a separatory funnel, cool the flask containing phenylmagnesium bromide solution briefly in an ice bath, remove the drying tube, and insert the stem of the separatory funnel into the top of the condenser. Run in the methyl benzoate solution *slowly* with only such cooling as is required to control the mildly exothermic reaction, which affords an intermediate addition compound which separates as a white solid. Replace the calcium chloride tube, swirl the flask until it is at room temperature and the reaction has subsided. The reaction is then completed by either refluxing the mixture for one-half hour, or stoppering the flask with the calcium chloride tube and letting the mixture stand overnight (subsequent refluxing is then unnecessary).[3]

This is a suitable stopping point

Pour the reaction mixture into a 250-ml Erlenmeyer flask containing 50 ml of 10% sulfuric acid and about 25 g of ice and use both ordinary ether and 10% sulfuric acid to rinse the flask. Swirl well to promote hydrolysis of the addition compound; basic magnesium salts are converted into water-soluble neutral salts and triphenylcarbinol is distributed into the ether layer. An additional amount of ether (ordinary) may be required. Pour the mixture into a separatory funnel (rinse the flask with ether), shake, and draw off the aqueous layer. Shake the ethereal solution with 10% sulfuric acid to further remove magnesium salts, wash it with saturated sodium chloride solution, and filter the solution through anhydrous sodium sulfate. If this solution is evaporated to dryness the residue will be found to be a solid (9.5 g) of low and wide melting range (110–115°), since it contains considerable biphenyl. The by-product can be removed by crystallization from ligroin, but instead of removing the ether and dissolving the residue in ligroin it is simpler to add the ligroin to the

---

[2] (a) Warm on the steam bath with swirling. Then see if boiling continues when the flask (condenser attached) is removed from the heating bath. (b) Try further mashing of the metal with a stirring rod. (c) Add a tiny crystal of iodine as a starter (in this case the ethereal solution of the final reaction product should be washed with sodium bisulfite solution in order to remove the yellow color). (d) Add a few drops of a solution of phenylmagnesium bromide or of methylmagnesium iodide (which can be made in a test tube). (e) Start afresh, taking greater care with respect to the dryness of apparatus and reagents.

[3] A rule of thumb for organic reactions: a 10° rise in temperature will double the rate of the reaction.

ethereal solution, distil, and so displace the ether by the less volatile crystallization solvent. Use 25 ml of 66–77° ligroin and concentrate the ether-ligroin solution (steam bath) in an Erlenmeyer flask under an aspirator tube. Evaporate slowly until crystals of triphenylcarbinol just begin to separate and then let crystallization proceed, first at room temperature and then at 0°. The product should be colorless and should melt not lower than 160°. Concentration of the mother liquor may yield a second crop of crystals. Total yield 5.0 g. Evaporate the mother liquors to dryness and save the residue for later isolation of the components by chromatography.

4. Reactions of Triphenylcarbinol

Prepare a stock solution of 4.0 g of triphenylcarbinol in 80 ml of warm (steam bath) glacial acetic acid for use in the first four reactions (a-d) which follow.

(**a**) To 10 ml of the stock solution add 2 ml of 47% hydrobromic acid, heat for 5 min on the steam bath, cool in ice, collect the product (0.8 g), and wash it with ligroin.

(**b**) To 10 ml of the stock solution add 2 ml of an acetic acid solution containing 5% of chloroacetic acid and 1% of sulfuric acid, heat 5 min, add water (5–6 ml) to produce a saturated solution, and let the mixture stand for crystallization. Collect the product by suction filtration and wash the crystals with 1:1 methanol-water; yield 0.8 g.

(**c**) To 10 ml of the stock solution add 2 ml of 47% hydriodic acid, heat for 1 hr, cool, add 1 g of sodium bisulfite in 20 ml of water, collect the precipitate, wash well with water, and crystallize the moist solid from methanol (15–20 ml).

(**d**) To 10 ml of the stock solution add a hot solution of 2 g of stannous chloride in 5 ml of concd hydrochloric acid, heat for 1 hr, cool, collect, and wash well with methanol; yield 0.8 g.

(**e**) Dissolve 1 g of the carbinol in 25 ml of methanol, add 1 ml of 47% hydrobromic acid, heat gently for 15 min, and let cool. Crystallization can be hastened by scratching. Wash the product (1.0 g) with methanol.

(**f**) Weigh 1 g of the carbinol and 2 g of malonic acid (crush any lumps) into a 25-ml Erlenmeyer flask, and heat the mixture for 7 min on the hot plate (about 150°) under an aspirator tube for removal of fumes. Cool, dissolve in 2 ml of benzene, add 10 ml of ligroin (60–90°) and set the solution aside. Wash the product with ligroin; yield 0.8 g.

	Mp
$(C_6H_5)_3CCl$	113°
$(C_6H_5)_3CBr$[4]	152°
$(C_6H_5)_3CI$	132°
$CH_3COOC(C_6H_5)_3$	97°
$CH_2ClCOOC$-$(C_6H_5)_3$	144°
$(C_6H_5)_3CH$	94°
$(C_6H_5)_3COOC$-$(C_6H_5)_3$	186°
9-Phenylfluorene	148°
p-$(C_6H_5)_2CHC_6H_4C$-$(C_6H_5)_3$	224°
$(C_6H_5)_3COCH_3$	97°
$(C_6H_5)_3COC$-$(C_6H_5)_3$	237°
$(C_6H_5)_3CCH_2CO_2H$	176°

[4] Put a small spatulaful of this compound in a 13 × 100-mm test tube, dissolve in benzene and cool, add a few mg of zinc dust, and stir with a rod to produce a yellow solution containing triphenylmethyl radical. Decant into a clean tube and shake with air; the cloudy precipitate is the peroxide, $(C_6H_5)_3COOC(C_6H_5)_3$.

5. Identification

Identify each product in reactions (a)-(f), and present some reason for your conclusion in addition to correspondence in mp. Some of the possible products are given in the table in the margin. Halogens can be detected by the **Beilstein test:** heat a 2–3 in. section of copper wire to redness in the oxidizing flame of a burner (nonluminous) until the flame is no longer colored, let the wire cool and touch it to a solid (or dip it in a liquid) and reheat; a volatile copper halide imparts a green color to the flame.

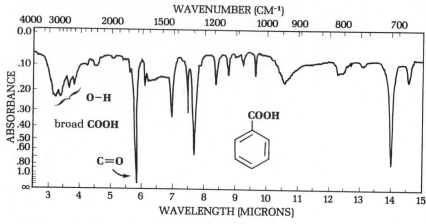

**Figure 21.3
Infrared spectrum of
benzoic acid in CS₂.**

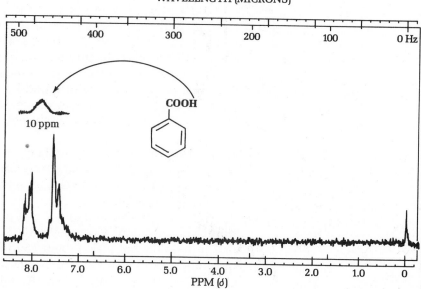

**Figure 21.4
Nmr spectrum of
benzoic acid.**

22 Aldehydes and Ketones

KEYWORDS

Phenylhydrazine
Phenylhydrazones
Semicarbazide
Semicarbazones
Bisulfite addition product
Iodoform test
Fluorene, fluorenone

Acid dichromate oxidation
Water-soluble carbonyl derivative
Girard reagent
Trimethylaminoacetohydrazide
  chloride
NMR and IR spectra

Sections 1–5 of this chapter present orientation experiments in preparation for the identification of unknowns (Section 6). First a test is described for recognizing a carbonyl compound by its reaction with phenylhydrazine in the presence of acetic acid as catalyst. All aldehydes and nearly all ketones rapidly form phenylhydrazones.

$$R_2C{=}O + H_2NNHC_6H_5 \rightarrow R_2C{=}NNHC_6H_5 + H_2O$$

Phenylhydrazine   **Phenylhydrazone**

The increase in molecular weight of 90 units attending the conversion renders the derivative so much less soluble than the starting material that it precipitates or crystallizes and thus gives evidence that a reaction has occurred. The phenylhydrazine test to be carried out later is a generally applicable method of distinguishing carbonyl compounds from substances of other classes.

Many phenylhydrazones are crystalline compounds suitable for characterization by melting point and mixed melting point determination, but some are oils and some are unstable. 2,4-Dinitrophenylhydrazones, prepared with 2,4-dinitrophenylhydrazine, $(NO_2)_2C_6H_3NHNH_2$, are often preferable for characterization because

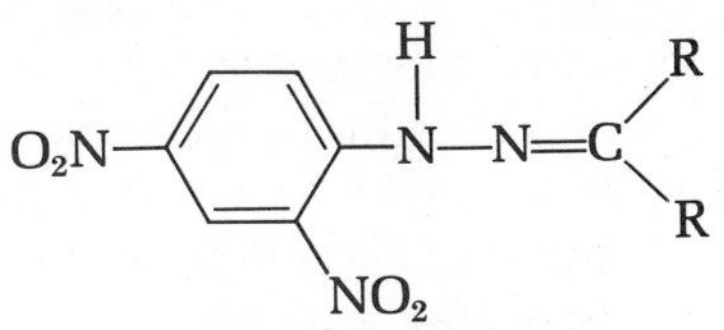

**2,4-Dinitrophenylhydrazone
of a ketone**

they are higher melting, less soluble, and more stable than the corresponding phenylhydrazones; the molecular weight change on conversion to the derivative is 180 units. Semicarbazones have comparable advantageous properties.

$$R_2C{=}O\ +\ H_2NNHCONH_2\ \to\ R_2C{=}NNHCONH_2$$

Semicarbazide   **Semicarbazone**

Oximes are less useful for characterization, for many are liquids at room temperature and those that are solids are often obtainable in crystalline form only with difficulty (the preparation of an oxime is described in Chapter 26). If a substance under study is to be compared with a compound that is known but not available for mixed melting point determination, probable identity can be established by showing that several derivatives of the substance in question all correspond in melting point with known derivatives. After the experience of preparing the phenylhydrazone, 2,4-dinitrophenylhydrazone, and semicarbazone derivatives of known compounds, you may find similar derivatives useful in characterization of your unknowns.

Aldehydes and ketones can be differentiated by the reaction with aqueous sodium bisulfite solution. Nearly all aldehydes, unhindered cyclic ketones, and most methyl ketones afford solid, water-soluble addition compounds, and the iodoform test distinguishes a methyl ketone from all aldehydes but acetaldehyde. Methyl aryl ketones such as acetophenone, $C_6H_5COCH_3$, fail to give bisulfite addition compounds but are recognizable by the iodoform test.

EXPERIMENTS

1. Phenylhydrazones

Reagent: $C_6H_5NHNH_2$
MW 108.14
Den 1.10

To prepare a stock solution of the phenylhydrazine reagent, measure 1.0 ml of phenylhydrazine with a calibrated capillary dropper into a 25-ml Erlenmeyer flask, add 3 ml of acetic acid from a 10-ml graduate, swirl, and note the heat of neutralization; dilute the solution with 4–5 ml of water, pour it into a 10-ml graduate, and make up the volume to 10.0 ml. Return the solution to the Erlenmeyer flask for storage. The molecular weight and specific gravity of phenylhydrazine are such that 1 ml of the stock solution contains 1 millimole of phenylhydrazine acetate, the amine salt of acetic acid. Excess acetic acid is used both as condensation catalyst and as solvent for the reagent.

A substance to be tested for the presence of a carbonyl group should first be examined to see whether or not it is soluble in water; if 8–10 micro drops dissolve in 1 ml of water at room temperature, the test is conducted by dissolving one millimole of sample in 1 ml of water and adding 1 ml of the stock solution (which contains 1 millimole of phenylhydrazine acetate); separation of an oil or solid constitutes a positive indication that the substance has a carbonyl group.   Test acetone (5 micro drops) in this way.

(a) Water-soluble
substances

Diethyl ketone (MW 86.13) is soluble to the extent of about 4 micro drops (47 mg) in 1 ml of water. Calibrate a capillary dropping tube for this substance by counting 30–40 micro drops delivered into a tared container and then weighing container and ketone. Calculate the number of micro drops required to give 1 millimole of the ketone (about 8; record the figure for future use); measure this amount into 1 ml of water in a small flask or test tube, and note that the first few drops dissolve but that an oily layer eventually appears. Add a few drops of methanol to just bring the oil into solution and then add 1 ml of stock phenylhydrazine solution. Rub the oil against the flask with a stirring rod briefly to see if it will crystallize easily and if not, discard it.

*(b) Water-insoluble
sample*

A typical water-insoluble liquid is acetophenone, $C_6H_5COCH_3$, MW 120.66. Calibrate the same capillary dropper for delivery of this substance, dissolve 1 millimole of substance in 1 ml of methanol, and add 1 ml of phenlhydrazine solution. Let the mixture stand for 5 min, collect the product, and see how the melting point corresponds with the value 105° reported for the pure derivative. A water-insoluble solid is likewise tested in methanol solution.

Save the residual stock solution for testing unknowns.

2. 2,4-Dinitrophenylhydrazones

Reagent:
$(NO_2)_2C_6H_3NHNH_2$
MW 198.14
Mp 197°

2,4-Dinitrophenylhydrazine[1] is a red solid of high melting point that has low solubility in ethanol. The yellow phosphate salt of this hydrazine (a base) is more soluble than the hydrazine itself, and a 0.1 M solution of the hydrazine in phosphoric acid-ethanol[2] is used in the following experiment and in the investigation of unknowns.

*Procedure for
unknowns*

To 10 ml (1 millimole) of the 0.1 M solution of 2,4-dinitrophenyl-hydrazine in phosphoric acid add 1 millimole of diethyl ketone (see calibration in Section 1), warm for a few minutes, and let crystallization proceed. The melting point reported is 156°; note the contrast between this derivative and the phenylhydrazone of the same ketone.

A second experiment illustrates an alternative procedure, applicable where the 2,4-dinitrophenylhydrazone is known to be sparingly soluble in ethanol. The test substance is cinnamaldehyde (C_6H_5CH=CHCHO, MW 132.15), an α,β-unsaturated aldehyde. Determine the number of micro drops corresponding to 1 millimole of the liquid. Measure 1 millimole of 2,4-dinitrophenylhydrazine into a 125-ml Erlenmeyer flask, add 30 ml of 95% ethanol, digest on the steam bath until all particles of solid are dissolved, and then add 1 millimole of

Notes for the instructor

[1] The reagent is available commercially but expensive. *Preparation.* Dissolve 100 g of 2,4-dinitrochlorobenzene (mp 50–52°) in 200 ml of triethylene glycol, stir mechanically in a salt-ice bath to 15° (or until crystallization starts), and add 28 ml of 64% hydrazine solution by drops at a temperature of 20 ± 3° (25 min). When the strongly exothermal reaction is over, digest the paste on the steam bath, add 100 ml of methanol and digest further, then cool, collect, and wash with methanol. Yield 98 g (100%), mp 190–192°.
[2] Dissolve 2.0 g of 2,4-dinitrophenylhydrazine in 50 ml of 85% phosphoric acid by heating, cool, add 50 ml of 95% ethanol, cool again, and clarify by suction filtration from a trace of residue.

cinnamaldehyde and continue warming. If there is no immediate change, add 6–8 micro drops of concd hydrochloric acid as catalyst and note the result. Warm for a few minutes, then cool and collect the product. Note that the derivative is more intensely colored than that of diethyl ketone; this is because it contains an α,β-double bond conjugated with the carbon-nitrogen double bond.

The alternative procedure strikingly demonstrates the catalytic effect of hydrochloric acid, but it is not applicable to a substance like diethyl ketone, whose 2,4-dinitrophenylhydrazone is much too soluble to crystallize from the large volume of ethanol. The first procedure is obviously the one to use for an unknown.

3. Semicarbazones

Reagent:
$H_2NCONHNH_3{}^+Cl^-$
MW 111.54

Semicarbazide (mp 96°) is not very stable in the free form and is used as the crystalline hydrochloride (mp 173°). Since this salt is insoluble in methanol or ethanol and does not react readily with typical carbonyl compounds in alcohol-water mixtures, a basic reagent is added to liberate free semicarbazide. In one procedure a suspension of semicarbazide hydrochloride and sodium carbonate in methanol is digested, filtered from the sodium chloride formed, and the carbonyl compound added to the filtrate. In another procedure a mixture of semicarbazide hydrochloride and sodium acetate in methanol-water is employed. The procedure that follows utilizes the aromatic amine pyridine as the basic reagent.

Prepare a stock solution by dissolving 1.11 g of semicarbazide hydrochloride in 5 ml of water; 0.5 ml of this solution contains 1 millimole of reagent. To 0.5 ml of the stock solution add 1 millimole of acetophenone and enough methanol (1 ml) to produce a clear solution; then add 10 micro drops of pyridine (a two-fold excess) and warm the solution gently on the steam bath for a few minutes until crystals begin to separate. The melting point of the pure product is 198°.

Repeat the reaction with cinnamaldehyde instead of acetophenone and see if you can observe a difference. Cinnamaldehyde semicarbazone melts at 215°.

4. Bisulfite Test

Prepare a stock solution by dissolving 5 g of sodium bisulfite in 20 ml of water with brief swirling. Put 1 ml of the solution into each of five 13 × 100-mm test tubes and to each tube add 5 micro drops of the following substances: benzaldehyde, 4-methylpentan-2-one, cinnamaldehyde, diethyl ketone, and acetophenone. Shake each tube occasionally during the next 10 min and note the results.

If the bisulfite test is applied to a liquid or solid that is very sparingly soluble in water, formation of the addition product is facilitated by adding a small amount of methanol before the addition of the bisulfite solution.

5. Iodoform Test

The reagent contains iodine in potassium iodide solution[3] at a concentration such that 2 ml of solution, on reaction with excess methyl ketone, will yield 174 mg of iodoform. If the substance to be tested is water-soluble, dissolve 4 micro drops of a liquid or an estimated 50 mg of a solid in 2 ml of water in a 20 × 150-mm test tube; add 2 ml of 10% sodium hydroxide and then slowly add 3 ml of the iodine solution. In a positive test the brown color of the reagent disappears and yellow iodoform separates. If the substance to be tested is insoluble in water, dissolve it in 2 ml of dioxane, proceed as above, and at the end dilute with 10 ml of water.

Test hexane-2,5-dione (water-soluble), n-butyraldehyde (water-soluble), and acetophenone (water-insoluble).

Iodoform can be recognized by its odor and yellow color and, more securely, from the melting point (119°). The substance can be isolated by suction filtration of the test suspension or by adding 2 ml of chloroform, shaking the stoppered test tube to extract the iodoform into the small lower layer, withdrawing the clear part of this layer with a capillary dropping tube, and evaporating it in a small tube on the steam bath. The crude solid is crystallized from methanol-water.

6. NMR Spectroscopy

Only the acidic protons of carboxylic acids, phenols, and enols appear farther downfield than aldehydic protons, which appear at *ca.* 9.6–10 ppm, and are thus easily recognized.

7. Infrared Spectroscopy

The carbon-oxygen stretching vibration of the carbonyl group provides one of the most intense and useful bands in infrared spectroscopy. Aldehydes generally absorb at 1725 cm^{-1} and aliphatic ketones at 1715 cm^{-1}, too close to each other to distinguish these functional groups, but the presence of the aldehydic C–H stretch at 2720 cm^{-1} and 2820 cm^{-1} serves to confirm the aldehyde. Aryl alkyl ketones, like acetophenone, can absorb anywhere between 1680 and 1700 cm^{-1}, depending on the nature of substituents on the aromatic ring. α,β-Unsaturated ketones fall in this same frequency range.

Open chain, saturated ketones have carbonyl stretching frequencies near 1715 cm^{-1}, such as six-membered ring ketones have. As the ring size decreases, the frequency changes to 1740 cm^{-1} for five-membered ring ketones and 1770 cm^{-1} for four-membered ring ketones.

Note for the instructor

[3] Dissolve 25 g of iodine in a solution of 50 g of potassium iodide in 200 ml of water.

8. Unknowns

Your unknown may be any of the aldehydes or ketones listed in Table 22.1 or it may be a noncarbonyl compound.[4] Hence, the first test should be that with phenylhydrazine reagent; if this test is negative, report accordingly and proceed to another unknown. At least one derivative of the unknown is to be submitted to the instructor, but if you first do the bisulfite and iodoform characterizing tests, the results may suggest derivatives whose melting points will be particularly revealing.

9. Citral

Prepare the 2,4-dinitrophenylhydrazone of the material you isolated from lemon grass oil (Chapter 9, Section 3) and so extend your previous characterization of the functional groups present.

10. Preparation of a Ketone-Hydrocarbon Mixture

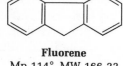

Fluorene
Mp 114°, MW 166.22

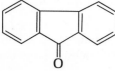

Fluorenone
Mp 83°, MW 180.21

One half hour unattended heating

This mixture of a ketone and a hydrocarbon (fluorenone and fluorene) is to be used in the separation experiment employing Girard's reagent (next section) and also in the separation by chromatography experiment (Chapter 23). The hydrocarbon component, fluorene, is inexpensive, but the yellow ketone derived from it by dichromate oxidation is very costly because of difficulty in pushing the oxidation to completion. Hence, you are to prepare your own mixture by brief, partial oxidation of fluorene.

In a 250-ml Erlenmeyer flask dissolve 5.0 g of practical grade fluorene in 25 ml of acetic acid by heating on the steam bath with occasional swirling. In a 125-ml Erlenmeyer dissolve 15 g of sodium dichromate dihydrate in 50 ml of acetic acid by swirling and heating on a hot plate. Adjust the temperature of the dichromate solution to 80°, transfer the thermometer and adjust the fluorene-acetic acid solution to 80°, and then, **under the hood,** pour in the dichromate solution. Note the time, the temperature of the solution, and heat on the steam bath for 30 min. Observe the maximum and final temperature, and then cool the solution and add 150 ml of water. Swirl the mixture for a full two minutes to coagulate the product and so promote rapid filtration, and collect the yellow solid in an 8.5-cm Büchner funnel (in case filtration is slow, empty the funnel and flask into a beaker and stir vigorously for a few minutes). Wash the filter cake well with water and then suck the filter cake as dry as possible. Either let the product dry overnight, or dry it quickly as follows: Put the moist solid into a 50-ml Erlenmeyer, add ether (20 ml) and swirl to dissolve, and add anhydrous sodium sulfate (10 g) to retain the water. Decant the ethereal solution through a cone of anhydrous sodium sulfate into a 125-ml Erlenmeyer, and use a Calcutta wash bottle for rinsing the flask and funnel with ether.

Note to Instructor

[4] See Tables 63.3 and 63.14 in Chapter 63 for additional aldehydes and ketones.

Table 22.1 Melting Points of Derivatives of Some Aldehydes and Ketones

Compound	Formula	MW	Den	Water solubility	Phenyl-hydrazone	2,4-DNP	Semi-carbazone
Acetone	CH_3COCH_3	58.08	0.79		42	126	187
n-Butanal	$CH_3CH_2CH_2CHO$	72.10	0.82	4 g/100 g	Oil	123	95(106)[a]
Diethyl ketone	$CH_3CH_2COCH_2CH_3$	86.13	0.81	4.7 g/100 g	Oil	156	138
Furfural	$C_4H_3O \cdot CHO$	96.08	1.16	9 g/100 g	97	212(230)[a]	202
Benzaldehyde	C_6H_5CHO	106.12	1.05	Insol.	158	237	222
Hexane-2,5-dione	$CH_3COCH_2CH_2COCH_3$	114.14	0.97	∞	120[b]	257[b]	224[b]
2-Heptanone	$CH_3(CH_2)_4COCH_3$	114.18	0.83	Insol.	Oil	89	123
3-Heptanone	$CH_3(CH_2)_3COCH_2CH_3$	114.18		Insol.	Oil	81	101
n-Heptanal	$n\text{-}C_6H_{13}CHO$	114.18	0.82	Insol.	Oil	108	109
Acetophenone	$C_6H_5COCH_3$	120.66	1.03	Insol.	105	238	198
2-Octanone	$CH_3(CH_2)_5COCH_3$	128.21	0.82	Insol.	Oil	58	122
Cinnamaldehyde	$C_6H_5CH{=}CHCHO$	132.15	1.10	Insol.	168	255	215
Propiophenone	$C_6H_5COCH_2CH_3$	134.17	1.01	Insol.	about 48°	191	182

[a] Both mps have been found, depending on crystalline form of derivative.
[b] Derivative.

Evaporate on the steam bath under an aspirator, heat until the ether is all removed, and pour the hot oil into a 50-ml beaker to cool and solidify. Scrape out the yellow solid; yield 4.0 g. Save for experiments in Section 9 of this chapter and Section 3 of Chapter 23.

11. Girard Separation of a Fluorene-Fluorenone Mixture

Girard reagent:
$Cl^-[(CH_3)_3\overset{+}{N}CH_2$-
$CONHNH_2]$
MW 167.64

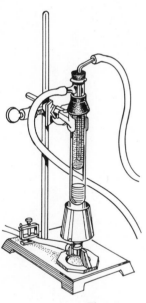

Figure 22.1
Cold finger reflux condenser.

Fluorene

Fluorenone

Girard's reagent, trimethylaminoacetohydrazide chloride, of structure similar to that of semicarbazide, condenses with carbonyl compounds in ethanol containing a little acetic acid as catalyst to give derivatives of the type $Cl^-[(CH_3)_3\overset{+}{N}CH_2CONHN=CR_2]$, which are soluble in water because of the presence of the dipolar ionic grouping. A Girard derivative, once formed and extracted from a water-ether mixture into the aqueous phase, is hydrolyzed easily with excess water under catalysis by mineral acid, with regeneration of the carbonyl compound.

In a 25×150-mm test tube, place 0.5 g of the fluorene-fluorenone mixture, 0.5 g of Girard reagent, 0.5 ml of acetic acid, and 5 ml of 95% ethanol. Reflux the mixture for 30 min. using a 10-ml round-bottomed flask and condenser or, better, a cold finger condenser, made from a 20×150-mm test tube with side arm and conveniently supported at any height desired by a No. 3 neoprene adapter (Fig. 22.1). Then cool the solution, pour it into a separatory funnel, and rinse the tube with 25 ml of ether and then with 10 ml of water. Add 25 ml of saturated sodium chloride solution (to avoid an emulsion), shake well and let the layers separate. Run the lower aqueous layer into a second separatory funnel, wash the ethereal layer with a little water (10 ml), add the wash water to the main water extract, and extract this once with ether. Then run the aqueous layer into a 125-ml Erlenmeyer, add 1 ml of concentrated hydrochloric acid, and heat on the steam bath for 10 min to hydrolyze the Girard derivative and drive off dissolved ether.

During the heating period, wash the ethereal solution in turn with 5% sodium bicarbonate solution and saturated sodium chloride solution and filter it through anhydrous sodium sulfate. This ethereal solution of nonketonic material may be colored from impurities; if so, it should be shaken briefly with decolorizing charcoal, filtered, and evaporated to dryness under an aspirator tube. Dissolve the residue in the least amount of methanol required (about 5 ml), let crystallization proceed at room temperature, and then at 0°; collect the product and wash it with chilled methanol. The resulting fluorene should be colorless and the melting point close to 114°; yield 140 mg.

Hydrolysis of the Girard derivative gives a yellow oil, which solidifies on cooling in ice to a bright yellow solid. Collect the solid (crude fluorenone) and let it dry (170 mg, mp 75–78°); crystallize it from a very small volume of 66–75° ligroin to purify. The fluorenone forms beautiful, large, bright yellow spars, mp 82–83°.

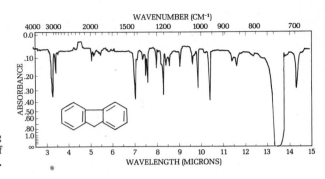

Figure 22.2 Infrared spectrum of fluorene in CS₂.

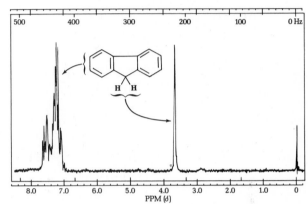

Figure 22.3 Nmr spectrum of fluorene.

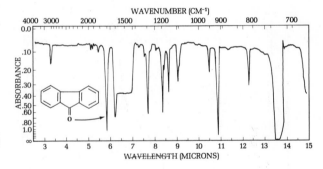

Figure 22.4 Infrared spectrum of fluorenone.

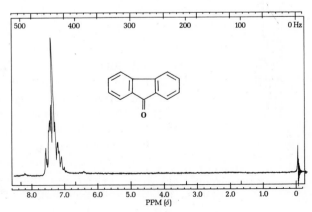

Figure 22.5 Nmr spectrum of fluorenone.

23 Column and Thin Layer Chromatography

KEYWORDS Adsorption
Alumina column
Eluant, elute
Packing the column
Solvent bubbler

Acetylation
Mother liquor
Fractions
Trituration

PART I.   Column Chromatography

Mixtures of organic substances often can be separated by adsorbing the substances from a small volume of solvent onto a column of alumina and then leaching the column with a sequence of solvents of increasing eluant power, that is, of increasing ability to dislodge the adsorbed substances from the adsorbate (Fig. 23.1). The eluant liquid is collected in numbered fractions, each of which is evaporated for examination of the residue. If the components of the mixture are of different chemical types, some will be eluted at an early stage of the process and others in intermediate or late fractions. The usual order of elution is:

Alkanes (*first*)
Alkenes
Dienes
Aromatic hydrocarbons
Ethers

Esters, acetates
Ketones
Alcohols
Diols
Acids (*last*)

Solvents employed for elution, listed in the order of increasing eluant power, are: petroleum ether, benzene, ether, methanol. In an investigation of an unknown mixture, elute first with petroleum ether, then with petroleum ether-benzene mixtures, varying from 4:1 to 1:4; then with benzene, benzene-ether mixtures, ether, and ether-methanol mixtures.

The directions which follow are for the preparation of a column and for three experiments involving applications of the technique. In the first experiment an alcohol is partially acetylated and the

Solvent
Sand

Alumina

Sand
Glass wool

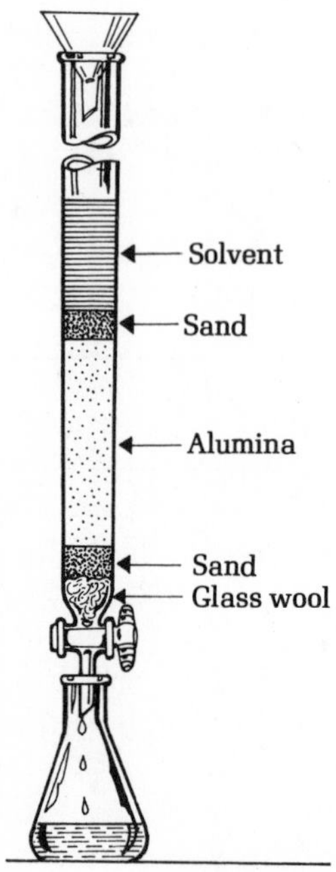

Figure 23.1
Chromatographic
column.

extent of reaction determined by separation of the easily eluted acetate from the strongly adsorbed alcohol; you should start the acetylation and let it run while you are preparing the column. The second experiment is separation of yellow fluorenone from the colorless hydrocarbon from which it is obtained by oxidation, and the third is an investigation of the triphenylcarbinol mother liquor of Chapter 21.3.

Ordinarily, 25 g of alumina is needed per gram of substance to be chromatographed, and this is the ratio specified in the first two experiments. In the third experiment, however, the difference in adsorbability is sufficient to permit a decrease in the proportion of alumina. In all three experiments the transition from solvents of low to high solvent power is done more rapidly than would be done if you were working with an unknown.

EXPERIMENTS

Figure 23.2 Chromatograph tube on ring stand.

Do not allow column to run dry during chromatography.

1. Preparation of the Chromatograph Column[1]

The column can be prepared using a burette, such as the one shown in Fig. 23.1, or using the less expensive and equally satisfactory chromatograph tube shown in Fig. 23.2, in which the flow of solvent is controlled by a screw pinchclamp. Weigh out the required amount of alumina (12.5 g in the first experiment), close the pinchclamp on the tube, and fill about half full with 30–60° petroleum ether. With a wooden dowel push a small plug of glass wool through the liquid to the bottom of the tube, dust in through a funnel enough sand to form a 1-cm layer over the glass wool, and level the surface by tapping the tube. Unclamp the tube and, with the right hand, grasp both the top of the tube and the funnel so that the whole assembly can be shaken to dislodge alumina that may stick to the walls, and with the left hand pour in the alumina. (Fig. 23.3). Do not grasp the part of the tube containing liquid, for this may cause the solvent to boil and disrupt the column. When the alumina has settled, add a little sand to provide a protective layer at the top. Open the pinchclamp, let the solvent level fall until it is just a little above the upper layer of sand, and then stop the flow. Alternatively the alumina can be added to the column (half filled with ligroin) by slurrying the alumina with ligroin in a beaker. The powder is stirred to suspend it in the solvent and *immediately* poured through a wide mouth funnel into the chromatograph tube. Add a protective layer of sand to the top. The column is now ready for use. Prepare 50-ml Erlenmeyer flasks as receivers by weighing each one carefully and marking them with numbers on the etched circle.

After use, the tube is conveniently emptied by pointing the open end into a beaker, opening the pinchclamp, and applying gentle air

[1] Aluminum oxide Merck (acid-washed) is suitable. Purified sand is supplied by J. T. Baker Chemical Co. A dowel ($^3/_{16}''$ or $^1/_4''$) is preferred to a fragile glass rod.

Figure 23.3
Technique useful when filling a chromatographic tube with alumina.

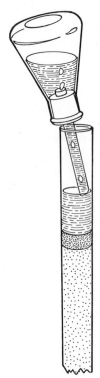

Figure 23.4
Bubbler for adding solvent automatically.

pressure to the tip. If the plug of glass wool remains in the tube, wet it with acetone and apply air pressure again.

2. Acetylation of Cholesterol

Cover 0.5 g of cholesterol with 5 ml of acetic acid in a small Erlenmeyer flask, swirl, and note that the initially thin slurry soon sets to a stiff paste of the molecular compound $C_{27}H_{45}OH \cdot CH_3CO_2H$. Add 1 ml of acetic anhydride and heat the mixture on the steam bath for any convenient period of time from 15 min to 1 hr; record the actual heating period. Cool, add water, and extract with ether. Wash the ethereal extract twice with water and once with 10% sodium hydroxide, dry, filter, and evaporate the ether. Dissolve the residue in 3-4 ml of benzene, transfer the solution with a capillary dropping tube onto a column of 12.5 g of alumina and rinse the flask with another small portion of benzene.[2] Open the pinchclamp, run the eluant solution into a 50-ml Erlenmeyer flask, and as soon as the solvent in the column has fallen to the level of the upper layer of sand fill the column with a part of a measured 50 ml of petroleum ether. When about 25 ml of eluate has collected in the flask (fraction 1), change to a fresh flask; add a boiling stone to the first flask and evaporate the solution to dryness on the steam bath. Evacuation using the aspirator helps to remove last traces of benzene (see Fig. 2.13). If this fraction 1 is negative (no residue), use the flask for collecting further fractions. Continue adding petroleum ether until the 50-ml portion is exhausted and then use 50 ml of benzene, followed by 75-100 ml of a 1:1 mixture of benzene and ether. A convenient bubbler (Fig. 23.4) made from a 125-ml Erlenmeyer flask, a short piece of 10-mm dia glass tubing, and a cork will automatically add solvent. Collect and evaporate successive 25-ml fractions of eluate. Evaporation of the solutions, particularly those rich in benzene, can be hastened by use of the aspirator vacuum. The ideal method for removal of solvents involves the use of a rotary evaporator (Fig. 23.5).

Cholesteryl acetate (mp 115°) and cholesterol (mp 149°) should appear, respectively, in early and late fractions with one or two negative fractions (no residue in between). If so, combine consecutive fractions of early and of late material and determine the weights and melting points. Calculate the percentage of the acetylated material to the total recovered and compare your result with those of others in your class employing different reaction periods. The total recovery is about 80%.

3. Separation of Fluorene and Fluorenone

Prepare a column of 12.5 g of alumina, run out excess solvent, and

---

[2] Ideally, the material to be adsorbed is dissolved in petroleum ether, the solvent of least eluant power. The present mixture is not soluble enough in petroleum ether and so benzene is used, but the volume is kept to a minimum.

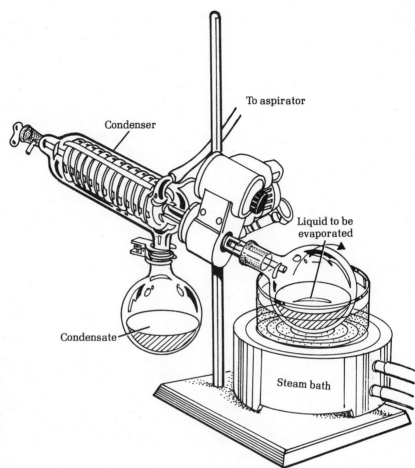

Figure 23.5
Rotary evaporator. The
rate of evaporation with
this apparatus is very
fast due to the thin film
of liquid over the entire
inner surface of the
rotating flask that is
heated under a vacuum.
Foaming and bumping
are also greatly reduced.

pour onto the column a solution of 0.5 g of the fluorene-fluorenone mixture previously separated with Girard reagent (Chapter 22, Section 9). See if you can better the yields by chromatography. Elute at first with petroleum ether and use tared 50-ml flasks as receivers. The yellow color of fluorenone provides one index of the course of the fractionation, and the appearance of solid around the delivery tip provides another. Wash the solid frequently into the receiver with ether delivered from a Calcutta wash bottle. When you think that one component has been eluted completely, change to another receiver until you judge that the second component is beginning to appear. Then, when you are sure the second component is being eluted, change to a 1:1 petroleum ether-benzene mixture and continue until the column is exhausted. It is possible to collect practically all of the two components in the two receiving flasks, with only a negligible intermediate fraction. After evaporation of solvent, pump out each flask at the suction pump and determine the weights and melting points of the products. Compare the outcome of the separation with that obtained with Girard reagent.

4. Examination of a Mother Liquor

Dissolve the residues from the purification of triphenylcarbinol in a little ether, pour the solution into a tared flask, evaporate and evacuate at the aspirator, and weigh. Digest 2 g of the oil or semi solid with 15–20 ml of 30–60° petroleum ether and prepare a column of 12.5 g of alumina. If solid has separated from the solution or sample, filter the mixture through a paper into the column and save the crystals for possible combination with more of the same material which might be isolated by chromatography. Since the early eluates are liable to contain mixtures, the first five fractions should be limited to 5 ml each and collected in small Erlenmeyer flasks (10- or 25-ml). When the fifth fraction has been collected, close the pinchclamp and defer further chromatography until the early eluates have been evaporated, pumped out, and examined (appearance and odor, particularly when warm). Let these fractions stand to see if, in time, crystals appear and continue with the chromatogram. If fraction 5 afforded only a trace of oil, elute with 20 ml of petroleum ether and collect the liquid in a 50-ml flask, as fraction 6. Then collect three fractions eluted by 20-ml portions of benzene, stop the chromatogram, and evaluate the results.

Late fractions consisting of oils usually crystallize readily when scratched. If so, add a little petroleum ether, digest briefly on the steam bath, rub the material off the walls with a stirring rod, cool, collect, and take the melting point. If one or more of the early fractions crystallize, the crystals will probably be contaminated with oil, which can be separated as follows: Make ready a Hirsch funnel and paper, chill a mixture of 2 ml of water and 8 ml of methanol in ice and cool the flask as well, introduce solvent and stir to loosen the crystals, and then collect the product and wash it with chilled solvent. This process is known as trituration.

Identify all crystalline products encountered and account for their presence.

PART II.   Thin Layer Chromatography

TLC

This microanalytical technique employs alumina as adsorbent, as in column chromatography, and silica gel as well but in the form of a thin layer on a glass plate. In this experiment the plate is a 3×1-inch microscope slide. A 1% solution of the substance to be examined is spotted onto the plate about 1 cm from the bottom end and the plate is inserted into a 4-oz wide-mouth bottle containing 4 ml of an organic solvent. The glass stopper is put in place and the time noted.

Colored compounds

The solvent travels rapidly up the thin layer by capillary action, and if the substance is a pure colored compound one soon sees a spot either traveling along with the solvent front or, more usually, at some distance behind the solvent front. One can remove the slide, quickly mark the front before the solvent evaporates, and calculate the Rf value (Chapter 34).

If two colored compounds are present and an appropriate solvent is selected, two spots will appear.

And colorless ones

Fortunately the method is not limited to colored substances. Any organic compound capable of being eluted from alumina will form a spot, which soon becomes visible when the solvent is let evaporate and the plate let stand in a stoppered 4-oz bottle containing a few crystals of iodine. Iodine vapor is adsorbed by the organic compound to form a brown spot. A spot should be outlined at once with a pencil, since it will soon disappear as the iodine sublimes away; brief return to the iodine chamber will regenerate the spot. The order of elution and the elution-power for solvents are the same as for column chromatography.

The adsorbent recommended for making TLC plates is a preparation of finely divided alumina containing plaster of paris as binder (Fluka[1]). A slurry of this material in water can be applied to microscope slides by simple techniques of dipping (A) or coating (B); for a small class, or for occasional preparation of a few slides by an individual, the more economical coating method is recommended. Either of two techniques of spraying (C) is applicable to the mass production of plates but requires special equipment.

(A) By Dipping.   Place 15 g of Fluka, or similar alumina formulated for TLC, in a 40 × 80 mm weighing bottle[2] and stir with a glass rod while gradually pouring in 75 ml of distilled water. Stir until lumps are eliminated and a completely homogeneous slurry results. Grasp a pair of clean slides[3] at one end with the thumb and forefinger, dip them in the slurry of adsorbent, withdraw with an even unhurried stroke, and touch a corner of the slides to the mouth of the weighing bottle to allow excess fluid to return to the container. Then dry the working surface by mounting the slides above a 70-watt hot plate, using as support two pairs of 1 × 7 cm strips of asbestos paper (or a pair of applicator sticks). Drying takes about a minute and a half and is evident by inspection. Remove the plate with a forceps as soon as it is dry, for cooling takes longer. If you dip and dry 8–10 slides in succession, finished plates will be ready for use when you are through. Alternatively, dry the slides in an oven at 110°. This heating activates the alumina. Coated slides dried at room temperature do not give as intense spots as those that have been heat-dried.

Keep the storage bottle of adsorbent closed when not in use. Note that this is not a stable emulsion but that the solid settles rapidly on standing. Stir thoroughly with a rod before each reuse.

Plates ready for use, as well as clean dry slides, are conveniently stored in a micro slide box (25 slides).

Figure 23.6 Preparation of TLC plates by dipping. The two microscope slides are grasped at one end, dipped into the silica gel slurry, removed and separated, and dried and activated by heating.

Notes for the instructor

[1] Aluminumoxid Fluka für Dünnschicht Chromatographie, Typ D5, Fluka AG, Buchs, S. G. Switzerland; Fluka Alumina is distributed by Gallard-Schlesinger Chemical Manufacturing Corp., Garden City, L. I., N. Y., 11530.
[2] Kimble Exax, 15146.
[3] The slides should be washed with water and a detergent, rinsed with tap water and then with distilled water, and placed on a clean towel to dry. Dry slides should be touched only on the sides, for a touch on the working surface may spoil a chromatogram.

(B) By Coating.   Place 2 g of Fluka alumina and 10 ml of distilled water in a 25-ml Erlenmeyer, stopper the flask and shake to produce an even slurry. Keep the flask stoppered when not in use. Place a clean, dry slide on a block of wood or box with the slide projecting about 1 cm on the left-hand side (if you are right handed) so that it can be grasped easily on the two sides. Swirl the flask to mix the slurry and draw a portion into a medicine dropper. Hold the dropper vertically and, starting at the right end of the slide, apply emulsion until the entire upper face is covered; make further applications to repair pin holes and eliminate bubbles. Grasp the left end of the slide with a forceps and even the emulsion layer by tilting the slide to the left to cause a flow, and then to the right; tilt again to the front and to the rear. Dry the slide on a hot plate as you did when preparing slides by dipping, procedure (A). The adsorbent should be about 0.25 mm thick.

(C) By Spraying.   Two techniques which enable an operator to spray a hundred small slides in a few minutes are of particular use in research, where large plates are often required. A batch of slides for student use can be prepared by spraying and let air-dry at room temperature, but each plate should then be activated by heating.[4]

One technique utilizes the Sutter spray gun of metal.[5] Place 15 g of Fluka alumina and 75 ml of distilled water in a 125-ml Erlenmeyer, stopper the flask and shake to produce an even suspension, and pour this into the chamber of the spray gun. Connect the gun with a piece of suction tubing to a source of gas pressure, preferably a cylinder of nitrogen or oxygen set at a pressure of 15 lbs/sq in. Place a batch of slides side by side on a piece of cardboard or paper. Hold the gun vertically with the nozzle about 2 ft. above the target, practice spraying a piece of paper, and then spray the slides evenly.

(D) Spotting Test Solutions.   This is done with micropipettes of either style shown in Fig. 23.7, made by drawing out a 5-mm soft glass tube. Satisfactory micropipettes can also be made by drawing open-end mp capillaries in a microburner flame. The bore should be of such a size that, when the pipette is dipped deep into benzene, the liquid flows in to form a tiny thread which, when the pipette is withdrawn, does not flow out to form a drop. Style (a) has the advantage that it can be cleaned by introducing solvent (benzene) with a capillary dropping tube at the top and then shaking the solvent on through the tube. To spot a test solution, let a 2–3 cm column of solution flow into the pipette, hold this vertically over a coated plate (cold), aim it at a point on the right side of the plate and about 1 cm from the bottom, and lower the pipette until the tip just touches the

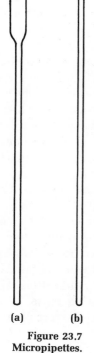

(a)   (b)

**Figure 23.7
Micropipettes.**

[4] Slides coated by dipping in a slurry of alumina in chloroform-methanol according to J. J. Peifer, *Microchimica Acta*, 529 (1962), dry very rapidly but still have to be steamed and heat-activated for the best performance.

[5] Sprühgerät nach B. Sutter, Hormuth-Vetter, Wiesloch (Baden), Germany.

adsorbent and liquid flows onto the plate; withdraw when the spot is about 2 cm in diameter. Make a second 2-mm spot on the left side of the plate, let it dry, and make two more applications of the same size (2-mm) at the same place. Determine whether the large or the small spot gives the better results.

EXPERIMENTS   **1. Dyes**

The first experiment utilizes a test solution[6] containing a green dye, a red dye, and a blue dye of the structures shown below. Or try mixtures of commercially available food colors. Chromatograms are to be run on the test solution, using the solvents listed:

   (a) Benzene (4 ml)
   (b) Benzene (3 ml)-Chloroform (1 ml)
   (c) Cyclohexane (4 ml)
   (d) Chloroform (4 ml)
   (e) Cyclohexane-Ethyl acetate. (Use 4 ml of a mixture of 8.5 ml of cyclohexane and 1.5 ml of ethyl acetate.)

For each chromatogram, (a)-(e), make two 2-mm spots of the test solution on the right and left sides of the plate and about 1 cm from the end, one of the two spots of single strength and the other of triple strength. For chromatogram (a) prepare a running chamber by pouring 4 ml of benzene into a 4-oz bottle, swirl to saturate the air with solvent, and put the stopper in place. With a forceps, grasp a plate that has been spotted and lower it into the bottle and rest it in a slanting position against the wall (Fig. 23.8). Replace the stopper, observe the plate carefully, and see if the dyes separate. Let chromatogram (a) continue for a time while starting others in the series. In each case, after a clear-cut pattern of separation or nonseparation has been achieved make a drawing to record the position, shape, and color of each spot.

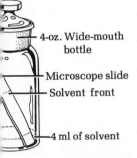

4-oz. Wide-mouth bottle

Microscope slide
Solvent front

4 ml of solvent

Figure 23.8
A method of developing thin layer chromatographic plates.

1,4-Di-*p*-toluidinoanthraquinone
(D & C Green 6)   **Sudan Red G**   **Indophenol**
(blue)

[6] Prepare an alcoholic solution containing 1% each of D & C Green 6, Sudan Red G, and Indophenol, available from Allied Chemical Corp., Specialty Chemicals Division, 40 Rector St., New York, N. Y., 10006.

Questions

(1) You probably have found that the dyes are separated effec-tively with benzene or with benzene-chloroform and that the rel-ative rate of travel indicates the same order of elution in each case. Examine the structures of the three dyes. Can you account for the order of travel? Think particularly about hydrogen bonding.

(2) If you noted a reversal in the order of elution from that in (1) above, can you account for the reversal? Again consider hydrogen bonding and remember that a hydrogen bond to oxygen is stronger than one to nitrogen. A solvent capable of hydrogen bonding might interfere with internal hydrogen bonding.

(3) Have you thought of an explanation for the results with chro-matograms (c) and (d)? If not, consider the fact that a liquid of poor solvent power may not move a substance spotted on a plate at all and that a liquid of high solvent power may move all the dyes along with the solvent front.

(4) Suppose the results of the thin layer experiments were to be used as a guide to running a column chromatographic separation of these pigments on alumina, what solvent or solvents would you select and how would you proceed?

2. Hydrocarbon Pigments

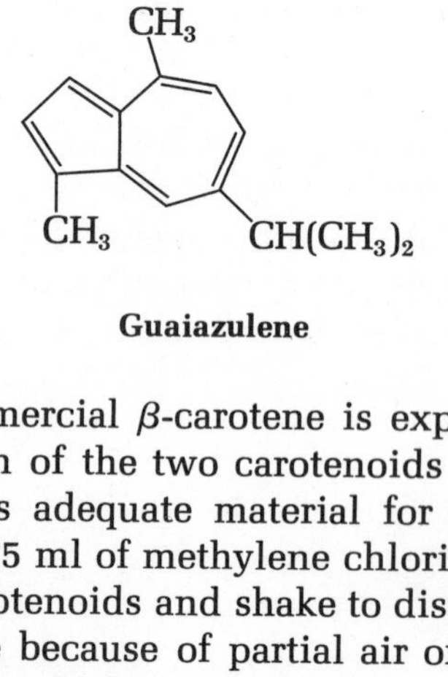

Guaiazulene 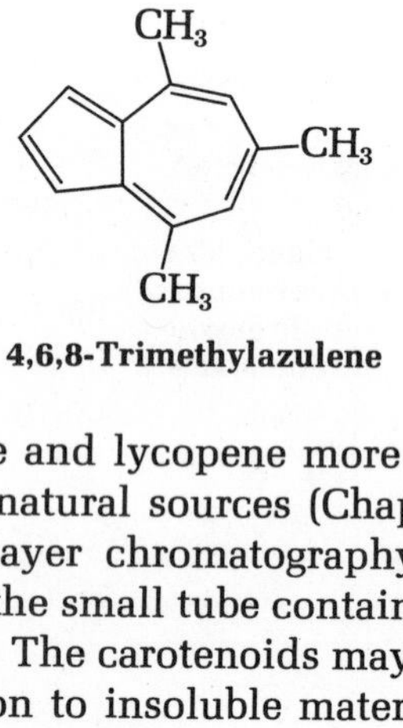   4,6,8-Trimethylazulene

Solubility in hexane at 0°, mg/ml: lycopene, 7.7 mg.; β-carotene, 109 mg.

Commercial β-carotene is expensive and lycopene more so, but isolation of the two carotenoids from natural sources (Chapter 58) provides adequate material for thin layer chromatography. Add about 0.5 ml of methylene chloride to the small tube containing the two carotenoids and shake to dissolve. The carotenoids may not all dissolve because of partial air oxidation to insoluble material, but there should be enough colored supernatant liquid for numerous chromatographic experiments. Keep the solutions out of the light, and when no more is required evaporate the solvent for safer storage.

If guaiazulene from guaiene is not available for testing, 4,6,8-tri-methylazulene[7] is recommended (1% solution in benzene, protected from light).

Autoxidation

The highly unsaturated hydrocarbons are subject to rapid photo-chemical autoxidation, but during a chromatogram run they are pro-tected by solvent vapor. However, on removal of a plate from the

---

[7] Aldrich Chemical Co. Aldrich stocks azulene itself, but at a higher price than 4,6,8-trimethyl-azulene.

chamber a spot may disappear rapidly and so should be outlined in pencil immediately. Save a plate from which outlined spots of the three hydrocarbons have disappeared and develop it in an iodine chamber as described in Experiment 3.

With care to make small spots, you should be able to run a chromatogram of the three hydrocarbons on one plate; put the blue hydrocarbon in the center lane. Spot two plates with the three hydrocarbons and run one in benzene and the other in cyclohexane.

Questions

Does either solvent effect a separation? If one of them fails, what is the explanation? The tomato is reported to contain one part of β-carotene to 10 parts of lycopene; do your chromatograms provide evidence for the presence or absence of β-carotene in the lycopene sample? Suggest a procedure for the separation of a mixture of 1 g of lycopene and 1 g of β-carotene, using a combination of crystallization and column chromatography. How would you control the separation? Which of the three hydrocarbons is the most sensitive to air oxidation?

3. Colorless Compounds

You are now to apply the thin layer technique to the more general case in which the chromatograms are developed in a 4-oz bottle containing crystals of iodine. During development, spots appear rapidly, but remember that they also disappear rapidly. Therefore, outline each spot with a pencil immediately on withdrawal of the plate from the iodine chamber. Solvents suggested are as follows:

Cyclohexane	Benzene (3 ml)–Chloroform (1 ml)
Benzene	9:1 Benzene-Methanol (use 4 ml)

Compounds for trial are to be selected from the following list (all 1% solutions in benzene):

Make your own selections

1. Anthracene*
2. Cholesterol
3. 2,7-Dimethyl-3,5-octadiyne-2,7-diol
4. Diphenylacetylene
5. *trans,trans*-1,4-Diphenyl-1,3-butadiene*
6. p-Di-*t*-butylbenzene
7. 1,4-Di-*t*-butyl-2,5-dimethoxybenzene
8. *trans*-Stilbene
9. 1,2,3,4-Tetraphenylnaphthalene*
10. Tetraphenylthiophene
11. p-Terphenyl*
12. Triphenylcarbinol
13. Triptycene

* Fluorescent under UV light.

It is up to you to make selections and to plan your own experiments. Do as many as time permits. One plan would be to select a
pair of compounds estimated to be separable and which have R_f
values determinable with the same solvent. One can assume that a
hydroxyl compound will travel less rapidly with a hydrocarbon solvent than a hydroxyl-free compound, and so you will know what to

Preliminary trials on
used plates

expect if the solvent contains a hydroxylic component. An aliphatic
solvent should carry along an aromatic compound with aliphatic
substituents better than one without such groups. However, instead
of relying on assumptions, you can do brief preliminary experiments
on used plates on which previous spots are visible or outlined. If
you spot a pair of compounds on such a plate and let the solvent rise
about 3 cm from the starting line before development, you may be
able to tell if a certain solvent is appropriate for a given sample.
If so, run a complete chromatogram on the two compounds on a
fresh plate. If separation of the two seems feasible, put two spots of
one compound on a plate, let the solvent evaporate, and put spots of
the second compound over the first ones. Run a chromatogram and
see if you can detect two spots in either lane (with colorless compounds, it is advisable not to attempt a three-lane chromatogram
until you have acquired considerable practice and skill).

4. Discussion

In case you have investigated hydroxylated compounds, you
doubtless have found that it is reasonably easy to separate a hydroxylated from a nonhydroxylated compound, or a diol from a mono-ol.
How, by a simple reaction followed by a thin layer chromatogram,
could you separate cholesterol from triphenylcarbinol? Heating a
sample of each with acetic anhydride–pyridine for 5 min on the
steam bath, followed by chromatography, should do it. Study of the
rate of acetylation of cholesterol by acetic acid (Chapter 23, Section 2) could be done more simply by the thin layer technique than
by column chromatography and with less material. On first trial of
a new reaction, one wonders if anything has happened and how
much, if any, starting material is present. A comparative chromatogram of reaction mixture with starting material may tell the story.
How crude is a crude reaction product? How many components are
present? The thin layer technique may give the answers to these
questions and suggest how best to process the product. A preparative
column chromatogram may afford a large number of fractions of
eluent (say 1 to 30). Some fractions probably contain nothing and
should be discarded, while others should be combined for evaporation and workup. How can you identify the good and the useless
fractions? Take a few used plates and put numbered circles on clean
places of each, spot samples of each of the fractions, and, without
any chromatography, develop the plates with iodine. Negative
fractions for discard will be obvious and the pattern alone of positive
fractions may allow you to infer which fractions can be combined.

Thin layer chromatograms of the first and last fractions of each suspected group would then show whether or not your inferences are correct.

5. Fluorescence

Four of the compounds listed in Section 3 are fluorescent under ultraviolet light and such compounds give colorless spots which can be picked up on a chromatogram by fluorescence (after removal from the UV-absorbing glass bottle). If a UV-light source is available, spot the four compounds on a used plate and observe the fluorescence. Take this opportunity to examine a white shirt or handkerchief under UV light to see if it contains a brightener, that is, a fluorescent white dye or optical bleach. Brighteners of the type of Calcofluor white MR, a sulfonated *trans*-stilbene derivative, are commonly used in detergent formulations for cotton, and the substituted coumarin derivative formulated is typical of brighteners used for nylon, acetate, and wool.

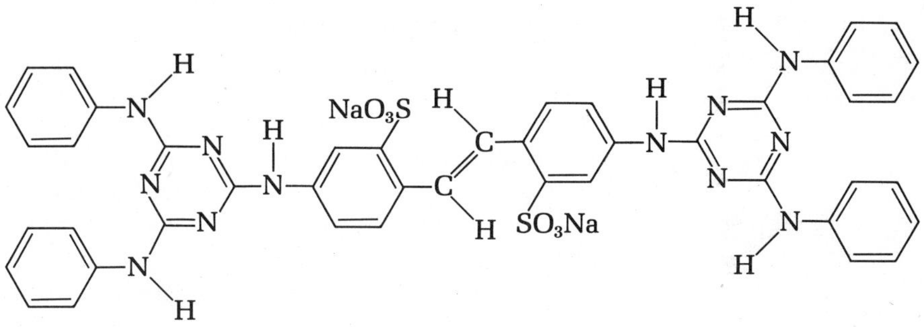

Do not look into a UV lamp

Calcofluor White MR

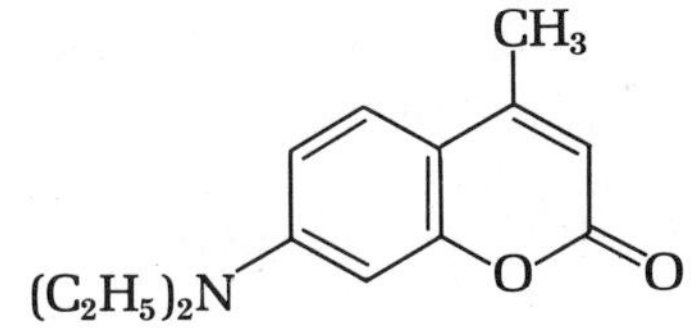

7-Diethylamino-4-methylcoumarin

Detergents normally contain 0.1–0.2% of optical bleach. The amount of dye on a freshly laundered shirt is approximately 0.01% of the weight of the fabric.

A technique of some use in the chromatography of nonfluorescent compounds is to incorporate an inorganic fluorescent indicator in the adsorbent. The whole plate is then fluorescent under a UV lamp and a dark spot appears where a migrating compound has quenched the fluorescence.

24 *Adipic Acid*

KEYWORDS Oxidation   Exothermic reaction
                          Acidic dichromate   Cyclohexanol, cyclohexanone
                          Alkaline permanganate   Manganese dioxide

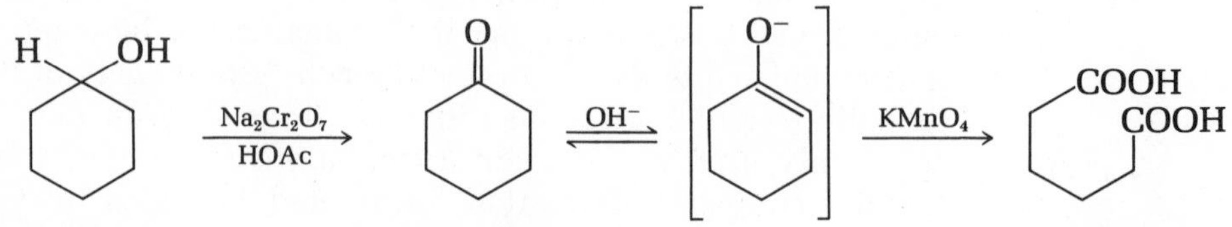

Cyclohexanol
Bp 161.5°, den 0.96,
MW 100.16

Cyclohexanone
Bp 157°, den 0.95,
MW 98.14, 1.5 g/100 ml $H_2O^{10°}$

Adipic acid
Mp 153°, 1.4 g/100 g $H_2O^{15°}$,
MW 146.14

Cyclohexanol can be oxidized directly to adipic acid by nitric acid in about 50% yield, but a stirring motor is required and toxic fumes have to be disposed of. Direct oxidation with permanganate is free from these limitations but takes several days. Since the first step in the permanganate reaction is the slow one, the overall reaction time is greatly reduced by oxidation of cyclohexanol to cyclohexanone with dichromate in acetic acid followed by permanganate oxidation of cyclohexanone catalyzed by alkali to promote formation of the intermediate enolate.

Before beginning the experiment write complete, *balanced* equations for both reactions in your laboratory notebook.

EXPERIMENTS   **1. Cyclohexanone**

In a 125-ml Erlenmeyer flask dissolve 15 g of sodium dichromate dihydrate in 25 ml of acetic acid by swirling the mixture on the hot plate (Chapter 2, Section 7) and then cool the solution with ice to 15°. In a second Erlenmeyer flask chill a mixture of 15.0 g of cyclohexanol and 10 ml of acetic acid in ice. After the first solution is cooled to 15°, transfer the thermometer and adjust the temperature in the second flask to 15°. Wipe the flask containing the dichromate solution, pour the solution into the cyclohexanol–acetic acid mixture, rinse the flask with a little solvent, note the time, and take the initially light orange solution from the ice bath, but keep the ice bath ready for use when required. The exothermic reaction that is soon evident can get out of hand unless controlled. When the temperature rises to 60° cool in ice just enough to prevent a further rise

Reaction time 45 min

152

and then, by intermittent brief cooling, keep the temperature close to 60° for 15 min. No further cooling is needed, but the flask should be swirled occasionally and the temperature watched. The usual maximal temperature is 65° (25–30 min). When the temperature begins to drop and the solution becomes pure green, the reaction is over. Allow 5–10 min more reaction time and then pour the green solution into a 250-ml round-bottomed flask, rinse the Erlenmeyer flask with 100 ml of water, and add the solution to the flask for steam distillation of the product (apparatus, Fig. 9.2). Distil as long as any oil passes over with the water and, since cyclohexanone is appreciably soluble in water, continue somewhat beyond this point.

To salt out the cyclohexanone add to the distillate 0.2 g of sodium chloride per ml of water present and swirl to dissolve the salt. Then pour the mixture into a separatory funnel, rinse the flask with ether, add more ether to a total volume of 25–30 ml, shake, and draw off the water layer. Then wash the ether layer with 25 ml of 10% sodium hydroxide solution to remove acetic acid, test a drop of the wash liquor to make sure it contains excess alkali, and draw off the aqueous layer. Wash with sodium chloride solution, filter the ethereal extract through a cone of anhydrous sodium sulfate into a tared 125-ml Erlenmeyer, and evaporate the ether on the steam bath under an aspirator tube. Cool to room temperature, evacuate the crude cyclohexanone at the water pump and weigh the product; yield 11–12.5 g.[1]

You have the choice of using the crude cyclohexanone in the following procedure for preparing adipic acid or of purifying it by distillation. To judge if the crude cyclohexanone is suitable for such use, prepare a carbonyl derivative such as the oxime (mp 89°), or 2,4-dinitrophenylhydrazone (mp 160°), or semicarbazone (mp 166°).

2. Adipic Acid

The reaction to prepare adipic acid is conducted with 10.0 g of cyclohexanone, 30.5 g of potassium permanganate, and amounts of water and alkali that can be adjusted to provide an attended reaction period of one-half hour, procedure (a), or an unattended overnight reaction, procedure (b).

Choice of controlling the temperature for 30 min or running an unattended overnight reaction

(a) For the short-term reaction, mix the cyclohexanone and permanganate with 250 ml of water in a 500-ml Erlenmeyer flask, adjust the temperature to 30°, note that there is no spontaneous temperature

[1] When the acetic acid solutions of cyclohexanol and dichromate were mixed at 25° rather than at 15° the yield of crude cyclohexanone was only 6.9 g. A clue to the evident importance of the initial temperature is suggested by an experiment in which the cyclohexanol was dissolved in 12.5 ml of benzene instead of in 10 ml of acetic acid and the two solutions were mixed at 15°. Within a few minutes orange-yellow crystals separated and soon filled the flask; the substance probably is the chromate ester, $(C_6H_{11}O)_2CrO_2$. When the crystal magma was let stand at room temperature the crystals soon dissolved, exothermic oxidation proceeded, and cyclohexanone was formed in high yield. Perhaps a low initial temperature insures complete conversion of the alcohol into the chromate ester before side reactions set in.

rise, and then add 2 ml of 10% sodium hydroxide solution. A temperature rise is soon registered by the thermometer. It may be of interest to determine the temperature at which you can just detect warmth, by holding the flask in the palm of the hand and by touching the flask to your cheek. When the temperature reaches 45° (15 min) slow the oxidation process by brief ice-cooling and keep the temperature at 45° for 20 min. Wait for a slight further rise (47°) and an eventual drop in temperature (25 min), and then heat the mixture by swirling it over a flame to complete the oxidation and to coagulate the precipitated manganese dioxide. Make a spot test by withdrawing reaction mixture on the tip of a stirring rod and touching it to a filter paper; permanganate, if present, will appear in a ring around the spot of manganese dioxide. If permanganate is still present, add small amounts of sodium bisulfite until the spot test is negative. Then filter the mixture by suction, wash the brown precipitate well with water, and evaporate the filtrate over a flame to a volume of 70 ml. If the solution is not clear and colorless, clarify it with decolorizing charcoal and evaporate again to 70 ml. Acidify the hot solution with concd hydrochloric acid to pH 1–2, add 10 ml acid in excess, and let the solution stand to crystallize. The yield of adipic acid, mp 152–153°, is 6.9 g.

(b) In the alternative procedure the weights of cyclohexanone and permanganate are the same but the amount of water is doubled (500 ml) to moderate the reaction and make temperature control unnecessary. The temperature is initially adjusted to 30°, 10 ml of 10% sodium hydroxide is added, the mixture is swirled briefly and let stand overnight (maximum temperature 45–46°). The work-up is the same as in procedure (a) and the yield of adipic acid is 8.3 g.

25 *Mandelic Acid*

KEYWORDS   Bisulfite addition compound   Mandelonitrile
            Cyanohydrin   Potassium cyanide

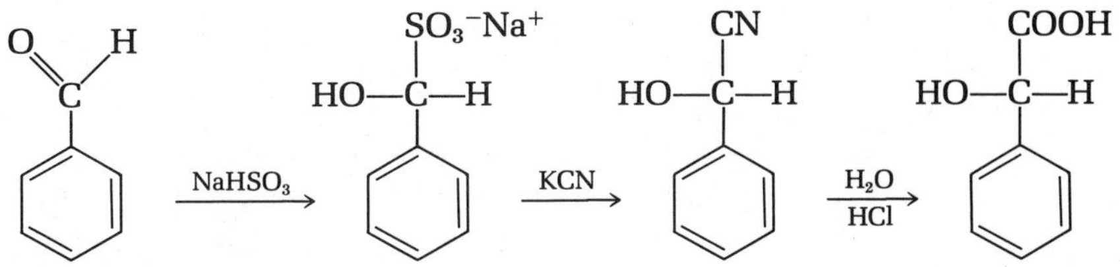

Benzaldehyde
MW 106.12
Den 1.05

Mandelonitrile

Mandelic acid
Mp 118.5°, MW 152.14
16.0 g/100 g $H_2O^{20°}$, pK_a 3.4

This α-hydroxy acid is prepared from benzaldehyde through the cyanohydrin, which is produced by interaction of the benzaldehyde bisulfite addition compound with potassium cyanide. This interchange reaction eliminates the hazard of working with volatile, toxic hydrogen cyanide, but even potassium cyanide is dangerous if safety measures are not observed. The interchange reaction is reversible, and excess cyanide is used to shift the equilibrium in favor of the cyanohydrin. Mandelonitrile, a liquid, undergoes changes on standing and hence should be processed further without delay; it is extracted with ether, the solution is carefully washed free of cyanide ion, and the nitrile is then hydrolyzed with hydrochloric acid. An intermediate ketimine hydrochloride, $C_6H_5CH(OH)C=NH\cdot HCl$, soon gives way to mandelic acid and ammonium chloride, both of which are soluble in the aqueous acid. The solution is cooled and extracted with ether, and the ether is then displaced by benzene for crystallization, since mandelic acid is much less soluble in this solvent than in ether.

EXPERIMENT

KCN is extremely toxic

Caution! Potassium cyanide is extremely poisonous if it gets into the blood through a cut, if transferred from the hand to the mouth, or if it comes into contact with acid and liberates hydrogen cyanide. Never touch the solid with the fingers; be very careful not to spill it and if you do, clean it up at once. Take care not to spill a solution or reaction mixture containing the substance, and discard the waste solution directly into the drain and wash the sink with excess water.

1. Procedure

In a 125-ml Erlenmeyer flask dissolve 11 g of sodium bisulfite i 30 ml of water by brief swirling, add 10 ml of benzaldehyde,[1] an swirl vigorously and stir until the oily aldehyde is all converted int the crystalline bisulfite addition compound. Cool to room tempe ature but not below, add 14 g of potassium cyanide (**caution**) an 25 ml of water (rinse down the walls), and take the mixture to a hooc Swirl and stir for about 10 min until all but a trace of solid has dis solved (break up lumps with a glass rod). Mandelonitrile separate as a thick oil. Pour the mixture into a separatory funnel, rinse th flask with small amounts of ether and water (*and at once wash th flask free of cyanide with water*), and then shake the mixture vigor ously for a full minute to insure complete reaction. Add 20 ml o ether, shake, and run the aqueous layer down the drain. Wash th ether extract with 25 ml of water and then with 25 ml of saturatec sodium chloride solution (note and account for the difference i appearance of the ether layer after the washing). Then run the solu tion into a 125-ml round-bottomed flask and rinse the funnel witl ether. Add 15 ml each of concentrated hydrochloric acid and water clamp the flask over a microburner, and fit the flask with a stillhead thermometer and a condenser-adapter for downward distillatio into an ice-cooled receiver. Distil off the ether and continue to i boiling point of 100°. Then remove the flame, disconnect the still head, and mount the condenser vertically for refluxing. On reflux

Reflux 15 min

ing for just a few minutes, the oily droplets of mandelonitrile shoulc disappear and give way to a clear yellow solution. Note the time and reflux for 15 minutes to ensure complete hydrolysis of th nitrile. Then cool the solution to room temperature and rinse intc a separatory funnel with a little ether. Add 20 ml more ether, shake well, let the layers separate, and draw off the aqueous layer into £ second separatory funnel or a dry flask, and wait a few minutes fo separation of a little more aqueous solution. The moist etherea] solution is to be run into a 250-ml round-bottomed flask, but firs measure 100 ml of benzene into this flask and make a mark at the level of the liquid. Then run in the ethereal extract and rinse the funnel with ether. Extract the aqueous solution with two furthe 20-ml portions of ether and add the extracts to the flask containing benzene. Add a boiling stone, put a stillhead with thermometer in place, and arrange for distillation from the steam bath into an ice-cooled receiver. Heat cautiously at first, since the flask is very full, and then more strongly. Water is eliminated gradually by azeotropic distillation and the boiling point rises as ether and water are elim-inated. Droplets of water adhere to the stillhead, but the solution in the boiling flask clears. Stop the distillation when the thermometer registers 91–92° and the volume has been reduced to the 100-ml

Note for the instructor

[1] As a safety measure, the benzaldehyde should be tested to make sure it does not contain an appreciable amount of benzoic acid.

mark. Disconnect the flask and inspect the bottom to see if a trace of ammonium chloride has separated, either as a film of solid or as a gum; make sure that no flames are near, and decant the hot solution into a 125-ml Erlenmeyer. Crystallization usually starts soon; yield 9.5–10.0 g, mp 118–119°.[2] Should no crystals appear, evaporate the solvent, scratch to induce crystallization, and digest the solid with benzene.

Note. When mandelic acid that has been crystallized from benzene is allowed to stand in contact with the mother liquor for several days, the needles gradually change into granular, sugarlike crystals of a molecular compound containing one molecule each of mandelic acid and benzene. These crystals are stable only in contact with benzene and at temperatures below 32.6°; on exposure to the air the benzene of crystallization evaporates and the mandelic acid is left as a white powder.

Note for the instructor

[2] A special experiment that provides starting material for use in Chapter 46 is reduction of mandelic acid to phenylacetic acid by the method of K. Miescher and J. R. Billeter, *Helv. Chim. Acta*, **22**, 601 (1939). The product can be distilled in vacuum in an apparatus such as that of Fig. 62.1.

26 *Pinacol and Pinacolone*

KEYWORDS Bimolecular reduction
Diradical
Pinacol-pinacolone rearrangement
Oxime

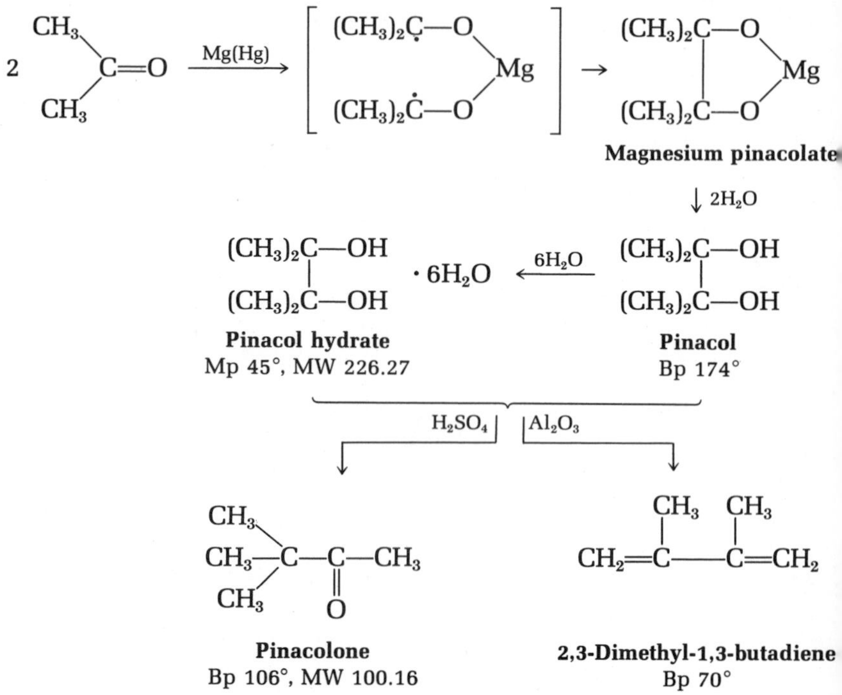
2
Mg(Hg)
Magnesium pinacolate
2H₂O
6H₂O
Pinacol hydrate
Mp 45°, MW 226.27
Pinacol
Bp 174°
H₂SO₄
Al₂O₃
Pinacolone
Bp 106°, MW 100.16
2,3-Dimethyl-1,3-butadiene
Bp 70°

Acetone on treatment with metallic reducing agents is converted to a considerable extent into the product of bimolecular reduction, pinacol. The diol is a liquid but can be isolated as the crystalline hexahydrate. In the following procedure dry acetone is treated under Grignard conditions with magnesium that has been amalgamated with 0.02 mole of mercury by reaction with mercuric chloride in the presence of a portion of the acetone: $HgCl_2 + Mg \rightarrow Hg + MgCl_2$. Bimolecular reduction probably proceeds through a diradical and affords magnesium pinacolate as a voluminous precipitate, which on treatment with water affords pinacol and then pinacol hydrate. This crystalline solid is easily separated from unreacted acetone and isopropyl alcohol, the product of normal, unimolecular reduction.

Pinacol on catalytic dehydration over alumina undergoes normal dehydration to 2,3-dimethyl-1,3-butadiene, but on dehydration with sulfuric acid the ditertiary glycol largely undergoes rearrangement and affords pinacolone in 70% yield. Fully purified pinacolone, as well as its oxime, has a fine camphorlike odor, and a similarity in structure is evident from the formulas.

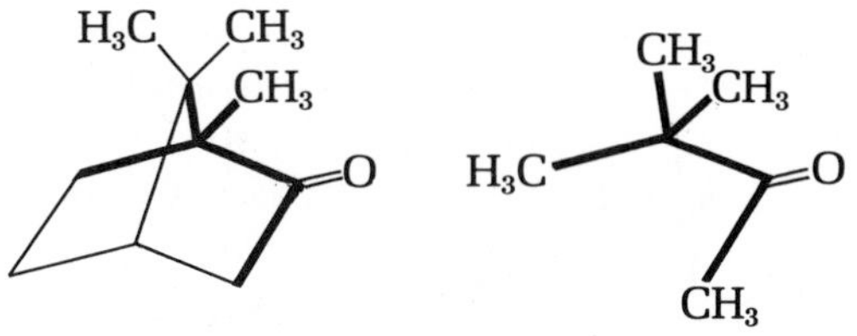

Camphor   **Pinacolone**

You are to prepare pinacol, purify it as the hydrate, and use this for preparation of pinacolone, which is to be characterized as the crystalline oxime. As a special experiment, pinacolone can be further converted by the haloform reaction into trimethylacetic acid.[1]

EXPERIMENTS   ## 1. Pinacol Hydrate

In a 500-ml round-bottomed flask equipped with a long reflux condenser, place 8 g of magnesium turnings and 100 ml of dry benzene. The apparatus should be thoroughly dry, as in a Grignard reaction. A solution of 9 g of mercuric chloride in 75 ml of dry acetone is placed in a small dropping funnel fitted into the top of the condenser by means of a grooved cork. Add about one fourth of this solution, and if the reaction does not start in a few minutes (vigorous ebullition) warm the mixture gently on the steam bath, but make ready to plunge the flask into cold water if moderation of the reaction becomes necessary. Once started, the reaction will proceed vigorously at this stage without further heating. Run in the remainder of the acetone solution at such a rate as to keep the reaction in progress. The boiling should be vigorous, but some cooling may be necessary in order to prevent escape of uncondensed acetone. When the reaction slows down (5–10 min), mount the flask on the steam bath with a single clamp placed as high as possible on a ring stand to keep the condenser upright. If the jaws of the clamp fit only loosely on the condenser, the flask can be grasped and swirled vigorously to dislodge the magnesium pinacolate that separates. Keep the mixture boiling briskly for one hour and swirl frequently during this period.

At the end of the 1-hr period pour 20 ml of water in through the

[1] *Organic Syntheses, Coll. Vol.* **1**, 526 (1941).

condenser and boil the mixture for 0.5 hr with frequent shaking. This converts the pinacolate into pinacol (soluble in benzene) and a precipitate of magnesium hydroxide. Filter the hot solution by suction, return the solid magnesium hydroxide to the flask and reflux it with 50 ml of ordinary benzene for 5–10 min, and filter this solution as before. Pour the combined filtrates into an Erlenmeyer flask and evaporate the solution on a steam bath under an aspirator tube to one third the original volume. Then add 15 ml of water, cool well in an ice bath, and scratch the walls of the flask with a stirring rod. The pinacol hydrate separates as an oil which soon solidifies; it should be collected only after thorough cooling and stirring for maximum crystallization. Scrape the crude material onto a suction funnel, wash it with a little cold benzene, press it well with a spatula, and let it drain for about 5 min in order to remove as much benzene as possible.

Extinguish flames

The crude material is purified by crystallization from water, in which it is very soluble. Transfer the product to a small Erlenmeyer flask, dissolve it in 25 ml of water by heating, and boil the solution gently for about 5 min in order to remove traces of benzene. If the solution is appreciably colored it should be clarified with decolorizing charcoal. Filter the hot solution through a rather large funnel, which has been warmed on the steam bath, into a second flask and set it aside to crystallize. Cool in ice before collecting the large crystals of pinacol hydrate. The purified product should be dried in a cool place since it sublimes easily. The yield is 18–20 g; mp 46–47°. Calculate the theoretical yield from the amount of magnesium employed, for acetone is taken in considerable excess.

2. Pinacolone

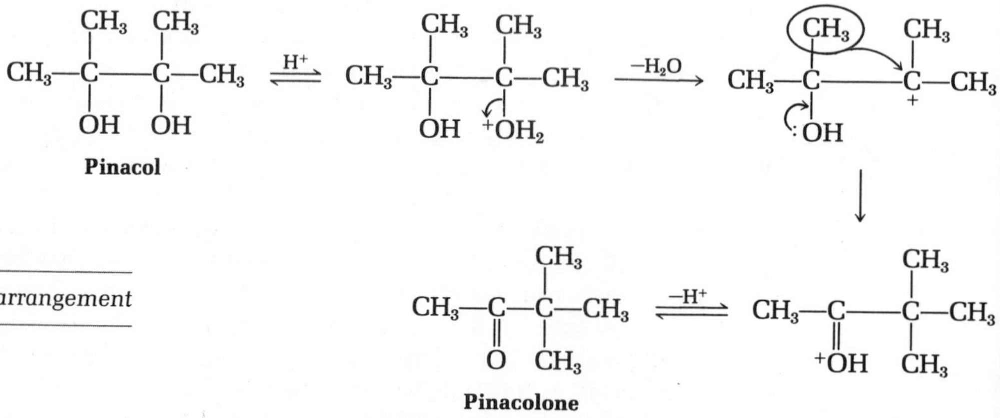

Pinacol rearrangement

Into a 250-ml round-bottomed flask pour 80 ml of water, then 20 ml of concentrated sulfuric acid, and dissolve 20 g of pinacol hydrate in the warm solution. Attach a reflux condenser, boil the mixture for 15 min, observe, and note carefully the changes that take place. Cool until the boiling ceases, attach a stillhead with a steam-inlet tube, and steam distil (Chapter 9) until no more pinacolone comes over. Separate the upper layer of crude pinacolone, dry it with calcium chloride, and purify the material by fractional distillation. Use a fractionating column and tetrachloroethane as chaser and see if you can detect and remove a small forerun containing dimethylbutadiene.

3. Pinacolone Oxime

$$(CH_3)_3C-\overset{\overset{\textstyle O}{\|}}{C}-CH_3 + H^+ \rightleftharpoons \left[(CH_3)_3C-\overset{\overset{\textstyle ^+OH}{\|}}{C}-CH_3 \leftrightarrow (CH_3)_3C-\overset{\overset{\textstyle O}{|}\diagdown H}{\underset{+}{C}}-CH_3 \right] \quad (1)$$

A conjugate acid

$$H_2N-OH + H^+ \rightleftharpoons H_3\overset{+}{N}-OH \quad (2)$$

A conjugate acid

$$(CH_3)_3C-\overset{\overset{\textstyle O}{\|}}{C}-CH_3 + H_3\overset{+}{N}\cdot OH\ \overset{Cl^-}{} \quad \xrightarrow[\text{Na}^+\text{OAc}^-]{\text{pH 5}}\ (CH_3)_3C-\overset{\overset{\textstyle N\diagup OH}{\|}}{C}-CH_3 + H_2O \quad (3)$$

Oxime

Protonation of pinacolone (eq. 1) makes it *more* electrophilic, and protonation of hydroxylamine (eq. 2) makes it *less* nucleophilic, as conjugate acids. A compromise is struck in the formation of the oxime (eq. 3) by adding sodium acetate to buffer the solution at pH 5 where pinacolone is converted to the electrophilic conjugate acid and yet there is a sufficient concentration of the nonprotonated nucleophile hydroxylamine, to allow the reaction to proceed.

Measure into a test tube provided with a cold finger condenser 1 ml of pinacolone and 3 ml each of 5 M hydroxylamine hydrochloride solution and 5 M sodium acetate solution, and add 5 ml of 95% ethanol in order to bring the oil completely into solution. Reflux the solution gently on the steam bath for 2 hr, at which time there should be sufficient oxime for identification. The oxime usually separates as an oily layer; on very thorough cooling and by rubbing the walls of the test tube with a stirring rod the material can be caused to solidify. Collect the product on a suction funnel, dry a portion on a filter paper, and determine the melting point. Pure pinacolone oxime melts at 77–78°. It is soluble in cold dilute hydrochloric acid, and the pleasant odor becomes particularly apparent on boiling the solution. The oxime evaporates rapidly, even at room temperature, and should not be left exposed to the air for more than a few hours.

QUESTION   1. Oximes can exhibit geometric isomerism; e.g., in pinacolone oxime the hydroxyl group can be cis or trans to the t-butyl group. How can you demonstrate that your product is one isomer and not a mixture?

27 Succinic Anhydride

KEYWORDS   Dehydration of a diacid   γ-Lactone
             Alcoholysis   Succinanilic acid
             Ammonolysis   Succinanil

Acetic anhydride can be prepared by the interaction of sodium acetate and acetyl chloride or by the addition of acetic acid to ketene:

$$CH_3COONa + CH_3COCl \rightarrow (CH_3CO)_2O + NaCl$$
$$CH_2{=}C{=}O + CH_3COOH \rightarrow (CH_3CO)_2O$$

The anhydride of an acid also can be prepared by treatment of the acid with a dehydrating agent, and in the case of a dibasic acid of the type of succinic acid this direct method is the only one applicable. This particular dehydration can be accomplished by use of either acetic anhydride or acetyl chloride:

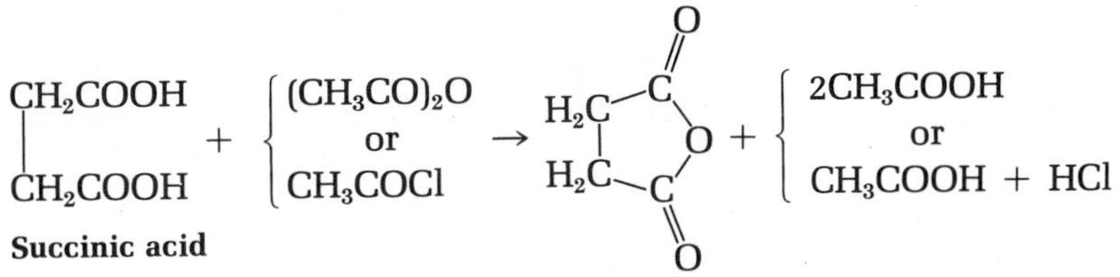

Succinic acid

Succinic anhydride
Mp 120°,
MW 100.07

The cyclic anhydride formed is not contaminated with inorganic reagents and the acetic acid produced serves as a solvent for crystallization. The procedure may be used for preparation of the anhydrides of glutaric, maleic, and phthalic acid.

Succinic anhydride is useful in affording routes to derivatives not available directly from the acid.

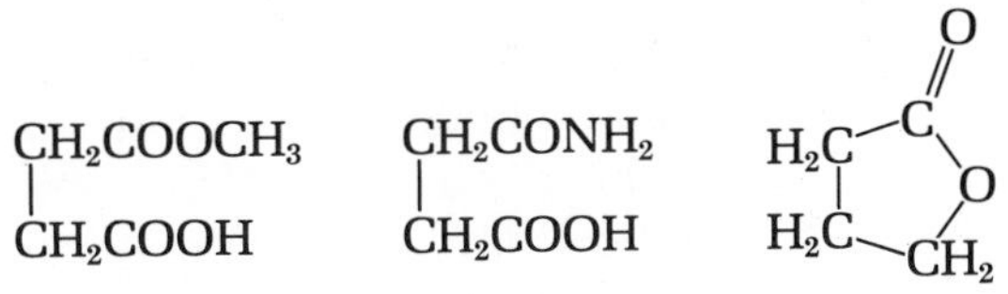

(1)   (2)   (3)

On alcoholysis succinic anhydride yields the monomethyl ester (**1**); the acid amide (**2**) is formed on ammonolysis of the anhydride; reduction gives butyrolactone (**3**), a typical γ-lactone. The reaction of the anhydride with aniline can be used for the identification of this primary amine, since the product is a crystalline substance of sharp melting point.

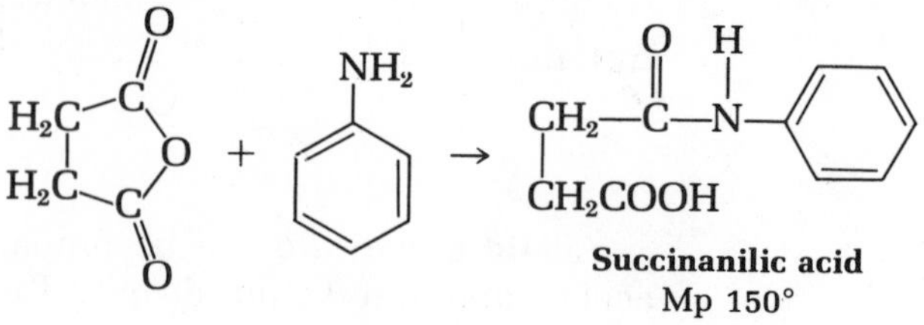

Succinanilic acid
Mp 150°

On treatment with acetyl chloride, succinanilic acid is cyclized to succinanil.

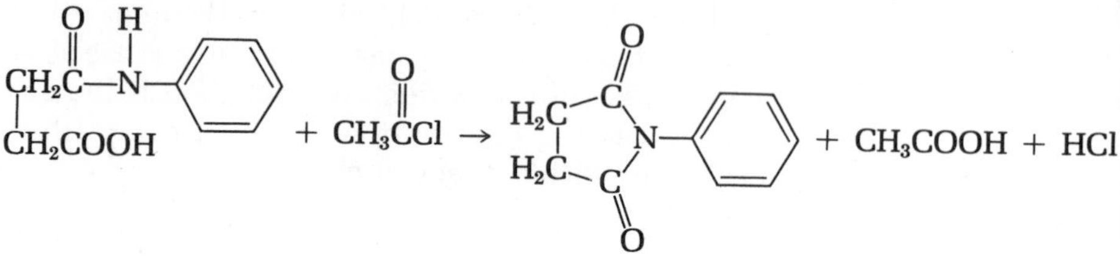

Succinanil
Mp 154°

EXPERIMENTS   1. Succinic Anhydride

In a 125-ml round-bottomed flask fitted with a reflux condenser, closed with a calcium chloride tube, place 15 g of succinic acid and 20 ml of acetic anhydride. Heat to the boiling point, noting that the crystals soon dissolve, and reflux gently for 15 min. Let the solution cool for a time undisturbed and observe the crystallization. Finally, cool in ice, collect the crystals on a dry suction funnel, and use several small portions of ether (Calcutta wash bottle) to rinse the reaction flask and wash the crystalline anhydride. The yield is 10–11 g; mp 119–120°. Test the product with cold sodium bicarbonate solution for the presence of unchanged acid.

Reaction time 15 min

2. Succinanilic Acid

Dissolve 1 g of succinic anhydride in 30 ml of benzene on the steam bath and to the boiling solution add all at once a solution of 0.9 ml of aniline in about 5 ml of benzene. The separation of the reaction product occurs in a striking manner. Cool the mixture, collect the crystalline product, and wash it with benzene. Determine the yield and melting point and note the action on a few mgs of the substance of cold sodium bicarbonate solution. Pure succinanilic acid melts at 150°.

The reaction of succinic anhydride with a primary amine can be carried out rapidly and with very small amounts of material. Dissolve a few small crystals of succinic anhydride in 1 ml of hot benzene in a 10 × 75-mm test tube, add one small crystal of p-toluidine (another primary amine), and observe the result.

3. Succinanil

Convert the remainder of the succinanilic acid to the cyclized form, succinanil (apparatus as in Section 1). Cover the succinanilic acid (about 1.8 g) with 5 ml of acetyl chloride and reflux until the reaction is complete (5 min). Allow crystallization to take place, collect the product, and wash it with ether as before. Test the solubility in bicarbonate solution. The pure material melts at 154°.

QUESTIONS   1. *How could succinic acid by converted into β-aminopropionic acid?*

2. Devise a method for the resolution of a dl-alcohol, assuming that an optically active derivative of aniline is available, and noting that the succinanilic acid, like other acids, can be esterified without difficulty.

28 trans,trans-1,4-Diphenyl-1,3-butadiene

KEYWORDS	Olefin synthesis	Phosphonium salt
	Wittig reaction	Phosphonate
	Ylide	Dimethylformamide, DMF

The Wittig reaction affords an invaluable method for the conversion of a carbonyl compound to an olefin, in this case the conversion of benzophenone to 1,1-diphenylethylene:

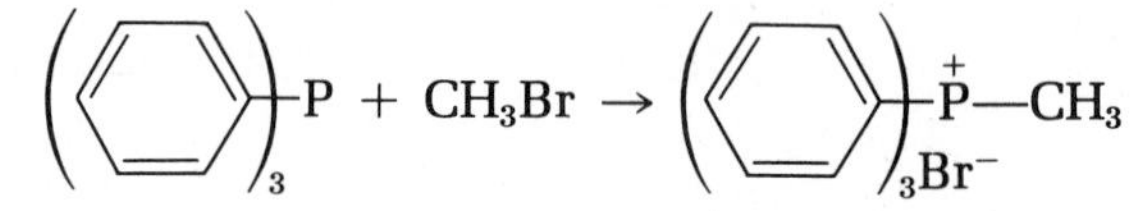

Triphenylphosphine

Methyl triphenyl-phosphonium bromide

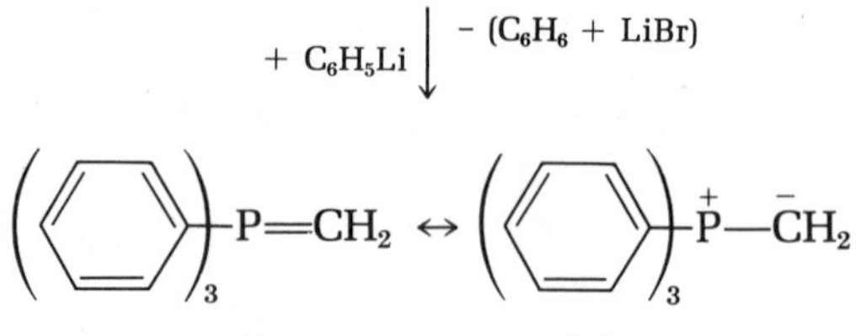

$+ C_6H_5Li$ $\Big| - (C_6H_6 + LiBr)$

Wittig reagent (an ylide)

Wittig reaction

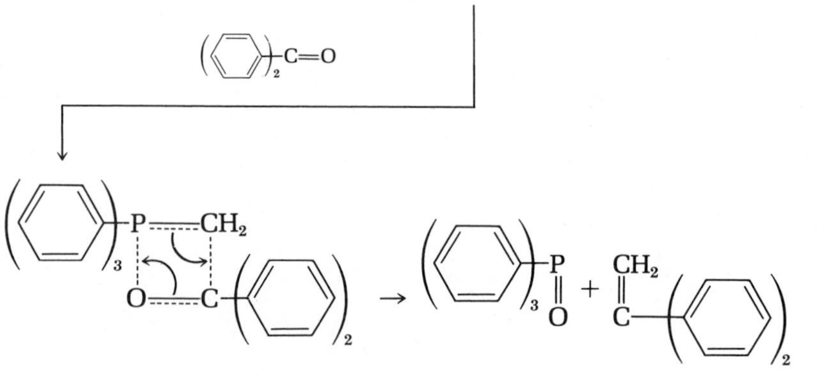

1,1-Diphenylethylene

Since the active reagent, an ylide, is unstable, it is generated in the presence of the carbonyl compound to be used by dehydro-halogenation of an alkyltriphenylphosphonium bromide with phenyllithium in dry ether in a nitrogen atmosphere.

When the halogen compound employed in the first step has an activated halogen atom ($RCH{=}CHCH_2X$, $C_6H_5CH_2X$, $XCH_2CO_2C_2H_5$) a simpler procedure known as the phosphonate modification is applicable. When benzyl chloride is heated with triethyl phosphite, ethyl chloride is eliminated from the initially formed phosphonium chloride with the production of diethyl benzylphosphonate. This phosphonate is stable, but in the presence of a strong base it condenses with a carbonyl component in the same way that a Wittig ylide condenses.

Simplified Wittig reaction

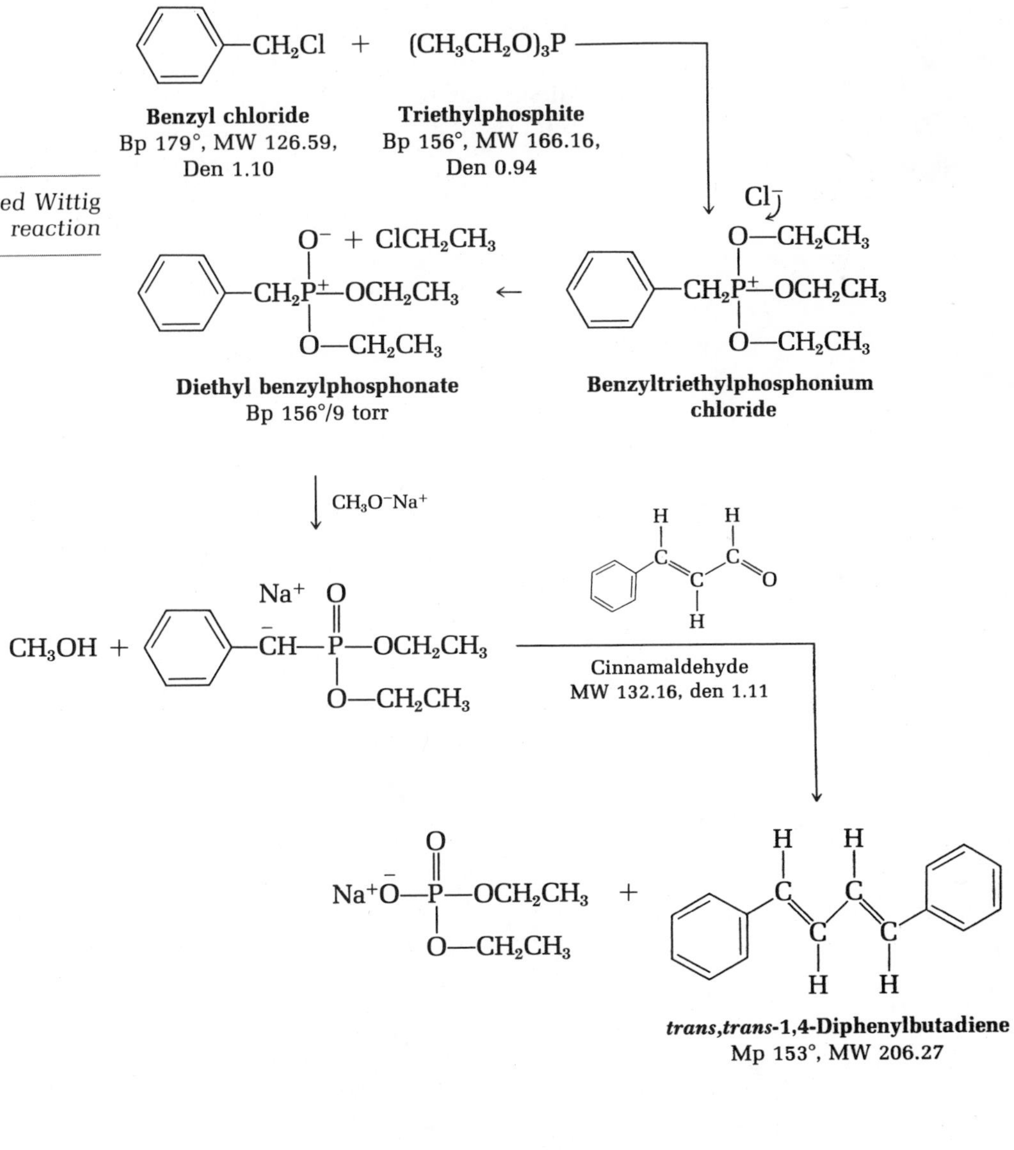

Benzyl chloride
Bp 179°, MW 126.59,
Den 1.10

Triethylphosphite
Bp 156°, MW 166.16,
Den 0.94

Diethyl benzylphosphonate
Bp 156°/9 torr

Benzyltriethylphosphonium chloride

$CH_3O^-Na^+$

$CH_3OH +$

Cinnamaldehyde
MW 132.16, den 1.11

***trans,trans*-1,4-Diphenylbutadiene**
Mp 153°, MW 206.27

Thus, it reacts with benzaldehyde to give *trans*-stilbene and with cinnamaldehyde to give *trans,trans*-1,4-diphenyl-1,3-butadiene.

Caution: Take care to keep organophosphorus compounds off the skin.

EXPERIMENT

Time for first step: 1 hr

Use freshly opened sodium methoxide

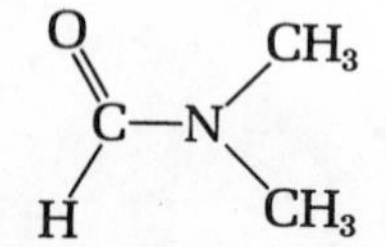

O   CH₃
‖
C—N
| \
H   CH₃

Dimethylformamide
MW 73.10, bp 153°

With the aid of pipettes and a pipetter or a suction pump, measure into a 25 × 150 mm test tube 5 ml of benzyl chloride and 7.7 ml of triethyl phosphite. Add a carborundum boiling stone, insert a cold finger condenser (Fig. 22.1), and with a small flame reflux the liquid gently for 1 hr. (Elimination of ethyl chloride starts at about 130°, and in the time period specified the temperature of the liquid rises to 190–200°.) Let the phosphonate ester cool to room temperature, pour it into a 125-ml Erlenmeyer flask containing 2.4 g of sodium methoxide, and add 40 ml of dimethylformamide, using a part of this solvent to rinse the test tube. Swirl the flask vigorously in a water-ice bath to thoroughly chill the contents and continue swirling while running in 5 ml of cinnamaldehyde by pipette. The mixture soon turns deep red and then crystalline hydrocarbon starts to separate. When there is no further change (about 2 min) remove the flask from the cooling bath and let it stand at room temperature for about 10 min. Then add 20 ml of water and 10 ml of methanol, swirl vigorously to dislodge crystals, and finally collect the product on a suction funnel using the red mother liquor to wash out the flask. Wash the product with water until the red color of the product is all replaced by yellow. Then wash with methanol to remove the yellow impurity, and continue until the wash liquor is colorless. The yield of the crude faintly yellow hydrocarbon, mp 150–151°, is 5.7 g. This material is satisfactory for use in the next experiment (1.5 g required). A good solvent for crystallization of the rest of the product is methylcyclohexane (bp 101°, 10 ml per g; use more if the solution requires filtration). Pure 1,4-diphenyl-1,3-butadiene melts at 153°.

29 p-Terphenyl

KEYWORDS Diels-Alder reaction   Decarboxylation
              Isomerization   Dimethyl acetylenedicarboxylate
              Hydrolysis   Lachrymator, vesicant

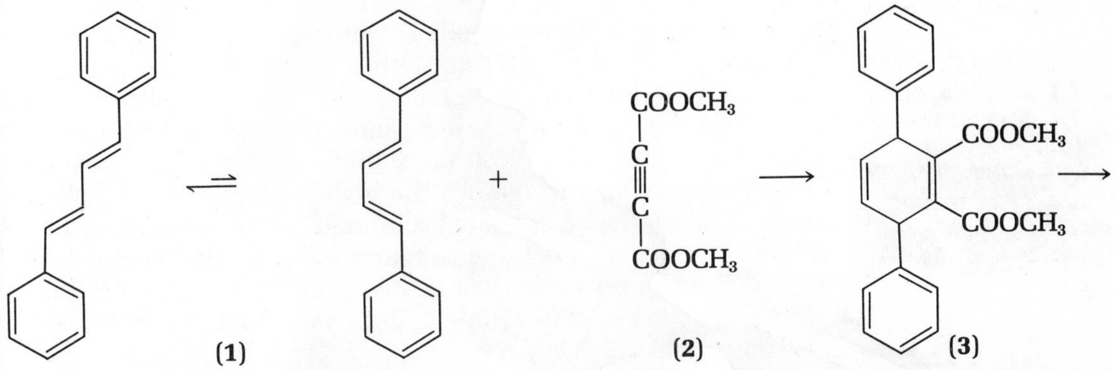

(1)
trans,trans-1,4-Diphenyl-1,3-butadiene
MW 206.27

(2)
MW 142.11

(3)
MW 348.38, Mp 98°

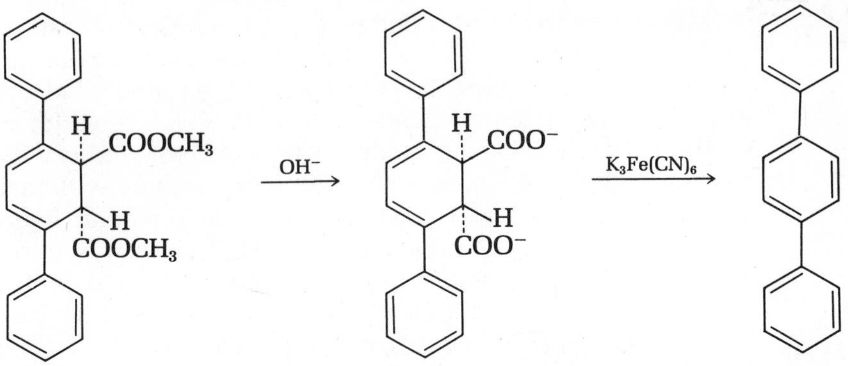

(4)
Mp 170°

(5)

(6)
p-Terphenyl
MW 230.29, Mp 211°

Diels-Alder reaction

trans,trans-1,4-Diphenyl-1,3-butadiene (1) is most stable in the transoid form, but at a suitably elevated temperature the cisoid form present in the equilibrium adds to dimethyl acetylenedicarboxylate (2) to give dimethyl 1,4-diphenyl-1,4-dihydrophthalate (3). This low-melting ester is obtained as an oil and when warmed briefly with methanolic potassium hydroxide is isomerized to the high-melting *trans*-ester (4). The free *trans*-acid can be obtained in

169

86% yield by refluxing the suspension of (3) for 4 hr, but in the recommended procedure the isomerized ester is collected, washed to remove dark mother liquor, and hydrolyzed by brief heating with potassium hydroxide in a high-boiling solvent. The final step, an oxidative decarboxylation, is rapid and nearly quantitative. It probably involves reaction of the oxidant with the dianion (5) with removal of two electrons and formation of a diradical, which loses carbon dioxide with formation of p-terphenyl (6).

EXPERIMENT

$CH_3OCH_2CH_2OCH_2$
|
$CH_3OCH_2CH_2OCH_2$
Triethylene glycol dimethyl ether (triglyme). Miscible with water. bp 222°.

Reaction time 30 min

Rapid hydrolysis of the hindered ester (4)

Place 1.5 g of *trans,trans*-1,4-diphenyl-1,3-butadiene and 1.0 ml (1.1 g) of dimethyl acetylenedicarboxylate (*caution, skin irritant*[1]) in a 25 × 150 mm test tube and rinse down the walls with 5 ml of triethylene glycol dimethyl ether (triglyme) (bp 222°). Clamp the test tube in a vertical position, introduce a cold finger condenser, and reflux the mixture gently for 30 min. Cool the yellowish solution under the tap, pour into a separatory funnel, and rinse out the tube with a total of about 50 ml of ether. Extract twice with water (50–75 ml portions) to remove the high-boiling solvent, shake the ethereal solution with saturated sodium chloride solution, and filter it through a paper containing anhydrous sodium sulfate into a tared 125-ml Erlenmeyer flask. Evaporate on the steam bath, eventually with evacuation at the aspirator, until the weight of yellow oil is constant; yield 2.5–2.8 g.

While evaporation is in progress, dissolve 0.5 g of potassium hydroxide (about 5 pellets) in 10 ml of methanol by heating and swirling; the process is greatly hastened by crushing the lumps with a stirring rod with a flattened-out head. Crystallization of the yellow oil containing (3) can be initiated by cooling and scratching; this provides assurance that the reaction has proceeded properly. Pour in the methanolic potassium hydroxide and heat with swirling on the hot plate for about 1 min until a stiff paste of crystals of the isomerized ester (4) appears. Cool, thin the mixture with methanol, collect the product and wash it free of dark mother liquor, and spread it out on a paper for rapid drying. The yield of pure, white ester (4) is 1.7–1.8 g. Solutions in methanol are strongly fluorescent.

Place the ester (4) in a 25 × 150 mm test tube, add 0.7 g of potassium hydroxide (7–8 pellets), and pour in 5 ml of triethyleneglycol. Stir the mixture with a thermometer and heat, raising the temperature to 140° in the course of about 5 min. By intermittent heating, keep the temperature close to 140° for 5 min longer and then cool the mixture under the tap. Pour into a 125-ml Erlenmeyer flask and rinse the tube with about 50 ml of water. Heat to boiling and, in case there is a small precipitate or the solution is cloudy, add a little decolorizing charcoal, swirl, and filter the alkaline solution by

---

[1] This ester is a powerful lachrymator (tear producer) and vesicant (blistering agent) and should be dispensed from a bottle provided with a pipet and a pipetter. Even a trace of ester on the skin should be washed off with methanol, followed by soap and water.

gravity. Then add 3.4 g of potassium ferricyanide and heat on the hot plate with swirling for about 5 min to dissolve the oxidant and to coagulate the white precipitate which soon separates. The product can be air-dried overnight or dried to constant weight by heating in an evacuated Erlenmeyer flask on the steam bath. The yield of colorless *p*-terphenyl, mp 209–210°, is 0.7–0.8 g. The *p*-terphenyl crystallizes well from dioxane.

1.0 ml dimethyl acetylenedicarboxylate
5.0 ml triethylene glycol dimethyl ether.

0-55 KOH, in 10 ml CH₃OH.

5 ml - triethylene glycol.
3.4 potassium ferricyanide.

30 Amines

KEYWORDS Organic bases   Benzenesulfonyl chloride
              Primary, secondary, tertiary   Sulfonamides
              Solubility in acid   Acidic hydrogen
              Hinsberg test   Acetyl and benzoyl derivatives

Amines, whether primary, secondary, or tertiary, differ from other classes of organic compounds, including other nitrogen-containing substances, in being basic. Some amines are not soluble enough in water to give a basic response to test paper, but nevertheless they combine with mineral acids to form salts. Hence their basic character can be recognized by a simple test with acid that distinguishes amines from neutral substances, such as amides ($RCONH_2$), N-acylamines ($RNHCOCH_3$), and nitriles (RCN). An amine salt can be recognized by its reaction with base. Procedures for the tests for basicity are given in Section 1. Section 2 presents a test for distinguishing between primary, secondary, and tertiary amines, Section 3 gives procedures for preparation of solid derivatives for melting point characterizations, and Section 4 gives spectral characteristics. After applying the procedures to known substances you are to identify a series of unknowns.

EXPERIMENTS **1. Basicity**

Substances to be tested:

Pyridine

Aniline, $C_6H_5NH_2$, bp 184°
p-Toluidine, $CH_3C_6H_4NH_2$, mp 43°
Pyridine, C_5H_5N, bp 115° (a tertiary amine)
Methylamine hydrochloride, dec. (salt of CH_3NH_2, bp −6.7°)
Aniline hydrochloride, dec. (salt of $C_6H_5NH_2$)
Aniline sulfate, dec.

First, see if the substance has a fishy, ammonia-like odor, for if it does it probably is an amine of low molecular weight. Then test the solubility in water by putting 2 micro drops if a liquid or an estimated 20 mg if a solid into a 10 × 75-mm test tube, adding 0.2 ml of water (a 5-mm column) and first seeing if the substance dissolves in the cold. If the substance is a solid rub it well with a stirring rod and break up any lumps before drawing a conclusion.

If the substance is *readily soluble in cold water* and if the odor is suggestive of an amine, test the solution with Hydrion paper and further determine if the odor disappears on addition of a few drops of 10% hydrochloric acid. If the properties are more like those of a salt, add a few drops of 10% sodium hydroxide solution. If the solution remains clear, addition of a little sodium chloride may cause separation of a liquid or solid amine.

If the substance is not soluble in cold water, see if it will dissolve on heating; be careful not to mistake the melting of a substance for dissolving. If it *dissolves in hot water*, add a few drops of 10% alkali and see if an amine precipitates. (If you are in doubt as to whether a salt has dissolved partially or not at all, pour off supernatant liquid and basify it.)

If the substance is *insoluble in hot water*, add 10% hydrochloric acid, heat if necessary, and see if it dissolves. If so, make the solution basic and see if an amine precipitates.

2. Hinsberg Test

The procedure for distinguishing amines with benzenesulfonyl chloride is to be run in parallel on the following substances:

Aniline, $C_6H_5NH_2$ (bp 184°)
N-Methylaniline, $C_6H_5NHCH_3$ (bp 194°)
Triethylamine, $(CH_3CH_2)_3N$ (bp 90°)

Primary and secondary amines react in the presence of alkali with benzenesulfonyl chloride, $C_6H_5SO_2Cl$, to give sulfonamides.

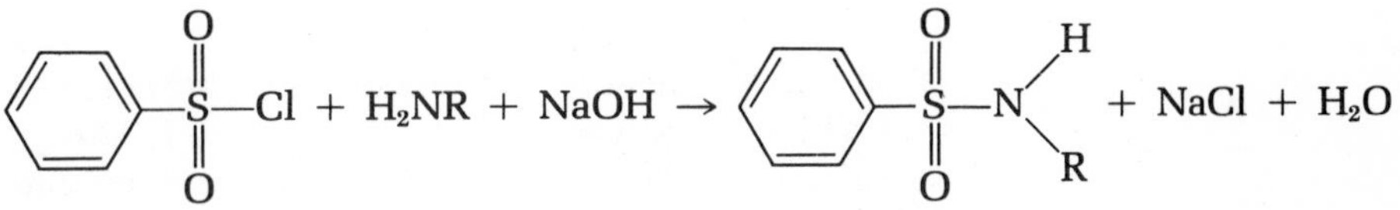

(Soluble)

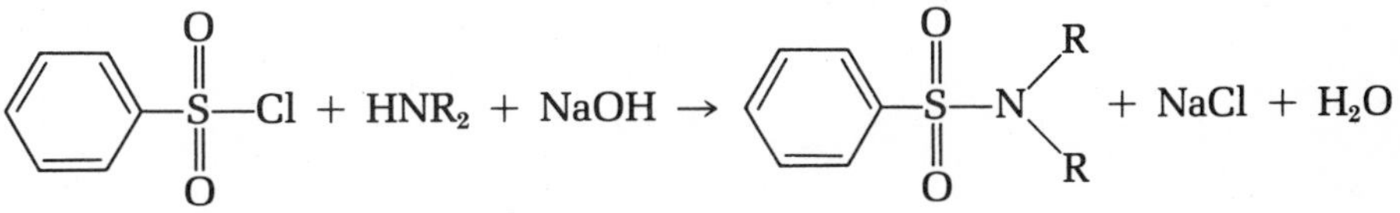

(Insoluble)

The sulfonamides are distinguishable, because the derivative from a primary amine has an acidic hydrogen available, rendering the product soluble in alkali (reaction 1), whereas the sulfonamide from a secondary amine is insoluble (reaction 2). Tertiary amines lack the necessary replaceable hydrogen for formation of benzenesulfonyl derivatives.

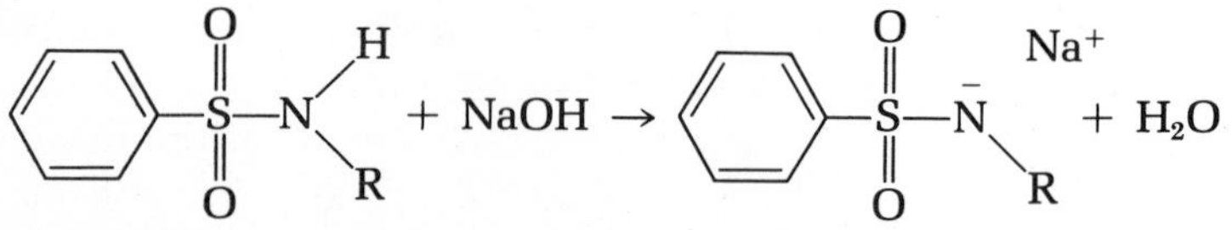

$$\text{C}_6\text{H}_5\text{—}\overset{\displaystyle O}{\underset{\displaystyle O}{\overset{\|}{\underset{\|}{S}}}}\text{—}\overset{H}{\underset{R}{N}} + \text{NaOH} \rightarrow \text{C}_6\text{H}_5\text{—}\overset{\displaystyle O}{\underset{\displaystyle O}{\overset{\|}{\underset{\|}{S}}}}\text{—}\overset{\text{Na}^+}{\underset{R}{\overset{-}{N}}} + \text{H}_2\text{O}$$

To about 4 ml of water in a 13 × 100-mm test tube (4-cm column) add either 4 micro drops of a liquid amine or an estimated 100 mg of an amine salt, 1 ml of 10% sodium hydroxide solution, and 7 micro drops of benzenesulfonyl chloride. Stopper the tube, shake for 5 min, note if there is any heat effect and if a solid separates. Warm the tube slightly, shake for a few minutes longer until the odor of benzenesulfonyl chloride is no longer apparent, and cool the mix-

*Distinguishing among
primary (1°),
secondary (2°) and
tertiary (3°) amines*

ture. If an oil separates at this point cool the tube in an ice bath, rub the oil against the walls with a stirring rod, and see if this is a product which will solidify or if it is unreacted amine. Make sure that the solution is still alkaline, and if there is a precipitate at this point it should be collected by suction filtration and a portion tested with dilute alkali to prove that it is actually insoluble in alkali. (An alkali-soluble product may precipitate at this point as the sodium salt if too much alkali is used.) If there is no precipitate the alkaline solution may contain an alkali-soluble sulfonamide. Acidify with concd hydrochloric acid and if a precipitate forms collect it and confirm the solubility in alkali.

3. Solid Derivatives

Acetyl derivatives of primary and secondary amines are usually solids suitable for melting point characterization and are readily prepared by reaction with acetic anhydride, even in the presence of water. Benzoyl and benzenesulfonyl derivatives are made by reaction of the amine with the acid chloride in the presence of alkali, as in Section 2 (the benzenesulfonamides of aniline and of N-methylaniline melt at 110° and 79°).

Solid derivatives suitable for characterization of tertiary amines are the methiodides and picrates:

$$\text{R}_3\text{N} + \text{CH}_3\text{I} \rightarrow \text{R}_3\overset{+}{\text{N}}\text{CH}_3 \;\; \text{I}^-$$

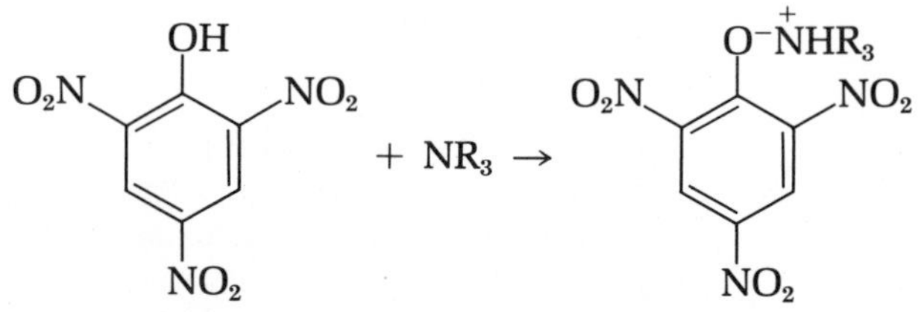

**2,4,6-Trinitrophenol
(Picric Acid)** + NR₃ → **Amine picrate**

Typical derivatives are to be prepared, and although determination of melting points is not necessary since the values are given, the products should be saved for possible identification of unknowns.

(a) Measure 4–5 micro drops (about 1 millimole) of aniline into a 13 × 100-mm test tube and add 5 micro drops of acetic anhydride. Note the heat effect, allow 2–3 min for completion of the reaction, and then cool and add water. The oily precipitate soon solidifies, mp 114°. Test its solubility in dilute HCl.

(b) Dissolve about 100 mg of aniline hydrochloride in 1 ml of water, add 5 micro drops of acetic anhydride, and then (at once) about 100 mg of sodium acetate.

(c) Dissolve about 100 mg of picric acid in 1 ml of methanol and to the warm solution add 4 micro drops of triethylamine and let the solution stand. The picrate melts at 171°.

4. NMR and IR Spectra of Amines

The proton bound to nitrogen can appear anywhere between 0.6 and 7.0 ppm on the nmr spectrum, depending upon solvent, concentration, and structure of the amine. The peak is sometimes extremely broad owing to slow exchange and interaction of the proton with the electric quadrapole of the nitrogen. If addition of a drop of D_2O to the sample causes the peak to disappear, this is evidence for an amine hydrogen, but alcohols, phenols, and enols will also exhibit this exchange behavior. See Fig. 30.1 for the nmr spectrum

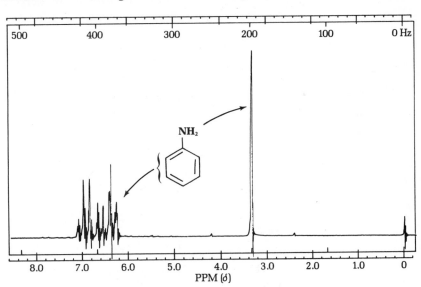

Figure 30.1
Nmr spectrum of aniline.

of aniline, in which the amine hydrogens appear as a sharp peak at 3.3 ppm. Infrared spectroscopy can also be very useful for identification purposes. Primary amines, both aromatic and aliphatic, show a weak doublet between 3300 and 3500 cm^{-1} and a strong absorption between 1560 and 1640 cm^{-1} due to NH bending (Fig 30.2). Secondary amines show a single peak between 3310 and

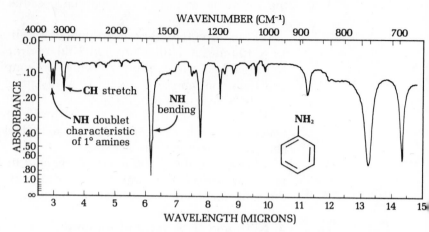

Figure 30.2
Infrared spectrum of aniline in CS$_2$.

3450 cm^{-1}. Tertiary amines have no useful infrared absorptions.

In Chapter 15 the characteristic ultraviolet absorption shifts of aromatic amines in the presence and absence of acids were discussed.

5. Unknowns (see Tables 63.5 and 63.6)

Determine first if the unknown supplied is an amine or an amine salt and then determine whether the amine is primary, secondary, or tertiary. Complete identification of your unknown may be required, depending on your instructor's wishes.

QUESTIONS

1. *How could you most easily distinguish between samples of β-naphthylamine and of acetanilide?*

2. *Would you expect the reaction product from benzenesulfonyl chloride and ammonia to be soluble or insoluble in alkali?*

3. *Is it safe to conclude that a substance is a tertiary amine because it forms a picrate?*

4. *Why is it usually true that amines that are insoluble in water are odorless?*

5. *Technical dimethylaniline contains traces of aniline and of methylaniline. Suggest a method of freeing it completely from these impurities.*

6. *How would you prepare aniline from aniline hydrochloride?*

31 Sugars

KEYWORDS

Mono-, di-, and oligosaccharides
Polysaccharides
Saccharin
Sodium cyclamate
Osazones
Phenylhydrazine

Fehling solution
Cupric ion
Tollens reagent
Red tetrazolium
Disaccharide hydrolysis
Trityl, triphenylmethyl

The term sugar applies to mono-, di-, and oligosaccharides, which are all soluble in water and thereby distinguished from polysaccharides. Many natural sugars are sweet, but data of Table 31.1

Table 31.1 Relative Sweetness of Sugars and Sugar Substitutes

Compound	Sweetness	
	To man	To bees
Monosaccharides		
D-Fructose	1.5	+
D-Glucose	0.55	+
D-Mannose	Sweet, then bitter	−
D-Galactose	0.55	−
D-Arabinose	0.70	−
Disaccharides		
Sucrose (glucose, fructose)	1	+
Maltose (2 glucose)	0.3	+
α-Lactose (glucose, galactose)	0.2	−
Cellobiose (2 glucose)	Indifferent	−
Gentiobiose (2 glucose)	Bitter	−
Synthetic sugar substitutes		
Aspartame	180	
Saccharin	550	
2-Amino-1-n-propoxy-4-nitrobenzene	4000	

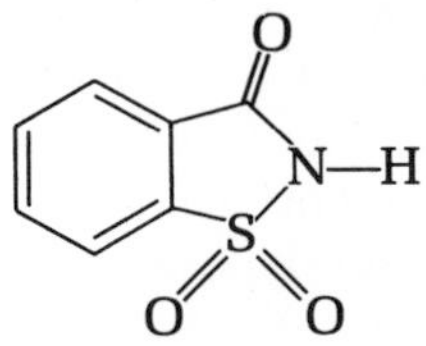

Saccharin

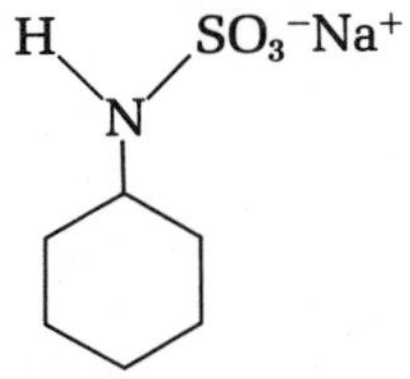

Sodium cyclamate

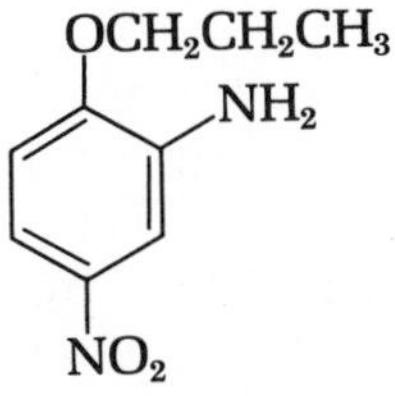

2-Amino-1-*n*-propoxy-4-nitrobenzene

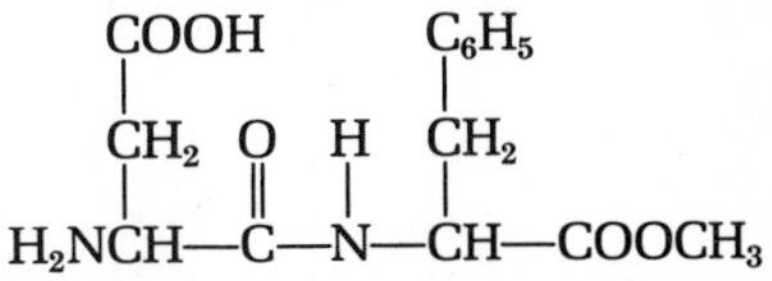

Aspartame
L-Aspartyl-L-phenylalanine
methyl ester

show that this form of physiological activity varies greatly with stereochemical configuration and that activity is exhibited by compounds of widely differing structural type.

Sugars are neutral, combustible substances, and these properties distinguish them from other water-soluble compounds. Some polycarboxylic acids and some lower amines are soluble in water, but the solutions are acidic or basic. Water-soluble amine salts react with alkali with liberation of the amine, and sodium salts of acids are noncombustible.

One gram of sucrose dissolves in 0.5 ml of water at 25° and in 0.2 ml at the boiling point, but the substance has marked, atypical crystallizing power. In spite of the high solubility it can be obtained in beautiful, large crystals (rock candy). More typical sugars are obtainable in crystalline form only with difficulty, particularly in the presence of a trace of impurity, and even then give small and not well-formed crystals. Alcohol is often added to a water solution to decrease solubility and thus to induce crystallization. The amounts of 95% ethanol required to dissolve 1-g samples at 25° are: sucrose, 170 ml; glucose, 60 ml; fructose, 15 ml. Some sugars have never been obtained in crystalline condition and are known only as viscous sirups. With phenylhydrazine many sugars form beautiful crystalline derivatives called osazones. Osazones are much less soluble in water than the parent sugars, since the molecular weight is increased by 178 units and the number of hydroxyl groups reduced by one. It is easier to isolate an osazone than to isolate the sugar, and sugars that are sirups often give crystalline osazones. Osazones of the more highly hydroxylic disaccharides are notably more soluble than those of monosaccharides.

Osazones

Some disaccharides do not form osazones, but a test for formation or nonformation of the osazone is ambiguous, because the glycosidic linkage may suffer hydrolysis in a boiling solution of phenylhydrazine and acetic acid, with formation of an osazone derived from a component sugar and not from the disaccharide. If a sugar has reducing properties it is also capable of osazone formation; hence an unknown sugar is tested for reducing properties before preparation of an osazone is attempted. Three tests for differentiation between reducing and nonreducing sugars are described below; two are classical and the third modern.

EXPERIMENTS   **1. Fehling Solution**[1]

The reagent is made just prior to use by mixing equal volumes of Fehling solution I, containing copper sulfate, with solution II, containing tartaric acid and alkali. The copper, present as a deep blue complex anion, if reduced by a sugar from the cupric to the cuprous state, precipitates as red cuprous oxide. If the initial step

---

Note for the instructor

[1] *Solution I:* 34.64 g of $CuSO_4 \cdot 5H_2O$ dissolved in water and diluted to 500 ml. *Solution II:* 173 g of sodium potassium tartrate (Rochelle salt) and 65 g of sodium hydroxide dissolved in water and diluted to 500 ml.

in the reaction involved oxidation of the aldehydic group of the aldose to a carboxyl group, a ketose should not reduce Fehling solution, or at least should react less rapidly than an aldose, but the comparative experiment that follows will show that this is not the case. Hence, attack by an alkaline oxidizing agent must be at the α-ketol grouping common to aldoses and hexoses, and perhaps proceeds through an enediol, the formation of which is favored by alkali. A new α-ketol grouping is produced, and thus oxidation proceeds down the carbon chain.

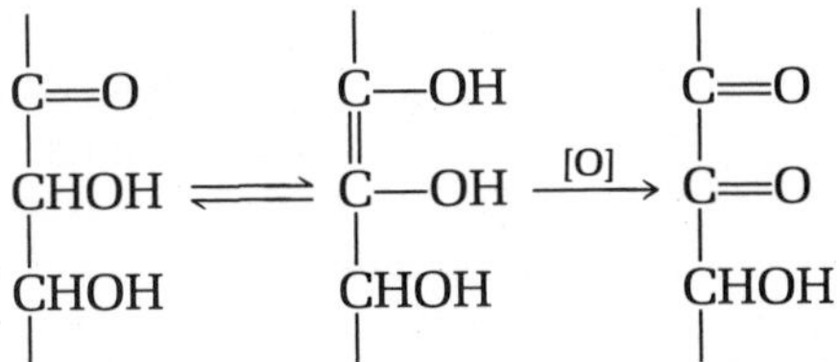

$$
\begin{array}{ccc}
\mid & \mid & \mid \\
C{=}O & C{-}OH & C{=}O \\
\mid & \parallel & \mid \\
CHOH & C{-}OH \xrightarrow{[O]} & C{=}O \\
\mid & \mid & \mid \\
CHOH & CHOH & CHOH \\
\mid & \mid & \mid \\
\textbf{α-Ketol} & \textbf{Enediol} & \textbf{α-Ketol}
\end{array}
$$

One milliliter of mixed solution will react with 5 mg of glucose; the empirically determined ratio is the basis for quantitative determination of the sugar. The Fehling test is not specific to reducing sugars, since ordinary aldehydes reduce the reagent although by a different mechanism and at a different rate.

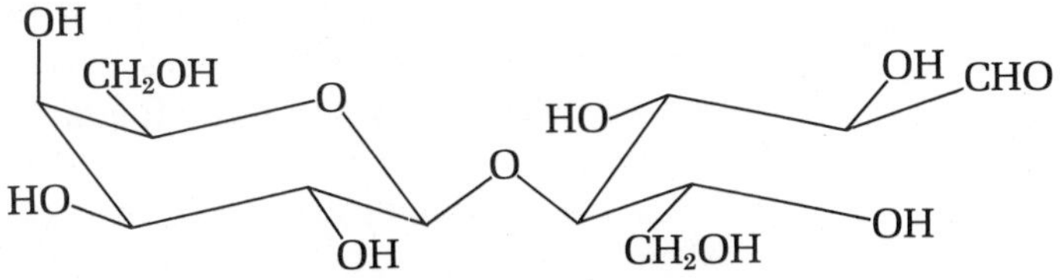

Lactose (aldehyde form)

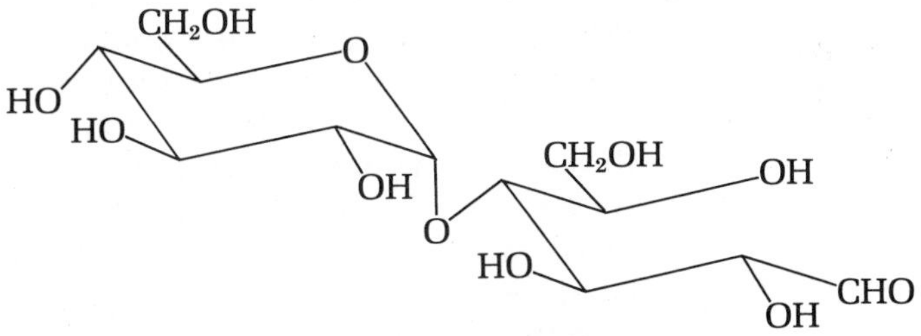

Maltose (aldehyde form)

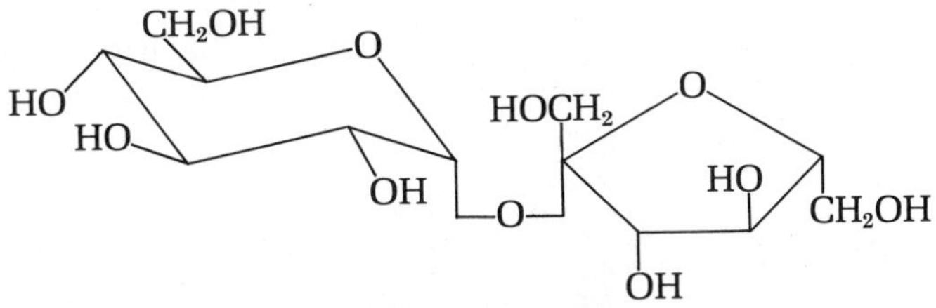

Sucrose

The following sugars are to be tested: 0.1 M solutions of glucose, fructose, lactose, maltose, sucrose (cane sugar).[2]

Introduce 10 micro drops of the 0.1 M solutions to be tested into each of five 13 × 100-mm test tubes carrying some form of serial numbers resistant to heat and water (rubber bands), and prepare a beaker of hot water in which all the tubes can be heated at once. Measure 5 ml of Fehling solution I into a small flask and wash the graduate before using it to measure 5 ml of solution II into the same flask. Mix until all precipitate dissolves, measure 2 ml of mixed solution into each of the five test tubes, shake, put the tubes in the heating bath, and observe the results. Empty, and wash the tubes with water and then with dilute acid (leave the markers in place and continue heating the beaker of water on a hot plate).

2. Tollens Reagent

$$R-C\underset{H}{\overset{\displaystyle O}{\big<}} + 2\ Ag(NH_3)_2OH \rightarrow 2\ AgO + RCOO^-NH_4^+ + H_2O + 3\ NH_3$$

Tollens reagent, a solution of silver ammonium hydroxide and sodium hydroxide, is reduced by aldoses and ketoses and by simple aldehydes, with deposition of metallic silver partly in the form of a mirror. The test is more sensitive than that of Fehling and better able to reveal small differences in reactivity, but it is less reliable in distinguishing between reducing and nonreducing sugars.

Add a few ml of 10% sodium hydroxide to each of the five marked test tubes and heat the tubes in the water bath (for thorough cleaning) while preparing a sufficient amount of Tollens reagent for five tests. Measure 2 ml of 5% silver nitrate solution and 1 ml of 10% sodium hydroxide into a test tube, and make a dilute solution of ammonia by mixing 1 ml of concd ammonia solution with 10 ml of distilled water. Add 0.5 ml of this solution to the precipitated silver oxide, stopper the tube, and shake. Repeat the process until almost all of the precipitate dissolves (3 ml, avoid an excess) and then dilute the solution to a volume of 10 ml. Empty the five tubes, rinse them with distilled water, and into each tube put one micro drop of a 0.1 M solution of glucose, fructose, lactose, maltose, and n-butanol.[3] Add 1 ml of Tollens reagent to each tube and let the reactions proceed at room temperature at first. Watch closely and try to define the order of reactivity as measured not by the color of a solution but by the time of appearance of the first precipitate of metal. After a few minutes put the tubes in the heating bath.

Notes for the instructor

[2] To prepare 0.1 M test solutions dissolve the following amounts of substance in 100 ml of water each: D-glucose monohydrate, 1.98 g; fructose, 1.80 g; α-lactose monohydrate, 3.60 g; maltose monohydrate, 3.60 g; sucrose, 3.42 g.
[3] Dissolve 0.72 g of n-butanol in 100 ml of water.

3. Red Tetrazolium[4]

The reagent (RT) is a nearly colorless, water-soluble substance that oxidizes aldoses and ketoses, as well as other α-ketols, and is thereby reduced. The reduced form is a water-insoluble, intensely colored pigment, a diformazan.

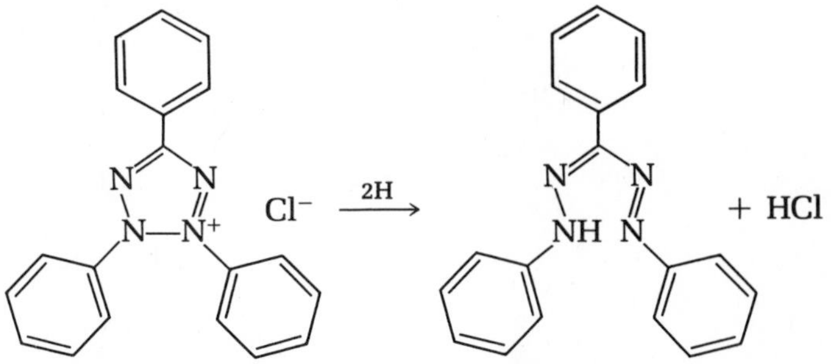

Red tetrazolium   **RT-Diformazan**

Red tetrazolium affords a highly sensitive test for reducing sugars and distinguishes between α-ketols and simple aldehydes more sharply than the Fehling and Tollens tests.

Put one micro drop of each of the five 0.1 M test solutions of Section 2 in the cleaned, marked test tubes, and to each tube add 1 ml of a 0.5% aqueous solution of red tetrazolium and one drop of 10% sodium hydroxide solution. Put the tubes in the beaker of hot water and note the order of development of color.

For estimation of the sensitivity of the test, use the substance that you regard as the most reactive of the five studied. Dilute 1 ml of the 0.1 M solution with water to a volume of 100 ml, and run a test with RT on 0.2 ml of the diluted solution.

4. Phenylosazones

Prepare 10 ml of stock phenylhydrazine reagent as in Chapter 22, Section 1, and put 1-ml (1-millimole) portions of it into four of the cleaned, numbered test tubes. Add 3.3-ml (0.3-millimole) portions of 0.1 M solutions of glucose, fructose, lactose, and maltose and heat the tubes in the beaker of hot water for 20 min. Shake the tubes occasionally to relieve supersaturation and note the times at which osazones separate. If at the end of 20 min no product has separated, cool and scratch to induce crystallization. (Save unused phenylhydrazine reagent.)

Collect and save the products for possible later use in identification of unknowns. Since osazones melt with decomposition, the bath in a mp determination should be heated at a standard rate (0.5° per sec).

Note for the instructor

[4] 2,3,5-Triphenyl-2H-tetrazolium chloride. Available from Aldrich, Dajac, Eastman, Fischer, MCB. Freshly prepared aqueous solutions should be used in tests. Any unused solution should be acidified and discarded.

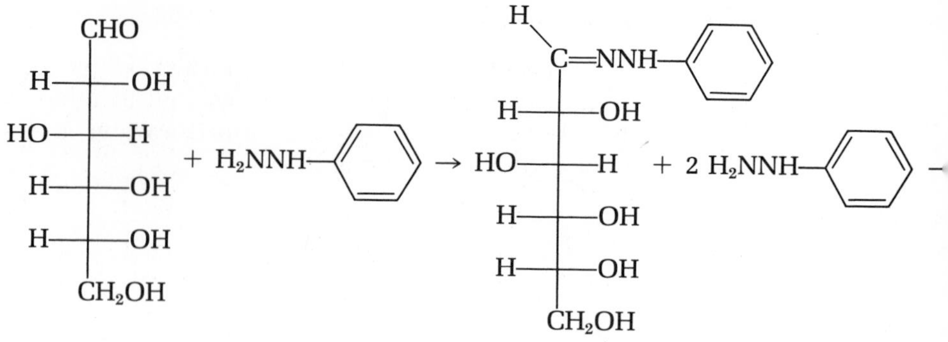

CHO
H——OH
HO——H   + H_2NNH—⟨benzene⟩   →
H——OH
H——OH
CH$_2$OH

H
  C=NNH—⟨benzene⟩
H——OH
HO——H   + 2 H_2NNH—⟨benzene⟩ —
H——OH
H——OH
CH$_2$OH

Glucose   **Phenylhydrazine**   **Glucose phenylhydrazone**

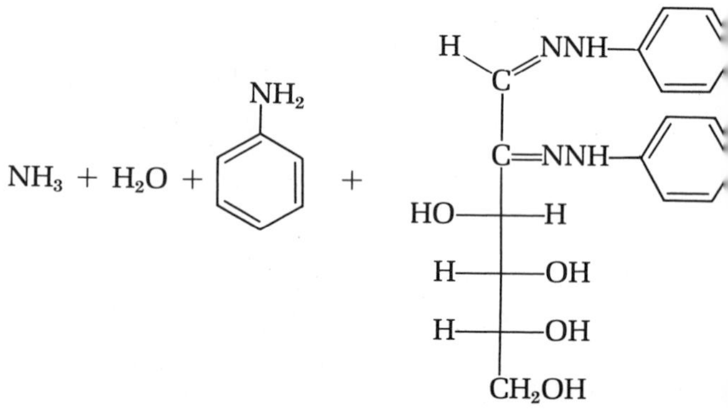

NH_3 + H_2O + ⟨benzene-NH_2⟩ +

H   NNH—⟨benzene⟩
  C
    C=NNH—⟨benzene⟩
HO——H
H——OH
H——OH
CH$_2$OH

Glucose phenylosazone

5. Hydrolysis of a Disaccharide

Sucrose is heated with dil HCl

The object of this experiment is to determine conditions suitable for hydrolysis of a typical disaccharide. Put 1-ml portions of a 0.1 M solution of sucrose in each of five numbered test tubes and add 5 micro drops of concd hydrochloric acid to tubes 2–5. Let tube 1 stand at room temperature and heat the other four tubes in the hot water bath for the following periods of time: tube 3, 2.5 min; tube 4, 5 min; tubes 1 and 5, 15 min. As each tube is removed from the bath it is cooled to room temperature, and if it contains acid, adjusted to approximate neutrality by addition of 15 micro drops of 10% sodium hydroxide. The neutral solution is then stored in a small test tube placed in a filter block adjacent to the numbered tube, which has been washed with water and returned to its place. At the end of the 15-min period, tube 2 is similarly neutralized and the solution stored in a small tube, adjacent to tube 2. Measure one micro drop of each neutral, stored solution into the appropriate numbered test tube, add 1 ml of red tetrazolium solution and a drop of 10% sodium hydroxide, and heat the five tubes together for 2 min and watch them closely.

In which of the tubes was hydrolysis negligible, incomplete, and extensive? Does the comparison indicate the minimum heating period required for complete hydrolysis? If not, return the stored solutions to the numbered test tubes, treat each with 1 ml of stock phenylhydrazine reagent and heat the tubes together for 5 min. On the basis of your results, decide upon a hydrolysis procedure to use in studying unknowns; the same method is applicable to the hydrolysis of methyl glycosides.

6. Evaporation Test

Few solid derivatives suitable for identification of sugars are available. Osazones are not suitable since the same osazone can form from more than one sugar. Acetylation, in the case of a reducing sugar, is complicated by the possibility of formation of the α- or the β-anomeric form, or a mixture of both. The unknown (next section) is supplied in 0.1 M aqueous solution, and if this is evaporated and the residue acetylated the ratio of α to β acetates may not be the same as that reported for a crystalline starting material. In the case of glucose, acetylation after evaporation and crystallization of the product from water gives β-glucose pentaacetate (two polymorphic forms, mp 111° and 135°), but the yield is only 36% (α-pentaacetate, mp 114°). The other sugars give less favorable results.

Triphenylmethyl, or trityl, ethers can be made by reaction of an anhydrous sugar in pyridine solution with trityl chloride, $(C_6H_5)_3CCl$. Primary alcoholic groups are attacked in preference to secondary, and hence the amount of reagent required varies: glucose requires one mole; fructose, lactose, and maltose require two; and sucrose three. 6-Tritylglucose forms solvated crystals of indefinite mp. Acetylation of this product gives tetraacetyl trityl-α-glucose, mp 130°, whereas if glucose is tritylated in pyridine, and the product then acetylated in the same solution without being isolated, the substance formed is tetraacetyl trityl-β-glucose, mp 163°.

Although preparation of the above derivatives is not recommended, the first step in the preparation, evaporation of the aqueous solution of the sugar, provides useful guidance in identification. On thorough evaporation of all the water from a solution of glucose, fructose, mannose, or galactose, the sugar is left as a sirup that appears as a glassy film on the walls of the container. Evaporation of solutions of lactose or maltose gives white solid products, which are distinguishable because the temperature ranges at which they decompose differ by about 100°.

Test for lactose, maltose

Measure 2 ml of a 0.1 M solution of either lactose or maltose into a 25 × 150-mm test tube, and add an equal volume of benzene (to hasten the evaporation). Connect the test tube through a filter trap to the suction pump with a rubber stopper that fits the test tube snugly. Make sure that the pressure-gauge of the filter trap is adjusted correctly so that it will show whether or not all connections are tight and if the suction pump is operating efficiently. Then turn the pump on at full force and rest the tube horizontally in the steam bath with all but the largest ring removed, so that the whole tube will be heated strongly. If evaporation does not occur rapidly, check the connections and trap to see what is wrong. If a water layer persists for a long time, disconnect and add 1–2 ml of benzene to hasten evaporation. When evaporation appears to be complete, disconnect, rinse the walls of the tube with 1–2 ml of methanol, and evaporate again, when a solid should separate on the walls. Rinse this down with methanol and evaporate again to produce a thoroughly anhydrous product. Then scrape out the solid and determine the melting point, or actually the temperature range of decomposition.

Anhydrous α-maltose decomposes at about 100–120°, and anhydrous α-lactose at 200–220°. Note that in the case of an unknown a point of decomposition in one range or the other is valid as an index of identity only if the substance has been characterized as a reducing sugar. Before applying the test to an unknown, perform a comparable evaporation of a 0.1 M solution of glucose, fructose, galactose, or mannose.

7. Unknowns

The unknown, supplied as a 0.1 M solution, may be any one of the following substances:

D-Glucose	Maltose
D-Fructose	Sucrose
D-Galactose	Methyl α-D-glucoside
Lactose	

You are to devise your own procedure of identification.

QUESTIONS   1. *What, do you conclude, is the order of relative reactivity in the RT test of the compounds studied?*

2. *Which test do you regard as the most reliable for distinguishing reducing from nonreducing sugars, and which for differentiating an α-ketol from a simple aldehyde?*

32 *Enzymic Resolution of DL-Alanine*

KEYWORDS   Resolution
            Enzyme
            Acylase, N-acyl derivatives

            N-acetyl-DL-alanine
            L-alanine
            Levorotatory
            Azlactone

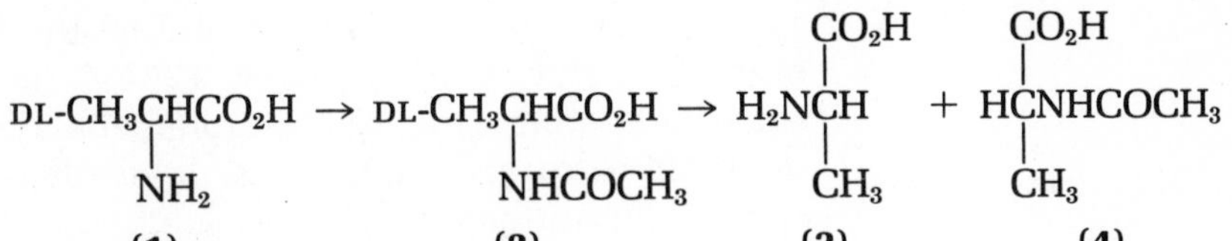

$$DL\text{-}CH_3CHCO_2H \rightarrow DL\text{-}CH_3CHCO_2H \rightarrow H_2NCH \quad + \quad HCNHCOCH_3$$

with pendant groups:

DL-CH₃CHCO₂H NH₂	DL-CH₃CHCO₂H NHCOCH₃	CO₂H H₂NCH CH₃	CO₂H HCNHCOCH₃ CH₃
(1)	**(2)**	**(3)**	**(4)**
Mp 295°, MW 89.10	Mp 137°, MW 131.13	Mp 297°, $\alpha_D + 14.4°$	Mp 125°, $\alpha_D + 66.5°$

Resolution of DL-alanine (**1**) is accomplished by subjecting the N-acetyl derivative (**2**) in weakly alkaline solution to the action of acylase, a proteinoid preparation from porcine kidney containing an enzyme that promotes rapid hydrolysis of N-acyl derivatives of natural L-amino acids but acts only immeasurably slowly on the unnatural D-isomers.[1] N-Acetyl-DL-alanine (**2**) can thus be converted into a mixture of L(+)-alanine (**3**) and N-acetyl-D-alanine (**4**). The mixture is easily separable into the components, since the free amino acid (**3**) is insoluble in ethanol and the N-acetyl derivative (**4**) is readily soluble in this solvent. Note that, in contrast to the weakly levorotatory D(−)-alanine (−14.4°), its acetyl derivative is strongly dextrorotatory.

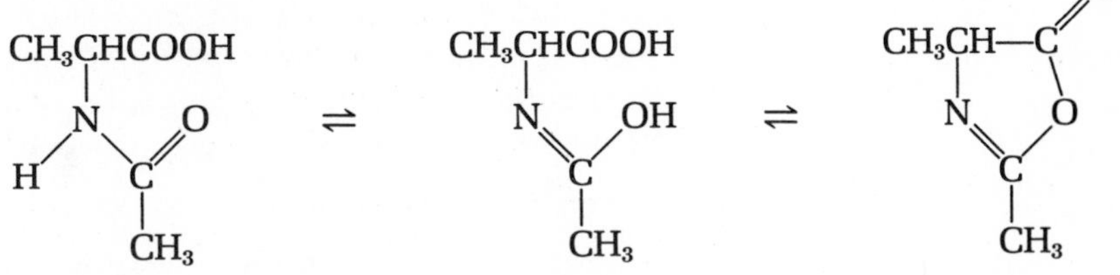

CH₃CHCOOH N—C=O H CH₃	⇌ CH₃CHCOOH N—C—OH CH₃	⇌ CH₃CH—C=O N O C—CH₃
(4)	**(5)**	**(6)**

[1] J. P. Greenstein, et al., *J. Biol. Chem.*, **175**, 969 (1948); **194**, 455 (1952).

The acetylation of an α-amino acid presents the difficulty that if the conditions are too drastic, the N-acetyl derivative (4) is con verted in part through the enol (5) to the azlactone (6).[2] However under critically controlled conditions of concentration, temperature and reaction time, N-acetyl-DL-alanine can be prepared easily ir high yield.

Place 2 g of DL-alanine and 5 ml of acetic acid in a 25 × 150-mm test tube, insert a thermometer, and clamp the tube in a vertical position. Measure 3 ml of acetic anhydride, which is to be added at exactly 100°. Heat the test tube with a small flame, with stirring until the temperature of the suspension has risen a little above 100° Stir the suspension, let the temperature gradually fall, and when it reaches 100° add the 3-ml portion of acetic anhydride and note the time. In the course of 1 min the temperature falls (91–95°, cooled by added reagent), rises (100–103°, the acetylation is exothermic) and begins to fall with the solid largely dissolving. Stir to facilitate reaction of a few particles of solid, let the temperature drop to 80°, pour the solution into a tared 125-ml round-bottomed flask, and rinse the thermometer and test tube with a little acetone. Add 10 ml

Check the pressure gauge

of water to react with excess anhydride, connect the flask to the suction pump operating at full force, put the flask *inside* the rings of the steam bath and wrap the flask with a towel. Evacuation and heating for about 5–10 minutes should remove most of the acetic acid and water and leave an oil or thick sirup. Add 10 ml of benzene and evacuate and heat as before for 5–10 min. If the product has not yet separated as a white solid or semisolid, determine the weight of the product, add 10 ml more benzene, and repeat the process. When the weight becomes constant, the yield of acetyl DL-alanine should be close to the theoretical amount. The product has a pronounced tendency to remain in supersaturated solution and hence does not crystallize.

Add 10 ml of distilled water[3] to the reaction flask, grasp this with a clamp, swirl the mixture over a free flame to bring all the product into solution, and cool under the tap. Remove a drop of the solution on a stirring rod, add it to 0.5 ml of a 0.3% solution of ninhydrin in water, and heat to boiling. If any unacetylated DL-alanine is present a purple color will develop. Pour the solution into a 20 × 150-mm test tube and rinse the flask with a little water. Add 1.5 ml of concentrated ammonia solution, stir to mix, check the pH with Hydrion paper, and if necessary adjust to pH 8 by addition of more ammonia with a capillary dropping tube. Add 100 mg of commercial acylase powder, or 2 ml of fresh acylase solution,[4] mix with a stirring

[2] The azlactone of DL-alanine is known only as a partially purified liquid.

[3] Tap water may contain sufficient heavy metal ion to deactivate the enzyme.

[4] The preparation of the acylase solution should be started on the day of the week that fresh pork kidneys are available at a slaughter house. The fat is sliced off two kidneys (about 150 g) and the kidneys cut into small pieces and either (a) ground in a mortar with sand and suspended in 300 ml of distilled water or (b) placed in a Waring blender with 300 ml of water and spun for 2 min. The homogenate prepared by either method is centrifuged in the cold until the supernatant liquid is clear (3000 × g for 30 min, or 2000 × g for 3 hr). An 18″-length of cellulose sausage casing (1″ in diameter) is wetted so that it can be opened and tied off at one end. The enzyme solution is carefully decanted into the sack, the other end is tied off, and the sack is let soak overnight in

**Figure 32.1
Filter paper identifi-
cation marker.**

Work-up time ½–¾ hr

rod, rinse the rod with distilled water and make up the volume until the tube is about half full. Then stopper the tube, mark it for identification (Fig. 32.1), and let the mixture stand at room temperature overnight, or at 37°[5] for 4 hrs.

At the end of the incubation period add 3 ml of acetic acid to deactivate (denature) the enzyme and if the solution is not as acidic as pH 5 add more acid. Rinse the cloudy solution into a 125-ml Erlenmeyer flask, add 100 mg of decolorizing carbon (0.5-cm column in a 13 × 100-mm test tube), heat and swirl over a free flame for a few moments to coagulate the protein, and filter the solution by suction. Transfer the solution to a 125-ml round-bottomed flask and evaporate on a rotary evaporator under vacuum, or add 20 ml of benzene (to prevent frothing) and a boiling stone and evaporate on the steam bath under vacuum to remove water and acetic acid as completely as possible. Remove the last traces of water and acid by adding 15 ml of benzene and evaporating again. The mixture of L-alanine and acetyl D-alanine separates as a white scum on the walls. Add 15 ml of 95% ethanol, digest on the steam bath, and dislodge some of the solid with a spatula. Cool well in ice for a few minutes, and then scrape as much of the crude L-alanine as possible onto a suction funnel, and wash it with ethanol. Save the ethanol mother liquor.[6] To recover the L-alanine retained by the flask, add 2 ml of water and warm on the steam bath until the solid is all dissolved, then transfer the solution to a 25-ml Erlenmeyer flask by means of a capillary dropping tube, rinse the flask with 2 ml more water, and transfer in the same way. Add the filtered L-alanine, dissolve by warming, and filter the solution by gravity into a 50-ml Erlenmeyer flask (use the dropping tube to effect the transfer of solution to filter). Rinse the flask and funnel with 1 ml of water and then with 5 ml of warm 95% ethanol. Then heat the filtrate on the steam bath and add more 95% ethanol (10–15 ml) in portions until crystals of L-alanine begin to separate from the hot solution. Let crystallization proceed. Collect the crystals and wash with ethanol. The yield of colorless needles of L-alanine, α_D +13.7 to +14.4°[7] (in 1 N hydrochloric acid) varies from 0.40 to 0.56 g, depending on the activity of the enzyme.

a pan of running tap water (dialysis removes soluble kidney components that would interfere with isolation of the amino acid; colored impurities are removed in the course of the isolation and do not interfere). The enzyme solution is centrifuged again to remove debris and stored at 5° until required; the volume is about 150 ml.

Commercial porcine kidney acylase is available from Schwarz/Mann, Division of Becton, Dickinson and Co., Orangeburg, N.Y. 10962.

[5] A reasonably constant heating device that will hold 15 tubes is made by filling a 1-liter beaker with water, adjusting to 37°, and maintaining this temperature by the heat of a 250-watt infrared drying lamp shining horizontally on the beaker from a distance of about 40 cm. The capacity can be tripled by placing other beakers on each side of the first one and a few cm closer to the lamp.

[6] In case the yield of L-alanine is low, evaporation of this mother liquor may reveal the reason. If the residue solidifies readily and crystallizes from acetone to give acetyl-DL-alanine, mp 130° or higher, the acylase preparation is recognized as inadequate in activity or amount. Acetyl-D-alanine is much more soluble and slow to crystallize.

[7] Determination of optical activity can be made in the student laboratory with a Zeiss Pocket Polarimeter, which requires no monochromatic light source and no light shield. For construction of a very inexpensive polarimeter, see W. H. R. Shaw, *J. Chem. Ed.*, **32**, 10 (1955).

33 Ninhydrin

KEYWORDS

Dimethyl phthalate
Ester interchange
β-Keto ester
Hydrolysis and
    decarboxylation

Indane-1,2,3-trione hydrate, ninhydrin
Blue color with amino acids, 570 nm
Preparation of powdered sodium
Pyridinium hydrobromide perbromide
Aspirator tube

Ninhydrin (**8**), p. 189, an expensive reagent used for identification of amino acids by paper-strip chromatography (Chapter 34), can be prepared by a six-step synthesis starting with a double ester condensation of dimethyl phthalate (**1**) with ethyl acetate under the influence of sodium to give the yellow sodium enolate (**2**). Ester interchange occurs during the process and the product is largely the methyl ester, sodio-2-carbomethoxyindane-1,3-dione (**2**). Dimethyl phthalate is preferable to the diethyl or dibutyl ester both because the reaction proceeds best with the lowest homolog and because, since the cost per pound is about the same for the three esters, the one of lowest molecular weight is the most economical. The yield is limited by the side reaction of ester condensation of ethyl acetate to ethyl acetoacetate, $CH_3COCH_2CO_2C_2H_5$.

In the next step the sodium enolate (**2**) is heated with dilute hydrochloric acid; the initially formed free enol of the β-keto ester readily undergoes hydrolysis and decarboxylation to indane-1,3-dione (**3**). This substance is largely enolic and highly sensitive; it has a doubly activated methylene group that is prone to enter into aldol-type condensation with a carbonyl group of a second molecule of dione. Nitration under critically controlled conditions affords 2-nitroindane-1,3-dione (**4**), a yellow substance that is strongly acidic and appears to exist largely in the acinitro form, $R_2CHNO_2 \rightarrow R_2C{=}N(O)OH$. It forms sparingly soluble salts with amines and has been suggested as a reagent for their characterization.

Bromination of 2-nitroindane-1,3-dione in aqueous solution affords the colorless 2-nitro-2-bromo-dione (**5**), which when heated in o-dichlorobenzene solution at 175° decomposes smoothly to dinitrogen trioxide and equal parts of indane-1,2,3-trione (**6**) and 2,2-dibromoindane-1,3-dione (**7**). Indane-1,2,3-trione, a beautifully crystalline red substance, is converted into the colorless 2-hydrate, ninhydrin (**8**), by crystallization from water.

The reaction of ninhydrin with amino acids of the type $RCH(NH_2)$-COOH involves initial formation of (**9**), which undergoes decarbox-

*The mother liquor of the dione (**3**), on standing in the dark for 1–7 days, may deposit yellow platelets of bindone, formed by self-condensation of (**3**)*

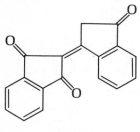

Bindone
Mp 208–210°

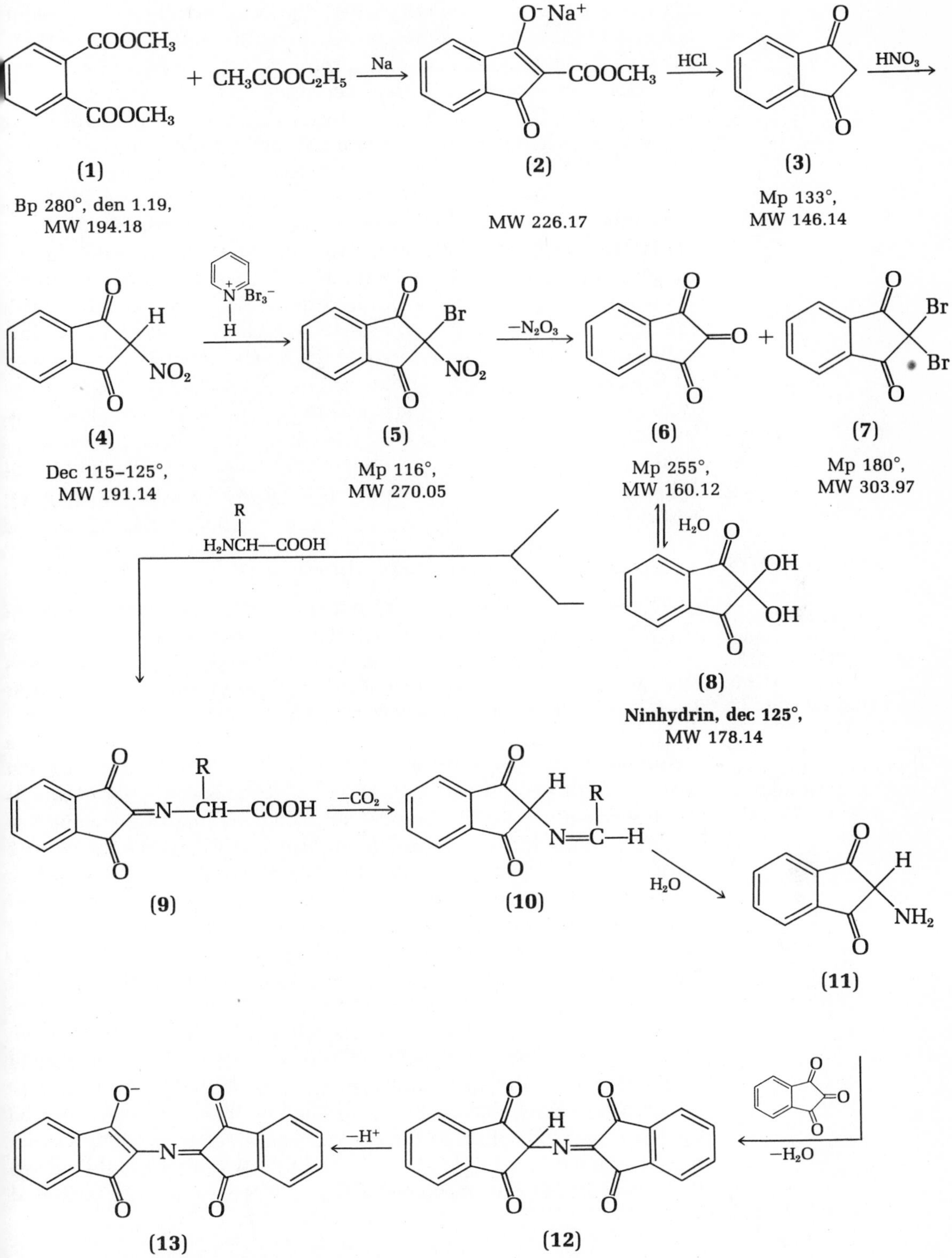
COOCH3
COOCH3
+ CH3COOC2H5
Na
O- Na+
COOCH3
HCl
HNO3
(1)
Bp 280°, den 1.19,
MW 194.18
(2)
MW 226.17
(3)
Mp 133°,
MW 146.14
H
NO2
Br3-
Br
NO2
−N2O3
+
Br
Br
(4)
Dec 115–125°,
MW 191.14
(5)
Mp 116°,
MW 270.05
(6)
Mp 255°,
MW 160.12
(7)
Mp 180°,
MW 303.97
R
H2NCH—COOH
H2O
OH
OH
(8)
Ninhydrin, dec 125°,
MW 178.14
N—CH—COOH
−CO2
N=C—H
H2O
NH2
(9)
(10)
(11)
O-
N
−H+
−H2O
(13)
(12)

ylation to the Schiff base (**10**), which in turn is readily hydrolyzed to the amine (**11**). This amine condenses with another mole of (**6**) to form (**12**), which, on loss of a proton, gives the blue anion (**13**). Therefore, most amino acids, according to this mechanism, will give the same final product (**13**). Since this reaction is quantitative and reproducible it serves as a precise method for amino acid analysis, in addition to the qualitative use made of it in the next chapter.

This color reaction of amino acids with ninhydrin lends itself to automation. In an automatic amino acid analyzer a mixture of amino acids (from hydrolysis of a protein or peptide) is chromatographed on a column of ion exchange resin using buffers to elute. The eluate is automatically mixed with ninhydrin, heated, and passed through the absorption cell of a spectrometer which operates in the visible wavelength range. The intensity of the blue color at 570 nm (as well as the peak at 440 nm from reaction of ninhydrin with proline and hydroxyproline, in which the α-amino group is secondary) is plotted by a chart recorder as a function of ml of eluate to give a series of peaks. Each amino acid is eluted by a characteristic volume of buffer; the relative amounts of the amino acids are given by the areas under the peaks.

EXPERIMENTS

1. Sodio-2-carbomethoxyindane-1,3-dione (2)

Preparation of powdered sodium. CAUTION! Read the whole of this paragraph and be prepared to do the shaking immediately after the sodium has been melted.

The ester condensation does not proceed satisfactorily if the sodium used is in chunks or slices. Hence, the first operation is preparation of powdered sodium, a mixture of fine particles and globules the size of bird shot. Remove a piece of sodium from its container (in which it is stored under kerosene) on the tip of a knife or with a forceps, press the piece in the folds of a *dry* towel to get rid of the kerosene (filter paper is less efficient), and by manipulating the material with a combination of knife and forceps and without touching the sodium with the fingers, cut off slices totalling a weight of 2 g (1.4 times the theoretical amount). Transfer these slices to a 20 × 150-mm test tube and add 5 ml of xylene. Clamp the tube in a vertical position at a height convenient for heating with a *very small* burner flame, insert an aspirator tube (dry) at the top for safety, heat the mixture gently until the xylene (bp 140°) just begins to boil. Turn off the flame when the rim of condensate rises about 1 inch above the surface of the liquid. The sodium is then in a molten state (mp 97°), even though the pieces roughly retain their original shape. The tube is to be stoppered and the contents shaken vigorously while the sodium is still molten, an operation that can be done efficiently and safely with use of a protective 25 × 150-mm test tube having a wad of cotton at the bottom and a section of 1¾″ Gooch tubing taped to the top (Fig. 33.1b; the rubber tubing should extend 6 cm beyond the rim of the test tube).[1] Before the sodium is melted

Note for the instructor

[1] A few "sodium stations," each equipped with a supply of sodium, xylene, 5-ml pipette, knife, tweezers, dry towel, protective tube, ring stand with clamp, burner, and a bottle for xylene-sodium residues should service a large class.

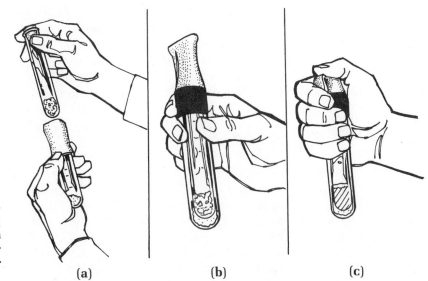

Figure 33.1
Using a safety tube in
the preparation of pow-
dered sodium.

(a)   (b)   (c)

under xylene, the Gooch rubber tubing of the protective tube should
be skinned back as in (a); the stoppered tube containing hot xylene
and sodium is then inserted, the Gooch tubing is extended as in (b),
folded over (c), and the tube grasped firmly in the hand with the
thumb pressing against the stopper under the protective rubber.
The tube assembly is then given a sharp whipping shake in the direction of its length, which should produce large globules of sodium.
Successive sharp shakes, made deliberately and each time followed
by inspection of the result, further reduce the particle size until after
five or six shakes the sodium is reduced to fine particles and small
globules. Shaking beyond this point may cause the particles to
coalesce. The rubber is then skinned back over the protective tube
and the smaller tube removed, let cool, and the bulk of the xylene
decanted into a xylene-sodium bottle.

Powdered sodium in this experiment can be replaced by a semisolid, 50% dispersion of sodium hydride in mineral oil.[2]

If the commercially available[3] dispersion of sodium in mineral oil
is employed, the mineral oil should be removed by adding xylene,
stirring, and decanting the xylene-mineral oil mixture from the
globules of sodium.

While the sodium-xylene mixture is cooling, measure into a
50-ml Erlenmeyer flask 10 ml of dimethyl phthalate and 20 ml of
ethyl acetate, and mix the liquids by swirling. Add 3 ml of the mixture to the test tube containing the powdered sodium, shake, and
pour the suspension into a dry 125-ml round-bottomed flask. Rinse
the test tube with another 3-ml portion of the ester mixture, and add
5 drops of 95% ethanol to the flask. Attach a short reflux condenser

[2] H. Gruen and B. E. Norcross, *J. Chem. Ed.*, **42**, 268 (1965).
[3] Sodium dispersion, 50% in mineral oil, SX232, Matheson, Coleman and Bell, 2909 Highland
Ave., Norwood, Ohio 45212.

and heat the mixture on the steam bath for 10 min.[4] Then add the rest of the ester mixture through the top of the condenser, and continue refluxing, either continuously, or on separate days for a total of 3–4 hr. The yellow sodium salt (2), usually begins to separate in about 1 hr; then the paste gradually thickens and may become an immobile mass. Even though there may only be time on the first day for refluxing for a few minutes, the operation is advantageous since the reaction will then proceed to a significant extent at room temperature in the course of 1–2 days.

Reflux time 3–4 hr either continuously or, more advantageously, on separate days

If the refluxing is interrupted, either attach a calcium chloride tube to the flask or fill the mouth with a plug of dry absorbent cotton.

At the end of the reflux period, break up the yellow cake with a spatula, add a little ethyl acetate for thinning, and collect the product on a Büchner funnel. For removal of dark-colored mother liquor, release the suction, cover the cake with ethyl acetate, scrape the solid and liquid together with a spatula to an even paste, and then apply suction. Since incomplete washing may lead to trouble in the next step, transfer the yellow salt to a beaker, stir it with ethyl acetate, and refilter. Press the cake well, applying suction as you do so, then spread the product out to dry (to constant weight). The yield of sodio-2-carbomethoxyindane-1,3-dione (2) varies from 5–10 g; the average is closer to the lower limit than to the upper. The variability is probably due to differences in the particle size of the sodium.[5] However, 5 g of product is ample for the next step.

While the salt is drying, dissolve a small sample (10 mg) in water, add a drop of acetic acid (pK_a 4.8), and then a drop of hydrochloric acid. Is the pK_a of the free enol larger or smaller than 4.8?

2. Indane-1,3-dione (3)

Place 7 g of the sodio-2-carbomethoxyindane-1,3-dione (2) just prepared (finely crushed) in one 250-ml Erlenmeyer flask and 100 ml of water in another. Pour 10 ml of concd hydrochloric acid into the water, heat the solution to 80°, pour it onto the salt, and keep the temperature close to 70° for 5–6 min. The salt is soon converted to the free enol (yellow) and this loses carbon dioxide and affords the dione 3, which separates as an almost colorless solid. Cool to 15–20°, collect the solid on a small filter, and wash with water weakly acidified with a little hydrochloric acid. To keep the drying time at a minimum, press the filter cake firmly under a spatula and let drain until the drip of filtrate has completely stopped. Note the weight of the wet cake (5–5.6 g), spread it out on a filter paper, scrape and respread it occasionally, and again note the weight after 1 hr (1.5 hr, 4.1 g; and 2 hr, 4.1 g). The crude light cream-colored

Swirl vigorously while heating

The substance stains the skin

[4] Addition of ethanol helps to get the reaction started, and the maximal effect is obtained when only a small amount of ester mixture is present.
[5] In industrial practice the reaction time can be shortened and the yield materially improved by use of highly reactive colloidally dispersed sodium.

For quick drying heat in an evacuated Erlenmeyer on the steam bath

indane-1,3-dione, mp 131–132°, can be nitrated directly if it is thoroughly dry. It crystallizes from benzene in long, fine needles, mp 132–133°.

3. 2-Nitroindane-1,3-dione (4)

Total time about 20 min

Indane-1,3-dione (3) is nitrated with 4 ml of acetic acid and 1 ml of fuming nitric acid per gram of dione at a temperature not to exceed 35°. Thus, place 3.5 g of indane-1,3-dione (3) in a 125-ml Erlenmeyer flask, add 14 ml of acetic acid, insert a thermometer, make an ice bath ready, and to the suspension at 25° add 3.5 ml of fuming nitric acid.[6] The temperature usually rises rapidly and should be checked at 35° by cooling sufficient to prevent a rise above 40° but not enough to cause much drop in temperature. The yellow nitro compound soon begins to separate, and after about 5 min at 40° the temperature begins to drop. After 5 min more the mixture can be cooled and the product collected and washed with a little ether (to remove acetic acid). The nitro compound (4) can be dried in a few minutes and is then ready for the next step; set aside an estimated 100 mg for the following tests.

Dissolve 100 mg of the nitroindanedione (4) in 10 ml of water and treat 1-ml portions as follows: (a) with 0.5 ml of concd hydrochloric acid and let stand undisturbed, (b) with 1 ml of 1% aniline hydrochloride and let stand (the salt melts at 209°); (c) with 5 ml of water and then 1 ml of 1% aniline hydrochloride solution, cool and scratch; (d) with one drop of 20% copper sulfate solution and scratch.

4. 2-Bromo-2-nitroindane-1,3-dione (5)

Total time about 30 min

Place a 50-ml Erlenmeyer flask containing 5 g of pyridinium hydrobromide perbromide and 25 ml of acetic acid inside the rings of a steam bath and let the mixture warm while dissolving 3 g of the nitroindanedione (4) in 50 ml of water in a 250-ml beaker by stirring. When the perbromide has all dissolved, grasp the flask with a towel and pour the hot solution into the aqueous solution with stirring. The color is discharged and the bromonitrodione separates at once as a white precipitate. Collect the precipitate, press the filter cake thoroughly with a spatula, and dry well by suction. Then dissolve the slightly moist product in ether, add 3–4 g of anhydrous sodium sulfate, swirl for a minute or two, filter, and evaporate the filtrate

Alternative procedure using concd HNO₃

[6] 90% HNO$_3$; den 1.48. The acid should be pipetted from a freshly opened bottle; use a pipetter. *Alternative procedure to the use of fuming nitric acid.* Mix 15 ml of acetic acid with 10 ml of acetic anhydride, make an ice bath ready, measure 5 ml of concd nitric acid and add it with a dropper in 1-ml portions. After each addition let the temperature rise (for reaction of the anhydride with the water in the acid) but not above 60°. When the heat effect is over, adjust to **exactly** 35° and pour the solution onto 3.5 g of dione 3. The temperature drops to 27–28° and then (2 min) begins to rise slowly. When it rises to 35° control to 34–35° by a quick dip in the ice bath. Control to 33–35° for a total of 20 min, cool to 5°, collect and wash the product well with ether. Yield 3.8 g.

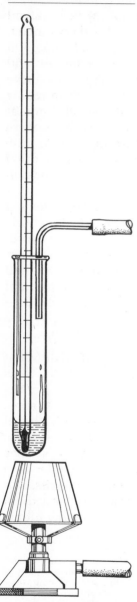

Figure 33.2
Aspirator tube used when heating reactants in the preparation of indane-1,2,3-trione.

to dryness. The solid residue, scraped out of the flask, is suitable for the next step; yield 3.1 g.[7]

5. Indane-1,2,3-trione (6)

Place 3 g of the bromonitroindanedione (5) and 3 ml of o-dichlorobenzene (bp 180°) in a 20 × 150-mm test tube, clamp the tube in a vertical position over a burner, insert a thermometer and an aspirator tube (Fig. 33.2), and heat over a small flame to a temperature of 170° when the solution turns green and oxides of nitrogen are evolved. By intermittent heating, maintain a temperature of 170–180° for 3 min, and then remove the thermometer and let the solution stand undisturbed to cool, when the indanetrione (6) separates as deep red needles. When crystallization appears to be complete at room temperature, cool the mixture well in an ice bath, which may cause separation of a little white hydrate. The crystals are collected on a Hirsch funnel, and since the trione and its hydrate are both insoluble in ether, this solvent is used to rinse the tube and wash the crystals free of o-dichlorobenzene (save the filtrate). The yield of crystalline product is 0.7 g.

If the o-dichlorobenzene-ether filtrate is warmed on the steam bath to evaporate the ether, and the residual solution is let stand for several hours, crystals of 2,2-dibromoindane-1,3-dione (7) separate.[8]

6. Ninhydrin (8)

In a 10-ml flask heat 0.7 g of indane-1,2,3-trione (6) with 2 ml of water until the solid is dissolved (with disappearance of the color), add 2 ml of concd hydrochloric acid, and place the flask in a small beaker of ice and water for crystallization. The hydrate separates as colorless prisms, which are collected, washed with a little ice-cold 1:1 hydrochloric acid-water, and dried; yield 0.5 g.

7. Tests

Dissolve 30 mg of ninhydrin in 3 ml of water and distribute the solution into three 13 × 100-mm test tubes. Add 1-ml portions of 1% solutions of glycine, DL-alanine, and DL-aspartic acid to the three tubes, heat the solutions together in the rings of a steam bath for 5 min, and note any differences in the speed of color development. Dilute 1 ml of the solution derived from glycine to 100 ml, note the shade and intensity of color, and test the effect of adding hydrochloric acid to one portion and alkali to another. Reaction with a protein can be demonstrated by pouring a few drops of the ninhydrin solution into a freshly opened eggshell, or, if your technique has been poor, by the appearance of your hands.

[7] Compound (5) crystallizes well from 1:1 benzene-ligroin. It is unstable in hydroxylic solvents and reacts with water with liberation of HOBr, which characterizes it as having positive bromine.
[8] By steam distillation of the mother liquor a total of 0.6 g of (7) can be recovered.

34 *Paper Chromatography of Amino Acids*

KEYWORDS   Partition chromatography
Stationary phase, adsorbed water
Moving phase, phenol
Lipophilic, hydrophilic
Rf values of amino acids
Molecular flow
Protein hydrolyzate
Filter paper strip

Applicator stick
Solvent front
Capillary action
Chromatogram
Ninhydrin and permanganate
   spot tests
Detection of proline

Filter paper under ordinary atmospheric conditions consists of cellulose containing 22% of adsorbed water (hydrogen-bonded; about two molecules per $C_6H_{10}O_5$ unit). When a folded strip of filter paper is inserted into a slightly slanting test tube (Fig. 34.1) with the lower end of the paper dipping into phenol, an organic solvent which is only partially miscible with water, the phenol ascends the strip by capillary flow without material disturbance of the adsorbed water. In paper partition chromatography, the adsorbed water is the stationary phase and the organic solvent the moving phase, just as silicone oil is the stationary phase and helium the moving phase in gas chromatography.

Before the folded strip is inserted, the paper is impregnated with minute amounts of an amino acid solution at the two ends of a starting line (Fig. 34.2) that will be a little above the level of the phenol. As phenol rises toward the top of the strip the amino acids are subjected to innumerable partitions between the moving lipid phase and the stationary water phase. Highly lipophilic amino acids travel almost as fast as the organic solvent, whereas very hydrophilic ones are largely retained by the adsorbed water and make little progress. When the easily discernible solvent front has reached the finish line, 10 cm from the starting line (Fig. 34.2b), the chromatogram is terminated and the strip removed and hung on a hook. Amino acids are all colorless and the strip bears no indication of their distribution until it is sprayed with a solution of ninhydrin. The oxidation-reduction reaction between ninhydrin and an amino

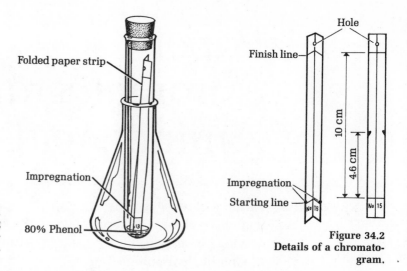

Figure 34.1
Apparatus for carrying out partition chromatography.

Figure 34.2
Details of a chromatogram.

acid (Chapter 33) produces a pigment that appears as a small spot (b). The position of the front (top) of the spot is noted and its distance from the starting line measured with a ruler. The rate of flow, or Rf value, of a particular amino acid is the ratio of the distance travelled by the acid to the distance travelled by the solvent. In the example of Fig. 34.2b, $Rf = (4.6/10) = 0.46$. The acids having primary α-amino groups give pink to purple spots when sprayed with ninhydrin; the secondary amines proline and hydroxyproline give yellow spots.

Rf values vary with the solvent system and type of filter paper used; those reported in Table 34.1 were determined by the procedure you will follow in this experiment. In the series glycine, alanine, valine, and leucine, Rf increases with increasing molecular weight; the larger the alkyl group the more the acid tends to move along with the organic solvent. Figure 34.3 includes Rf values for three rarer, straight-chain acids that are the higher n-alkyl homologs of alanine; that the five points for the n-alkyl series fall on a smooth curve demonstrates a regular relationship, if not a strict proportionality, between Rf and molecular weight. The structure of the alkyl group is of minor but noticeable influence, since the isoalkyl compounds valine and leucine do not travel quite as fast, and hence are slightly more hydrophilic than the n-isomers.

The balance between the hydrophilic and lipophilic character of a given acid is expressed by a quantity defined as the molecular flow (Mf), calculated thus:

$$Mf = MW/100 \, Rf$$

Inspection of the fifth column of the table shows that structurally related compounds have Mf values in a range characteristic of the acid type. For the seven n- and i-alkyl acids plotted in Fig. 34.3, the average Mf is 1.6. That the Mf values for methionine (1.9) and

Table 34.1 Rf Values of Amino Acids

Acid	Formula	Rf	MW	$Mf = \dfrac{MW}{100\,Rf}$	Cost[a]	Yield: mg/100 mg of		
						Hair	Silk	Gelatin[b]
Cystine (CyS-SCy)	$HO_2CCH(NH_2)CH_2SSCH_2CH(NH_2)CO_2H$	.16	240.30	15	low	18.0	42.3	25.5
Glycine (Gly)	$CH_2(NH_2)CO_2H$	.42	75.07	1.8	v. low	4.1	24.5	11.4
Alanine (Ala)	$CH_3CH(NH_2)CO_2H$	.59	89.09	1.5	v. low	2.8	3.2	2.0
Valine (Val)	$(CH_3)_2CHCH(NH_2)CO_2H$	.75	117.15	1.6	low	5.5	0.8	2.5
Leucine (Leu)	$(CH_3)_2CHCH_2CH(NH_2)CO_2H$	.79	131.18	1.7	low	11.2	—	2.5
Isoleucine (Ile)	$CH_3CH(CH_2CH_3)CH(NH_2)COOH$	—	131.18	—	v. high	—	—	2.0
Methionine (Met)	$CH_3SCH_2CH_2CH(NH_2)CO_2H$	.77	149.21	1.9	v. low	0.7	—	2.5
Phenylalanine (Phe)	$C_6H_5CH_2CH(NH_2)CO_2H$	.82	165.19	2.0	high	2.4	—	14.1
Proline (Pro)	$HNCH_2CH_2CH_2CHCO_2H$	.85	115.13	1.4	high	4.3	1.5	
Ornithine (Orn)	$H_2NCH_2CH_2CH_2CH(NH_2)CO_2H$	.67	132.16	2.0	high			
cation	$^+H_3NCH_2CH_2CH_2CH(NH_2)CO_2H$	.47	133.17	2.8	—			
Lysine (Lys)	$H_2NCH_2CH_2CH_2CH_2CH(NH_2)CO_2H$	.71	146.19	2.1	v. high	1.9	0.4	3.6
cation	$^+H_3NCH_2CH_2CH_2CH_2CH(NH_2)CO_2H$	.53	147.20	2.8	—			
Arginine (Arg)	$HN=C(NH_2)NHCH_2CH_2CH_2CH(NH_2)CO_2H$	.76	174.20	2.3	low	8.9	1.1	9.0
cation	$^+H_2N=C(NH_2)NHCH_2CH_2CH_2CH(NH_2)CO_2H$	.60	175.21	2.9	—			
Histidine (His)	$HNCH=NCHCH_2CH(NH_2)COOH$	—	155.16	—	high			
Tryptophan (Try)	$HN—C_6H_4—CH=C—CH_2CH(NH_2)COOH$	—	204.23	—	high			
Serine (Ser)	$HOCH_2CH(NH_2)CO_2H$	.43	105.09	2.4	high	10.6	12.6	4.4
Threonine (Thr)	$HOCH(CH_3)CH(NH_2)CO_2H$	.51	119.12	2.3	high	8.5	1.5	2.9
Tyrosine (Tyr)	$HOC_6H_4CH_2CH(NH_2)CO_2H$	.62	181.19	2.9	low	2.2	10.6	1.7
Aspartic acid (Asp)	$HO_2CCH_2CH(NH_2)CO_2H$	.32	133.10	4.2	v. low	3.9	—	6.1
anion	$HO_2CCH_2CH(NH_2)CO_2^-$	.25	132.09	5.3	—			
Glutamic acid (Glu)	$HO_2CCH_2CH_2CH(NH_2)CO_2H$	.40	147.13	3.7	v. low	13.1	—	12.1
anion	$HO_2CCH_2CH_2CH(NH_2)CO_2^-$	.35	146.12	4.0	—			

[a] Per 100 g of L- or DL-form: v. low < $5.00, low < $10.00, high $15–$20, v. high > $30.00 (1973 prices).
[b] Also 9.7 mg of hydroxyproline, unique to gelatin.

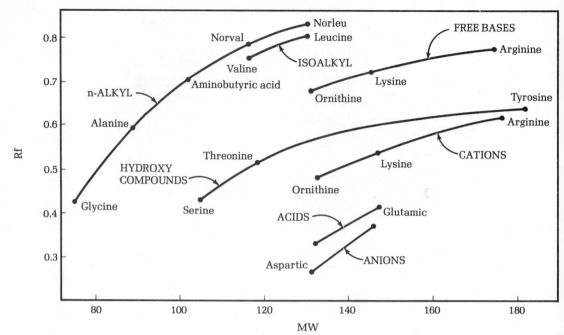

Figure 34.3
Rf values of amino acids
as a function of MW.
Lines connect similar
amino acids.

phenylalanine (2.0) are slightly higher means that the sulfur atoms and the phenyl group render the substances a little more hydrophilic, weight for weight, than the alkyl acids. Proline is distinctive in being a cyclic secondary amine; it has the lowest *Mf* value (1.4) of all the acids. Cystine is anomalous in both *Mf* and solubility. The high molecular weight would seem to be offset by the presence of two hydrophilic dipolar ion groupings, but the *Rf* is very low and the solubility in water (25°) is only 0.112 g/l., as compared to 24.3 g/l. for leucine. The anomaly may be associated with the existence of the substance in a chelated endocyclic structure.

Introduction of polar groups into the side chains of the alkyl acids (av. *Mf* = 1.6) produces successive hydrophilic shifts as follows: basic group, *Mf* = 2.1; hydroxyl group, *Mf* = 2.5; carboxyl group, *Mf* = 4.0. Figure 34.3 shows that the curves are of similar types but displaced to lower *Rf* values. The average *Mf* value of 2.6 found for dipeptides (Fig. 34.4) indicates that the lipophilic effect of the peptide link is intermediate between those of the hydroxyl and carboxyl groups. If ornithine, lysine, or glutamic acid is chromatographed not as the free base but as the monohydrochloride, the ionic group provides better retention by the stationary water phase and in each case *Rf* is about 0.2 unit lower than found for the free base. Since lysine perchlorate gives the same *Rf* as the hydrochloride, the rate of flow is determined by the lysine cation alone. A smaller hydrophilic

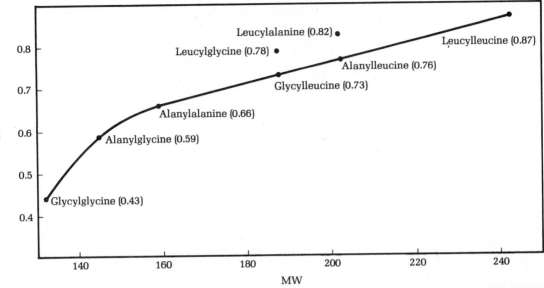

Figure 34.4
Rf values of dipeptides
as a function of MW.
Average *mf* = 2.6.

displacement results when aspartic and glutamic acid are chromatographed as anions. The displacements are sometimes useful aids in the identification of unknowns, which is the major objective of the present experiment.

Sections 1–3 of the following experiments describe general procedures, and Section 4 lists five chromatograms that are to be run on known acids and mixtures; these should be started at the very beginning of the period so that they can be carried to a stopping point by the end of the day. As soon as the chromatograms are under way the two protein hydrolysis experiments should be started; during the 1-hr reflux period the characterizing tests of Sections 5–8 can be carried out. The protein hydrolyzate should be worked up on the first day and an overnight chromatogram started.

<table>
<tr><td>EXPERIMENTS</td><td>

1. Procedure of Chromatography[1]

</td></tr>
</table>

A strip of paper 13 cm long and 1.3 cm wide is to be cut from a roll of half inch, Eaton-Dikeman No. 613 filter paper,[2] marked, punched, folded, and impregnated as shown in Fig. 34.5 without being touched at the critical part with the fingers, since contact with skin protein can give rise to false spots. This can be done by initially cutting a strip about 4 cm longer than required and using the 2-cm section at

Notes for the instructor

[1] Wilkens-Anderson supplies the filter paper recommended, the strip dispenser (Fig. 34.6), the amino acid dispenser, and stock amino acid solutions.
[2] Supplied in rolls $7\frac{1}{4}$ and $9\frac{5}{8}$ in. in diameter. Half-inch Whatman No. 1 paper ($8\frac{3}{8}$ in. roll, 600 ft) can be used but gives slightly lower *Rf* values.

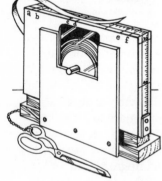

Figure 34.5
Details of chromatograms
for the partition of
proteins.

Handle

Not
to be
touched

Serial No — No. 15

Handle

each end as a handle that is to be cut off after all other operations have been completed. Figure 34.5 shows such a strip marked with starting and finish lines (c and b, 10 cm apart) and boundary lines (a and d); it is also punched and marked with a serial number. A strip dispenser (Fig. 34.6) protects the paper and expedites the preparation of strips. Grasp the end of the paper tape and pull out enough to give a 10-cm (or a 25-cm) strip, mark pencil lines at a, b, c, and d (or, for a 25-cm strip at a, b, e, and f), punch a hole centered between a and b with the point of a sharp pencil, and then cut off the strip cleanly at a right angle at a point about 2 cm above a. Record the serial number of the experiment in the space between c and d, and then fold the strip down the center. A convenient technique of folding (without contact of the fingers in the vital area) is to place the strip with the markings down on a clean, smooth surface, such as the back of a notebook or writing pad, place a transparent 30-60° triangle so that it covers exactly one half of the paper (Fig. 34.7),

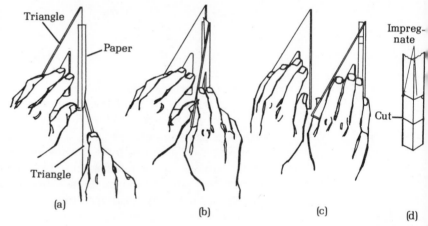

Figure 34.7
A technique for folding
a chromatogram.

Figure 34.6
Strip dispenser. This
dispenser, which will
hold any of the rolls
mentioned in Footnote 2,
can be made from a
11/4" × 9" × 12" card-
board box. For accom-
modation of the 9 5/8"
roll, the box is mounted
on blocks as shown and
its middle section cut
away. The slot is 8 cm
from the top; marking-
line (a) is 1.3 cm from
the edge, and the other
lines are at the follow-
ing distances from (a):
(b), 1.5 cm; (c), 11.5 cm;
(d), 13 cm; (e), 26.5 cm,
(f), 28 cm. A small hole,
centered between (a) and
(b), is punched with an
awl through both layers
of cardboard.

press it down firmly, insert under the paper the tip of a second triangle, fold the paper up against and over the first one (b), and then withdraw the first triangle while flattening the strip on the crease (c). Note that the triangle for drawing formulas (Fig. 34.8) serves also for folding paper strips.

Now grasp the folded strip by the lower handle and apply a tiny spot of amino acid solution to each edge of the strip at the ends of the starting line (d). The wetted area should be no more than 1–2 mm in diameter and may be visible only from the under side. The application is conveniently made with the smaller end of a toothpick that has been soaked in the solution until the wood is saturated.[3] To remove excess solution, touch the end first to the neck of the bottle and then to the strip handle; make practice spots on the handle and

[3] An amino acid dispenser can be made by inserting a toothpick through a hole drilled in the plastic cap of a specimen vial containing each amino acid solution. Matching colors on a base board, the vial, and the cap for each amino acid help to ensure return to the proper place.

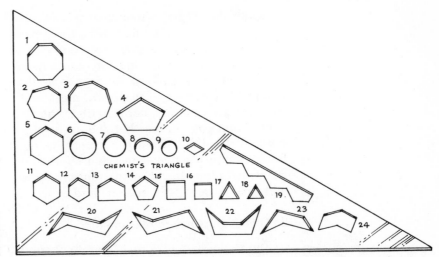

Figure 34.8
Chemist's triangle (Distributor: Reinhold Publishing Corp.).

then make the actual applications and let the spots dry (1–2 min). About 0.2 ml of 80% phenol is to be withdrawn from a bottle in which the solution is overlayered with 90–120° ligroin to prevent oxidation[4] and introduced to the bottom of a 20 × 150-mm test tube without getting it on the walls of the tube. This can be done by inserting the capillary tip of an 8-mm dropping tube through the ligroin layer while exerting a slight pressure, drawing up phenol solution, inserting the dropper carefully into the test tube until the tip is at the bottom, and running in a 2-cm column of liquid. Mount the tube in a stand, an Erlenmeyer flask, or beaker in a slightly slanting position, cut off both handles of the impregnated strip, and with a forceps, lower the strip carefully into the test tube so that it touches only the top and bottom, as in Fig. 34.1. Make sure that the hole is up and the amino acid spots down,[5] cork the tube, and note the time. It takes 2–3 hours for the solvent front to complete the travel of 10 cm and reach the finish line. When this point is reached (note the time) grasp the strip with a forceps and withdraw it while rinsing it with a stream of acetone (Calcutta wash bottle) to remove phenol and promote rapid drying. Hang the strip on either an applicator stick (Fig. 34.9), a hook, or a pin. When it is dry, grasp the lower end with a forceps and spray the strip uniformly but lightly with ninhydrin solution[6] (the paper should become moist but not dripping).

Caution: Phenol is toxic and caustic. If it gets on the skin wash the area well with soap and water and soak it in dilute alcohol

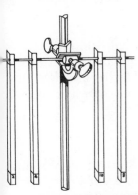

Figure 34.9
Chromatograph strips hung on an applicator stick for drying. The ends of the stick can be pointed in a pencil sharpener.

[4] Either (a) warm a mixture of 80 g of analytical grade phenol with 20 g of distilled water until dissolved and cool, or (b) pour 114 ml of boiling water into a 1-lb bottle of phenol and shake for a few minutes; the solid soon dissolves (and the temperature drops to about 13°). The solution is at once covered with a generous layer of 90–120° ligroin and shaken to effect saturation (very little ligroin dissolves). If the solution is not protected from air it soon turns pink and then brown and eventually acquires capacity for destroying amino acids by oxidation.

[5] If you notice soon enough that the strip is upside down you may be able to withdraw it with a forceps, cut off the wetted section, and insert it in the correct way.

[6] Dissolve 400 mg of ninhydrin and 1.5 ml of s-collidine in 100 ml of 95% ethanol. A basic solution is required since the test solutions are acidic; pyridine can be substituted for collidine but gives less range in the color of the spots. Ethanol is superior to the frequently used butanol because it evaporates more rapidly and promotes quick development of spots at room temperature. A perfume atomizer works well as a sprayer.

Notes for the instructor

Spots usually begin to appear within 5–10 min; with aspartic acid there may be an induction period of several hours. Each application of a test solution should give rise to a separate spot, not more than 3–4 mm in diameter, on each side of the strip, but one may appear better defined and more reliable than the other. Outline the boundary of each spot in pencil, measure the distance of spot front from the starting line, and calculate the Rf.

It is not imperative that the chromatogram be extended to the point where the solvent has exactly reached the finish line. If you wish to stop it either somewhat short of or somewhat beyond this point, withdraw the strip and *mark the position of the solvent front,* and then rinse it with acetone and proceed as above.

2. Overnight Chromatograms

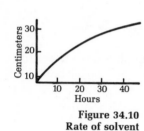

Figure 34.10
Rate of solvent travel.

Figure 34.10 shows that the rate of solvent flow decreases markedly with time.[7] Hence, use of a longer strip gives much greater flexibility in the time of terminating the chromatogram. With a 25-cm strip, the chromatograms can be run for 3–5 hrs, 15 hrs, or any time up to 24 hrs. Such a strip can be measured on the same dispenser (Fig. 34.7), marked with lines at *a, b, e,* and *f,* and folded as before. The amino acid sample is doubled by making one pair of applications, drying, and repeating the process. The amount of phenol is increased to 1 ml (4-cm column in the dropper); the phenol is introduced as before into a 20 × 150-mm test tube, and this is slid into a 250-ml graduate and adjusted to a slanting position. The strip is lowered into place and the graduate stoppered.

3. Basification

For conversion of either lysine or arginine hydrochloride to the free base or conversion of either aspartic or glutamic acid to the anion, the test solution is spotted on a strip and let dry and then 10% sodium hydroxide is applied at the same spots (use a fresh toothpick and discard it) and let dry.

4. Knowns

Prepare five 10-cm strips; impregnate them with 0.1 M solutions[8] of each of the following acids or pairs. In the case of a pair, apply one acid in two spots, dry, and then apply the other acid over the first in the same two spots.

---

[7] The curve conforms to the equation:

$$\log T \text{ (min)} = \sqrt{0.42 \times \text{Distance (cm)}}$$

[8] The solutions are made by dissolving 0.01 mole of a hydrochloride in 100 ml of water, or dissolving 0.01 mole of a free amino acid and 1 ml of concd hydrochloric acid in 100 ml of water (cystine and tyrosine require 4 ml of acid). In the case of the compounds of higher molecular weight the acid is required to effect solution; in other cases it is required to prevent growth of microorganisms.

Notes for the instructor

Arginine hydrochloride   Phenylalanine and glycine
Arginine (see Section 3)   Alanine and aspartic acid
Leucine and glutamic acid

See if you can effect good separation of spots, and compare your
Rf values with those of Table 34.1. You may note a general displacement of values, which will be of guidance in studying unknowns.

5. Detection of Basic Amino Acids

Put a tiny drop of arginine hydrochloride solution on the bottom
of an inverted beaker. With a fresh toothpick (that is to be discarded
after the test) add a tiny drop of 2% phosphotungstic acid solution
and see if a white precipitate forms. Repeat with lysine hydrochloride and, for comparison, with some one of the nonbasic acids.

6. Ninhydrin Spot Test

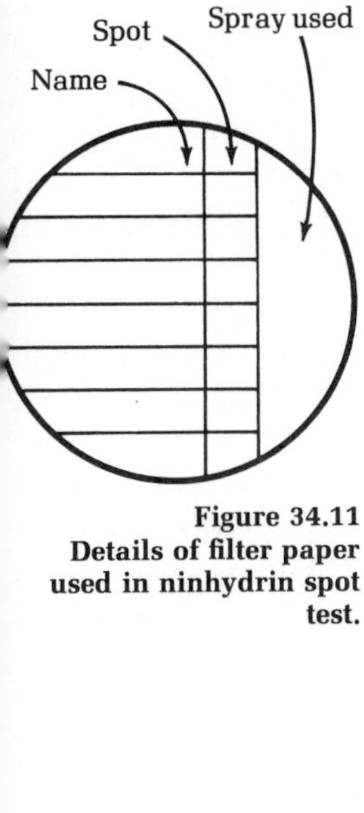

Figure 34.11
Details of filter paper
used in ninhydrin spot
test.

Mark a piece of filter paper as in Fig. 34.11 and enter the names:
alanine, proline (make two applications), methionine, aspartic acid,
glutamic acid, arginine. In the box opposite each name make an
appropriate spot of any size desired, let the spots dry, spray with
ninhydrin-collidine, and let dry. Repeat, using ninhydrin-pyridine
spray. The papers can be entered in the notebook, but the initial
colors should be recorded; application of label lacquer helps to
retard fading.

7. Permanganate Spot Test

Spot a paper as in Section 6 with alanine, methionine, cystine,
tyrosine, and proline (two applications); dry, and spray lightly with
an aqueous solution containing 1% potassium permanganate and
2% sodium carbonate. Put a check mark beside any spots that at
once appear yellow on a pink background, and without delay bleach
the paper by holding it briefly under a watch glass covering a beaker
(hood) in which sulfur dioxide is generated by occasional addition of
sodium bisulfite and hydrochloric acid. Let the excess dioxide
evaporate, spray with ninhydrin-collidine, and let dry.

Permanganate oxidizes the sulfur atom of methionine and cystine
and the phenolic group of tyrosine, but the recovery test shows that
original acid is still present. The nature of the yellow pigment is not
known.[9]

8. Differentiation of Tyrosine from Sulfur Amino Acids

This test is based upon the fact that the phenolic tyrosine couples

---

[9] Proline gives a weak positive response in the permanganate test if the spray is left on for 1/2–1
min but not if it is bleached at once as specified; proline is adequately identified by the yellow
ninhydrin spot test.

with diazotized amines, whereas methionine and cystine do not.

Dip a 10-cm strip of filter paper halfway into a 1% solution of 2,5-dichlorobenzenediazonium chloride[10] and hang it up to dry. Mark one-half of the wetted area T and the other M, and put two spots each of tyrosine and methionine in the appropriate areas. Then dip the strip into 1% aqueous sodium carbonate solution and hang the strip up and again let dry; record the initial result (note that both solutions contain hydrochloric acid) and the final appearance (5–10 min).

9. Protein Hydrolyzate

1-Hr reflux; 20-min workup; overnight chromatogram

In a 25 × 150-mm test tube clamped so that it can be heated over a free flame, place 100 mg of either hair, silk, or gelatin (cooking gelatin serves well), 3 ml of constant boiling (20%) hydrochloric acid,[11] and a boiling stone. Insert a cold finger and let its side-tube rest on the rim of the test tube,[12] and reflux for 1 hr. Remove the condenser, introduce an aspirator tube, and boil the solution down to a volume of about 1 ml. Then rinse down the walls with 3 ml of acetone, add 3 ml of benzene, and heat the tube in an open steam bath while evacuating at the full force of the aspirator to remove the hydrochloric acid and solvents (azeotropic distillation) and leave a film of amino acid hydrochlorides. Add 1 ml of water, warm briefly, and then cool.

Try to identify the major component acids of the mixture (see Table 34.1) by applying such of the tests (Sections 5–8) as seem appropriate and by running an overnight chromatogram (Section 2). In some instances it may be possible to identify a group of components of comparable Rf values but not to differentiate between the members of the group. Treatment of a small sample with phosphotungstic acid and chromatography of the filtrate may be helpful.

10. Unknowns

Test solutions (containing hydrochloric acid) of the following acids (0.1 M) serve as satisfactory unknowns:

Cystine	Methionine	Tyrosine
Glycine	Proline	Aspartic acid
Alanine	Lysine	Glutamic acid
Leucine	Arginine	

---

[10] Du Pont Naphthanil Diazo Scarlet GG.
[11] Mix 2.5 ml of concd hydrochloric acid with 2 ml of water.
[12] The cooling tube close to the boiling liquid breaks up an otherwise troublesome foam.

A 0.1-ml sample (2 drops delivered by a medicine dropper into a specimen vial) should suffice for identification by a combination of spot tests (Sections 5–8) and chromatography, either on the material as supplied or after basification (Section 3). Estimate the weight of sample that you actually used for a successful identification.

35 *Oleic Acid from Olive Oil*

KEYWORDS   Saponification, olive oil   Urea inclusion compounds
               Oleic acid, linoleic acid   Host-guest ratio
               High bp solvent, triethylene glycol   Filter thimble

Saponification is the name given to alkaline hydrolysis of fats and oils to give glycerol and the alkali metal salt of a long chain fatty acid (a soap). In this experiment saponification of olive oil is accomplished in a few minutes by use of a solvent permitting operation at 160°.[1] Of the five acids found in olive oil, listed in Table 35.1, two

Table 35.1   Acids of Split Olive Oil (Typical Composition)

Acid	Formula	%	MW	Mp
Oleic	$C_{18}H_{34}O_2$	64	282.45	13°, 16°
Linoleic	$C_{18}H_{32}O_2$	16	280.44	−5°
Linolenic	$C_{18}H_{30}O_2$	2	278.42	liquid
Stearic	$C_{18}H_{36}O_2$	4	284.07	69.9°
Palmitic	$C_{18}H_{32}O_2$	14	256.42	62.9°

are unsaturated and three are saturated. Oleic acid and linoleic acid are considerably lower melting and more soluble in organic solvents than the saturated components, and when a solution of the hydrolyzate in acetone is cooled to −15°, about half of the material separates as a crystallizate containing the two saturated compounds, stearic acid and palmitic acid.

Urea inclusion complexes

The unsaturated acid fraction is then recovered from the filtrate and treated with urea in methanol to form the urea inclusion complex. Normal alkanes having seven or more carbon atoms form complexes with urea in which hydrogen-bonded urea molecules are oriented in a helical crystal lattice in such a way as to leave a cylindrical channel in which a straight-chain hydrocarbon fits. The guest molecule (hydrocarbon) is not bonded to the host (urea), but merely trapped in the channel. The cylindrical channel is of such a diameter (5.3 Å) as to accommodate a normal alkane, but not a thick, branched-chain hydrocarbon such as 2,2,4-trimethylpentane. In Fig. 35.1 a model of *n*-nonane to the scale 0.2 cm = 1 Å con-

Note for the instructor

[1] The experiment gains in interest if students are given the choice of olive oil from two sources, for example Italy and Israel, and instructed to compare notes.

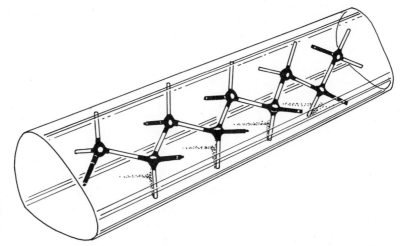

Figure 35.1
Model of *n*-nonane
molecule in a plastic
cylinder, 14.3 cm in
diameter.

structed from plastic-metal atoms[2] is inserted into a tightly fitting
cellulose acetate cylinder, 14.3 cm in diameter, which defines the
space occupied by the resting guest molecule. Table 35.2 shows that
as the hydrocarbon chain is lengthened, more urea molecules are

Table 35.2 Molecules of Urea per Molecule of
            n-Alkane

Alkane	Ratio	Alkane	Ratio
C_6	No complex	C_{11}	8.7
C_7	6.1	C_{12}	9.7
C_8	7.0	C_{16}	12.3
C_9	7.3	C_{24}	18.0
C_{10}	8.3	C_{28}	21.2

required to extend the channel but that the host-guest relationship
is not stoichiometric. The higher saturated fatty acids form urea-
inclusion complexes in which the ratio of host to guest molecules
is about the same as for the corresponding alkanes:

$$\left.\begin{array}{l}\text{Myristic acid } (C_{14})\\ \text{Palmitic acid } (C_{16})\\ \text{Stearic acid } (C_{18})\end{array}\right\} \text{requires} \left\{\begin{array}{l}11.3\\12.8\\14.2\end{array}\right\} \text{moles of urea}$$

With the elimination of saturated acids from the olive oil hydrolyzate by crystallization from acetone, the problem remaining in
obtaining oleic acid is to remove the doubly unsaturated linoleic
acid (see p. 208). Models and cylinders show that the introduction
of just one *cis* double bond is enough to widen the molecule to the
extent that it can no longer be inserted into the 14.3-cm wide channel
which accommodates n-alkanes (Fig. 35.1). However, a model of
3-nonyne likewise fails to fit into the 14.3-cm channel, and the fact

[2] L. F. Fieser, *J. Chem. Ed.*, **40**, 457 (1963); *ibid.*, **42**, 408 (1965).

H   H
⌄   ⌄

18 〜〜〜〜〜〜〜 $\overset{10}{}\overset{9}{}$ 〜〜〜〜〜〜 COOH
                                                        1

Oleic acid

H   H H   H
⌄   ⌄ ⌄   ⌄

18 〜〜〜 $\overset{13}{}\overset{12}{}\overset{10}{}\overset{9}{}$ 〜〜〜〜〜 COOH
                                                        1

Linoleic acid

that this acetylenic hydrocarbon nevertheless forms a urea complex
indicates that the channel is subject to some stretching, namely to
diameter of 16.2 cm, as in Fig. 35.2. From examination of the olei

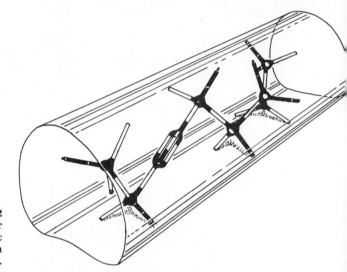

Figure 35.2
3-nonyne molecular
model in a plastic
cylinder, 16.2 cm in
diameter.

acid model together with the 16.2-cm cylinder, it is evident tha
when the carbon atoms are arranged in the particular manner showr
in Fig. 35.3 the oleic acid molecule can be accommodated in this
channel. Careful study of the drawing in Fig. 35.4 will show that ar
attempt to insert the linoleic acid model into the same 16.2-cm
cylinder meets with failure; the carboxy half of the molecule is
accommodated, including the 9,10-double bond, but the 12,13
double bond imposes a stoppage and leaves a five-carbon tail pro
jecting. Thus, when a solution of the two acids in hot methanol is
treated with urea and let cool, crystals of the oleic acid complex
separate and the linoleic acid is retained in the mother liquor.

Predict the host-guest ratio

The experiment has two objectives. One is to isolate pure oleic
acid; the other is to determine the number of molecules of urea in
the inclusion complex per molecule of the fatty acid. Make a prediction in advance.

[3] J. Radell, J. W. Connolly, and L. D. Yuhas, *J. Org. Chem.*, **26**, 2022 (1961).

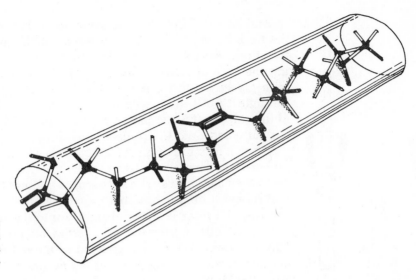

**Figure 35.3
Oleic acid model in a
plastic cylinder, 16.2 cm
in diameter.**

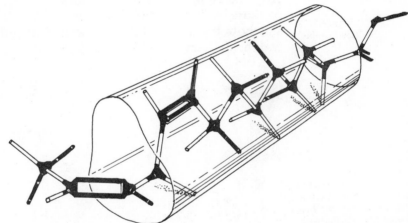

**Figure 35.4
Model of linoleic acid
molecule does not have
adequate space in the
16.2-cm cylinder.**

EXPERIMENT   Pour about 10 g of olive oil into a tared 125-ml Erlenmeyer flask
and adjust the weight to exactly 10.0 g using a capillary dropping
tube. Add 2.3 g of potassium hydroxide pellets and 20 ml of triethylene glycol, insert a thermometer, and bring the temperature to 160°
by heating over a free flame at first and then on a hot plate; the two
Rapid saponification   layers initially observed soon merge. Then, by removing the flask
from the hot plate and replacing it as required, keep the tempera-
ture at 160° for 5 min to insure complete hydrolysis, and cool the
thick yellow sirup to room temperature. Add 50 ml of water, using
it to rinse the thermometer, and acidify the soapy solution with 10
ml of concentrated hydrochloric acid. Cool to room temperature,
extract the oil with ether, wash the ether with saturated sodium
chloride solution, and filter the mixture through anhydrous sodium
sulfate into a tared 125-ml Erlenmeyer filter flask. Evaporate the
solution on the steam bath and evacuate at the aspirator until the
weight is constant (9–10 g of acid mixture).

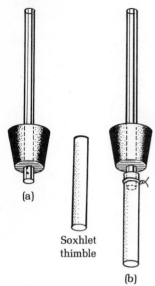

(a)

Soxhlet
thimble

(b)

**Figure 35.5
Filter thimble.**

Add 75 ml of acetone to dissolve the oil in the filter flask and place the flask in an ice bath to cool while making preparations for quick filtration. An internal filter for the filter flask is made by moistening the hole of a No. 4 one-hole fusiform rubber stopper (Fig. 2.6) with glycerol and thrusting a 10-cm section of 9-mm glass tubing through it until the tube projects about 8 mm from the smaller end of the stopper (Fig. 35.5a). A 10×50-mm Soxhlet extraction thimble is then wired onto the projecting end (b).

When a mixture of oleic acid crystals and mother liquor has been prepared as described below, the internal filtering operation is performed as follows: The side arm of a filter flask is connected with $^3/_{16}"$ rubber tubing to a rubber pressure bulb with valve,[4] the stopper carrying the filter is inserted tightly into the flask, and a receiving flask (125 ml) is inverted and rested on the rubber stopper (Fig. 35.6a). The two flasks are grasped in the left hand, with the thumb and forefinger pressing down firmly on the rim of the empty flask to keep the stopper of the other flask in place. While operating the pressure bulb constantly with the right hand, turn the assembly slowly to the left until the side arm of the filter flask is slanting up and the delivery tube is slanting down (Fig. 35.6b). Squeeze the bulb

**Figure 35.6
Apparatus for filtering
crystals from mother
liquor. (a) Flasks with
pressure bulb and filter
thimble assembly;
(b) position of flasks
when filtering.**

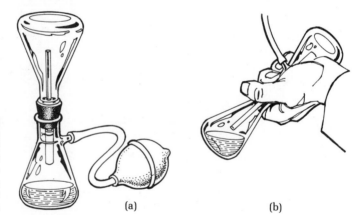

(a)

(b)

constantly until the bulk of the liquid has been filtered and then stop. Remove the filtrate and place the filter flask in the normal position.

Crystallization of the saturated acids requires cooling in a dry ice-acetone bath at −15° for about 15 min and this is done most conveniently in a beaker half-filled with acetone and mounted on a magnetic stirrer and provided with a toluene low temperature thermometer. Add crushed dry ice a little at a time to bring the bath temperature to −15°. The filter flask containing the acetone solution of acids is fitted with the rubber pressure bulb, but the stopper carrying the filter is not put in place until later. Place the filter flask in the cooling bath and swirl the mixture occasionally. Add more

*Quick filtration at
−15°*

[4] Will Scientific, Inc. 23276.

dry ice as required to maintain a bath temperature of $-15°$. After the first crystals appear, as a white powder, swirl and cool the solution an additional 10 min to let the whole mixture acquire the temperature of the bath. Then introduce the filter and inverted receiver and filter quickly by the technique described previously. Stop the process as soon as the bulk of liquid has been collected, for the solid will melt rapidly.[5]

Evaporate the solvent from the filtrate on the steam bath and evacuate the residual fraction of unsaturated acids at the aspirator until the weight is constant. The yield of a mixture of stearic and palmitic acids is 5–7 g. In a second 125-ml Erlenmeyer flask dissolve 11 g. of urea in 50 ml of methanol and pour the solution onto the unsaturated acid fraction. Reheat to dissolve any material that separates and let the solution stand until a large crop of needles has separated. Then cool thoroughly in an ice bath with swirling, collect the product, and rinse the flask with filtrate. Press down the crystals and drain to promote rapid drying and spread out the needles on a large filter paper. After a few minutes, transfer the crystals to a fresh paper. The yield of colorless oleic acid-urea inclusion complex is 10–12 g. The complex does not have a characteristic melting point.

Determine the host/guest ratio

When the complex is fully dry, bottle a small sample, note carefully the weight of the remainder and place it in a separatory funnel. Add 25 ml of water, swirl, and note the result. Then extract with ether for recovery of oleic acid. Wash the extract with saturated sodium chloride solution, filter the solution through sodium sulfate into a tared 125-ml Erlenmeyer flask, evaporate the solution on the steam bath, and pump out to constant weight on the steam bath. The yield of pure oleic acid is 2–3 g. From the weights of the complex and of the acid, calculate the number of moles of urea per mole of acid in the complex.

[5] Alternative technique: Do the crystallization in an Erlenmeyer cooled in a salt-ice bath to $-15°$ and filter by suction on a small Büchner funnel that has been chilled outdoors in winter weather or in a refrigerator freezing compartment.

36 Sulfonation

KEYWORDS   Fuming sulfuric acid (H$_2$SO$_4$ + 7–30% SO$_3$)
p-Toluenesulfonic acid
2-Naphthalenesulfonic acid
p-Toluidine salt

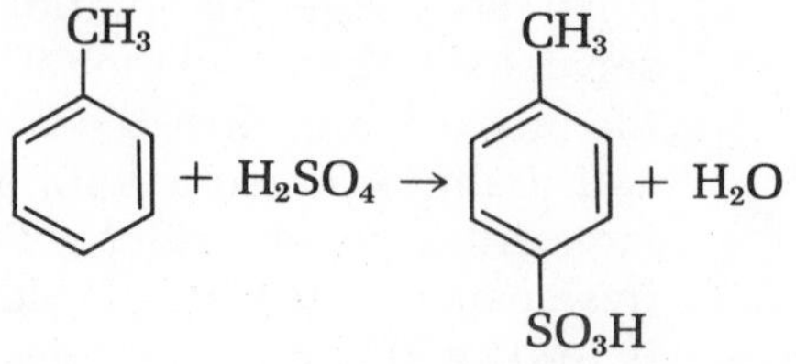

$$\text{CH}_3\text{-benzene} + \text{H}_2\text{SO}_4 \rightarrow \text{CH}_3\text{-benzene-SO}_3\text{H} + \text{H}_2\text{O}$$

p-Toluenesulfonic acid

Phenol is so prone to enter into substitution reactions that it can be sulfonated with dilute sulfuric acid. Naphthalene can be α sulfonated with concd sulfuric acid at 0–60°. Benzene, less reactive than naphthalene, on reaction with fuming sulfuric acid (oleum 7–30% SO$_3$) at 40° affords benzenesulfonic acid, and at 200° the product is benzene-m-disulfonic acid. Toluene, more reactive than benzene and comparable to naphthalene, is sulfonated by concd sulfuric acid even at 0°. The proportion of the three isomeric products formed varies with the temperature as follows:

Reaction Temperature	Toluenesulfonic Acids (% Formed)		
	Ortho	Meta	Para
0°	45.2	2.5	52.3
35°	33.3	5.3	61.4
100°	17.4	10.1	72.5

In the present experiment the reaction is conducted at a still higher temperature (about 175°) to further favor formation of the para isomer and to shorten the reaction time. Intimate mixing of the two layers is accomplished by drawing a stream of dry air through the reaction mixture. The amount of sulfuric acid is kept at a minimum in order to avoid disulfonation; that some toluene remains unreacted is unimportant, since the hydrocarbon is inexpensive. The chief reaction product, p-toluenesulfonic acid, is isolated as the sodium salt, produced by partial neutralization of the acid mixture and addition of sodium chloride:

$$p\text{-}CH_3C_6H_4SO_3H + NaCl \rightleftharpoons p\text{-}CH_3C_6H_4SO_3{}^-Na^+ + HCl$$

Sodium p-toluenesulfonate is infusible, but can be characterized as to purity and identified by conversion to the p-toluidine salt, which has a characteristic melting point:

$$p\text{-}CH_3C_6H_4SO_3{}^-Na^+ + p\text{-}H_2NC_6H_4CH_3 + HCl \rightarrow$$

$$p\text{-}CH_3C_6H_4SO_3{}^-\overset{+}{H_3N}C_6H_4CH_3\text{-}p + NaCl$$

EXPERIMENTS   **1. p-Toluenesulfonic Acid**

Measure 32 ml (0.3 mole) of toluene[1] and 19 ml of concd sulfuric acid into a 125-ml round-bottomed flask mounted over a micro-burner and fitted with a short reflux condenser. Air stirring is accomplished with the arrangement shown in Fig. 36.1. The 6-mm

**Figure 36.1
Stirring with air drawn
in by suction applied
through the side tube.**

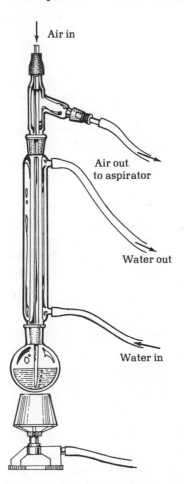

Air in

Air out
to aspirator

Water out

Water in

Note for the instructor   [1] Merck reagent grade toluene is satisfactory; thiophene derivatives in inadequately purified toluene cause blackening of the reaction mixture.

glass tube secured at the top to the stillhead extends to the bottom of the reaction flask. Draw a very slow stream of air through the two phase reaction mixture and apply heat gently until the toluene begins to boil and to drip back from the condenser. Note the time, adjust the flame to provide for very gentle boiling, and continue heating for 15–20 min, when only a thin layer of toluene should persist (a little toluene may condense in the side tube).

Reaction time:
15–20 min

The reaction mixture while still warm is poured into 100 ml of distilled water in a 400-ml beaker and the flask is rinsed with a little water from a wash bottle. (NOTE: If the mixture is allowed to cool before pouring into water the mixture sets to a stiff paste of crystals of p-toluenesulfonic acid in the form of a monohydrate. This can be dissolved by warming and treated as above.) The acid solution is partly neutralized by adding carefully and in small portions with vigorous stirring 15 g of sodium bicarbonate; 40 g sodium chloride is then added and the mixture is stirred until the granular salt has all given way to pearly plates of sodium p-toluenesulfonate (probe for granular solid at the bottom with the stirring rod, or hold up the beaker and inspect the bottom). Then cool thoroughly in an ice water bath and collect the product on a 75 mm (ID) Büchner funnel.

Crude sulfonate

Use only gentle suction at first, since the liquor is strongly enough acidic to weaken the paper. Use filtrate to rinse the beaker, since the sulfonate is extremely soluble in water in the absence of sodium chloride. Press down the solid with a spatula, and then release the suction and cover the filter cake with 20 ml of saturated sodium chloride solution, using it to rinse spatula and funnel. Apply full suction and let the cake drain thoroughly.

Transfer the crude sulfonate to a paper and thence to a 250-ml Erlenmeyer and dissolve it in 100 ml of distilled water at the bp. Weigh out 15 g of sodium chloride and add this in small portions with swirling to bring it into solution (a thin layer of toluene, if present, will be removed later). When the addition is complete, rinse down any salt adhering to the walls, make sure that all solid is dissolved, and then swirl the flask vigorously in ice water until crystallization begins. After standing for thorough cooling, collect the product and rinse the flask with filtrate. Dry the cake with suction and then release the suction and pour in 20 ml of methanol to remove a trace of toluene and speed up drying. After applying full suction to drain off the wash liquor, transfer the crystals to a tared 400-ml beaker and dry on the steam bath. The yield of pure, colorless sodium p-toluenesulfonate is 16–20 g.

2. Characterization

Dissolve 1 g of the sodium sulfonate in the minimum quantity of boiling water, add 0.5 g of p-toluidine and 1 ml of concentrated hydrochloric acid, and bring the material into solution by heating and by adding more water if required. (If there are any oily drops of the amine, more HCl should be added.) The solution is cooled and

CH$_3$

NH$_2$

p-Toluidine

once in an ice bath and the walls of the flask are scratched until the
p-toluidine salt crystallizes. The product is collected, washed with
a very small quantity of water, and a few crystals are saved for seed.
The rest is recrystallized from water, with clarification if necessary.
If no crystals separate, the solution is inoculated with a seed crystal.
The pure salt melts at 197°. A product melting within two or three
degrees of this temperature is satisfactory.

3. 2-Naphthalenesulfonic acid[2] (Special Experiment)

$$+ H_2SO_4 \rightarrow \qquad SO_3H \qquad + H_2O$$

**2-Naphthalene-
sulfonic acid**

A 200-ml three-necked round-bottomed flask is clamped in a
position to be heated over a microburner and a mechanically driven
stirrer is mounted in the middle opening. A suitable stirrer is a
glass rod bent through an angle of 45° about 2 cm from the end, with
the shaft turning in a bearing made by inserting a short glass tube
in a cork stopper of suitable size. The bearing may require lubri-
cation with glycerol. A thermometer is inserted with a cork in one of
the side tubulatures in such a way that the bulb will be immersed
when the flask is one-third filled. Sulfuric acid is to be run in from a
dropping funnel, which should be clamped in such a position that
it will deliver into the third tubulature of the flask.

In the flask melt 50 g of naphthalene, start the stirrer, and adjust
the flame until a steady temperature of 160° is maintained. In the
course of 3 min run in from the dropping funnel 45 ml of concentrated sulfuric acid, keeping the temperature at 160° (the flame may
be removed). After stirring for 3 min longer pour the solution into
400 ml of water. In a well conducted operation there will be no
precipitate of naphthalene, but there may be 1–2 g of the water-
insoluble di-β-naphthyl sulfone, formed thus:

$$C_{10}H_7SO_3H + C_{10}H_8 \rightarrow C_{10}H_7SO_2C_{10}H_7 + H_2O$$

To remove it boil the solution with decolorizing charcoal and filter
under gentle suction.

Partially neutralize the clarified solution by adding cautiously
20 g of sodium bicarbonate in small portions. Saturate the solution
at the boiling point by adding sodium chloride (30–35 g) until
crystals persists in the hot solution, and then allow crystallization to
take place. Recrystallize the material, using this time only a small

---

[2] A stirring motor and a three-necked flask are required. Two students can work together on the
sulfonation and divide the aqueous solution of the sulfonic acid.

amount of sodium chloride. Test the purity of the product by exam
ining a sample of the p-toluidine salt which, when pure, melts a
217–218°. Heat the salt in a beaker on the steam bath until the weigh
is constant. The yield is 70–75 g.

37 Sulfanilamide from Benzene

KEYWORDS

Sulfa drug, sulfanilamide
Anti-bacterial agent
Nitronium ion, nitrobenzene (toxic)
Reduction, tin and hydrochloric acid
Aniline double salt
Steam distillation

Acetylation in aqueous solution, acetic anhydride
Acetanilide, p-acetaminobenzene-sulfonyl chloride
Chlorosulfonic acid
Ammonolysis
Speed, high yield, pure product

$$HNO_3 + 2\ H_2SO_4 \rightleftharpoons NO_2^+ + H_3O^+ + 2\ HSO_4^-$$

Nitronium ion

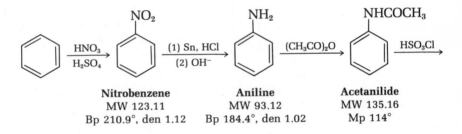

Nitrobenzene	Aniline	Acetanilide
MW 123.11	MW 93.12	MW 135.16
Bp 210.9°, den 1.12	Bp 184.4°, den 1.02	Mp 114°

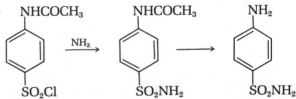

p-Acetaminobenzene-sulfonyl chloride
MW 233.68

p-Acetaminobenzene-sulfonamide
MW 214.25

Sulfanilamide
(*p*-Aminobenzene-sulfonamide)
MW 172.20, mp 163–164°

This experiment is a six-step synthesis of the first known sulfa drug, sulfanilamide. A number of N-substituted sulfanilamides are still important anti-bacterial agents. These drugs compete with *p*-aminobenzoic acid in bacterial enzyme systems and thus inhibit bacterial growth. Higher organisms, such as man, do not require p-aminobenzoic acid in their enzyme systems and are therefore not affected by the sulfanilamides.

Benzene is nitrated with a mixture of nitric and sulfuric acid, the latter acid promoting formation of the nitronium ion. The nitrobenzene formed is reduced to aniline by tin and hydrochloric acid. A double salt of tin having the formula $(C_6H_5NH_3)_2SnCl_4$ separates partially during the reaction, and at the end it is decomposed by addition of excess alkali, which converts the tin into water-soluble stannite or stannate (Na_2SnO_2 or Na_2SnO_3). The aniline liberated is separated from inorganic salts and the insoluble impurities derived from the tin by steam distillation and is then dried, distilled, and acetylated, either with acetic anhydride in aqueous solution or by refluxing with acetic acid. Treatment of the resulting acetanilide with excess chlorosulfonic acid effects substitution of the chlorosulfonyl group and affords p-acetaminobenzensulfonyl chloride. The alternative route to this intermediate via sulfanilic acid is unsatisfactory, because sulfanilic acid being dipolar is difficult to acetylate. In both processes the amino group must be protected by acetylation to permit formation of the acid chloride group. The next step in the synthesis is ammonolysis of the sulfonyl chloride and the terminal step is removal of the protective acetyl group.

Use the total product obtained at each step as starting material for the next step and adjust the amounts of reagents accordingly. You are to choose between alternative procedures for the acetylation of aniline and decide whether to purify an intermediate or use it as such to avoid purification losses. Aim both for a high overall yield of pure final product and also for its speedy production. Keep a record of your actual working time. Study the procedures carefully beforehand so that your work will be efficient. A combination of consecutive steps that avoids a needless isolation saves time and increases the yield.

EXPERIMENTS   1. Nitrobenzene

Explanation of procedure

The nitration of benzene is conducted with a mixture of nitric and sulfuric acids diluted with a small amount of water to inhibit dinitration. The reaction is exothermic at the start, when the concentrations of benzene and nitric acid are maximal, but becomes sluggish toward the end as the concentrations of reactants decrease. Consequently, the mixture is cooled during the early stages to prevent the reaction from getting out of hand and, later, is heated to force the diluted components to react. Since both benzene and nitrobenzene are insoluble in the acid mixture and form an upper layer, thorough agitation must be maintained by swirling throughout the entire reaction period, in order to obtain a successful result in the time indicated.

If this reaction were carried out on a large scale, the mixed acid would be added from a pressure-compensated dropping funnel

(Fig. 37.1), and the reaction mixture would be stirred with a motor-driven stirrer (Fig. 37.2).

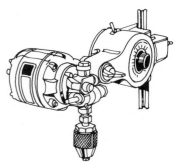

Figure 37.2
Stirring motor.

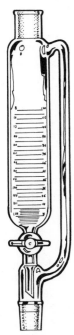

Figure 37.1
Pressure compensated
dropping funnel.

20 min of continuous swirling and careful control of temperature

Before mixing the reagents, prepare an ice-water bath and plug in a hot plate. Measure 5 ml of water into a 125-ml Erlenmeyer, add 25 ml of concentrated sulfuric acid and cool. Add 15 ml of concd (71%) nitric acid and cool again. Then add 17 ml (15 g) of benzene, insert a thermometer, note the time, and swirl vigorously to promote interaction of the immiscible layers. Watch the temperature closely and when it approaches 60° plunge the flask briefly into the ice bath to check a further rise. Swirl constantly and, by very brief cooling in the ice bath as required, try to maintain the temperature close to 60°, but not above, and not below 55°. Within 7–8 min the temperature should begin to drop, and after 10 min, if the initial strongly exothermic reaction seems to be over, heat the flask occasionally on the hot plate and maintain a temperature close to 60° for another 10 min. Then cool in ice to 25°, add 75 ml of water, cool again, and pour the solution (or a part of it) into a small separatory funnel containing 30 ml of ether. Shake, draw off, and discard the lower aqueous layer. Pour in and extract any further reaction mixture (rinse the flask with a little ether). Wash the ethereal extract once with water and then shake it with 25 ml of 10% sodium hydroxide solution. In case the ethereal extract is yellow shake it with two or three further portions of alkali until the pigmented by-product is all removed. Then shake the ethereal solution with saturated salt solution, drain off the salt solution, filter the ethereal solution through anhydrous sodium sulfate, and evaporate the ether on the steam bath under an aspirator tube. Pour the residual oil into a 50-ml round-bottomed flask, attach a stillhead fitted with an air condenser (an empty fractionating column), and distil. Remove a small fore-run and collect nitrobenzene boiling at 205–207°. *Do not distil to dryness.* If m-dinitrobenzene is present in the residue it may decompose explosively should the temperature rise much above 207°. Yield 18.0 g. Nitrobenzene is toxic in contact with the skin or if the vapor is inhaled.

Can you identify, by inference, the yellow pigment extracted from water into ether and extracted from ether by alkali? It is formed from thiophene-free as well as from thiophene-containing benzene and is produced as a minor reaction product whether the nitric acid contains oxides of nitrogen (yellow) or is colorless and kept free of oxides of nitrogen by addition of urea. Suggestion: acidify the yellow alkaline extract, introduce a piece of silk, and heat.

2. Aniline

The reduction of the nitrobenzene is carried out in a 500-ml round-bottomed flask suitable for steam distillation of the reaction product. Put 25 g of granulated tin and 12.0 g of nitrobenzene in the flask, make an ice-water bath ready, add 55 ml of concd hydrochloric acid, insert a thermometer, and swirl well to promote reaction in the three-phase system. Let the mixture react until the temperature reaches 60° and then cool briefly in ice just enough to prevent a rise above 60°, so the reaction will not get out of hand. Continue to swirl, cool as required, and maintain the temperature in the range 55–60° for 15 min. Remove the thermometer and rinse it with water, fit the flask with a reflux condenser (to catch any nitrobenzene that may steam distil), and heat on the steam bath with frequent swirling until droplets of nitrobenzene are absent and the color due to an intermediate reduction product is gone (about 15 min). During this period dissolve 40 g of sodium hydroxide in 100 ml of water and cool to room temperature.

Reaction time: 0.5 hr

At the end of the reduction reaction, cool the acid solution in ice (to prevent volatilization of aniline) during gradual addition of the solution of alkali. Then attach a stillhead with steam-inlet tube, condenser, adapter, and receiving Erlenmeyer (Fig. 9.2); heat the flask with a microburner to prevent flask from filling with water from condensed steam and proceed to steam distil. Since aniline is fairly soluble in water (3.6 g/100 g$^{18°}$) distillation should be continued somewhat beyond the point where the distillate has lost its original turbidity (50–60 ml more). Make an accurate estimate of the volume of distillate by filling a second flask with water to the level of liquid in the receiver and measuring the volume of water.

Suitable point of interruption

At this point consider the conversion of aniline into acetanilide and make a choice between alternative procedures. You could reduce the solubility of aniline by dissolving in the steam distillate 0.2 g of sodium chloride per ml, extract the aniline with 2–3 portions of methylene chloride, dry the extract, distil off the methylene chloride (bp 41°), and then distil the aniline (bp 184°). You can then select either of the two following procedures for the acetylation of aniline. The first (a) requires pure, dry aniline, as you would obtain it by extraction and distillation, but both processes might be attended with losses; furthermore a 4-hr reflux period is required. Procedure (b) also calls for pure, dry aniline, but note that the first step is to dissolve the aniline in water and hydrochloric acid. Your steam distillate is a mixture of aniline and water, both of which have been distilled. Are they not both water-white and presumably pure? Hence, an attractive procedure would be to assume that the steam

Alternative choices

distillate contains the theoretical amount of aniline and to add to it, in turn, appropriate amounts of hydrochloric acid, acetic anhydride, and sodium acetate, calculated from the quantities given in (b).

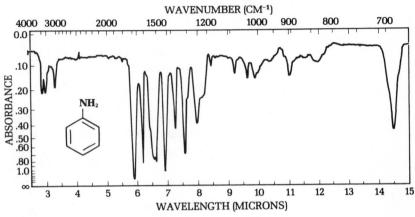

Figure 37.3
Infrared spectrum of aniline.

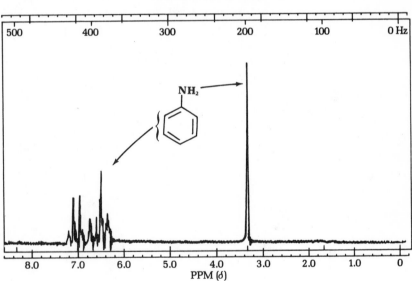

Figure 37.4
Nmr spectrum of aniline.

3. Acetanilide

(a) Acetylation in Aqueous Solution

Choice of procedures
(a) and (b)

Dissolve 5.0 g (0.054 mole) of aniline in 135 ml of water and 4.5 ml (0.054 mole) of concd hydrochloric acid, and, if the solution is colored, filter it by suction through a pad of decolorizing charcoal. Measure out 6.2 ml (0.065 mole) of acetic anhydride, and also prepare a solution of 5.3 g (0.065 mole) of anhydrous sodium acetate in 30 ml of water. Add the acetic anhydride to the solution of aniline hydrochloride with stirring and at once add the sodium acetate solution. Stir, cool in ice, and collect the product. It should be colorless and the mp close to 114°. Since the acetanilide *must be*

completely dry for use in the next step, it is advisable to put th
material in a tared 125-ml Erlenmeyer flask and to heat this on th
steam bath under evacuation until the weight is constant.

(b) Refluxing with Acetic Acid

In a 125-ml flask equipped with an air condenser set for reflux
place 5.0 g of aniline and 20 ml of acetic acid; in case the aniline i
discolored add a small pinch of zinc dust. Adjust the flame so tha
the ring of condensate rises to about 15 cm from the top of the con
denser. Reflux for at least 4 hr and pour the hot reaction mixture i
a thin stream into 200 ml of cold water. Collect the product and
wash it with water. Pure acetanilide is colorless and melts at 114°

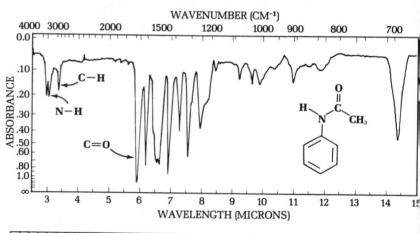

Figure 37.5
Infrared spectrum of
acetanilide in CHCl₃.

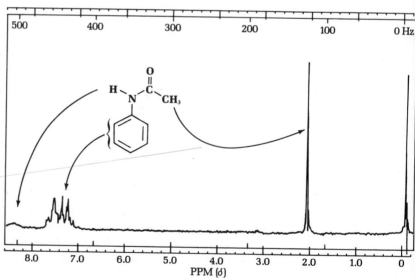

Figure 37.6
Nmr spectrum of acet-
anilide. The amide
proton shows a charac-
teristically broad peak.

4. Sulfanilamide

The chlorosulfonation of acetanilide in the preparation of sulfanilamide is conducted without solvent in the 125-ml Erlenmeyer

used for drying the precipitated acetanilide from procedure 3(a). Since the reaction is most easily controlled when the acetanilide is in the form of a hard cake, the dried solid is melted by heating the flask over a free flame; as the melt cools, the flask is swirled to distribute the material as it solidifies over the lower walls of the flask. Let the flask cool while making provision for entraining the hydrogen chloride evolved in the chlorosulfonation. Fit the Erlenmeyer with a stopper connected by a section of rubber tubing to a glass tube fitted with a cork into the neck of a 250-ml filtering flask half-filled with water. The tube should be about 1 cm above the surface of the water and *must not dip into the water.* Cool the flask containing the acetanilide thoroughly in an ice-water bath, and for 5.0 g of acetanilide measure 12.5 ml of chlorosulfonic acid in a graduate supplied with the reagent and kept away from water. Add the reagent in 1–2 ml portions with a capillary dropping tube, and connect the flask to the gas trap. The flask is now removed from the ice bath and swirled until a part of the solid has dissolved and the evolution of hydrogen chloride is proceeding at a rapid rate. Occasional cooling in ice may be required to prevent too brisk a reaction. In 5–10 min the reaction subsides and only a few lumps of acetanilide remain undissolved. When this point has been reached, heat the mixture on the steam bath for 10 min to complete the reaction, cool the flask under the tap, and deliver the oil by drops with a capillary dropper and stirring into 75 ml of ice water contained in a beaker cooled in an ice bath (hood). Rinse the flask with cold water and stir the precipitated p-acetaminobenzenesulfonyl chloride for a few minutes until an even suspension of granular white solid is obtained. Collect and wash the solid on a Büchner funnel. After pressing and draining the filter cake, transfer the solid to the rinsed reaction flask, add (for 5 g of aniline) 15 ml of concd aqueous ammonia solution and 15 ml of water, and heat the mixture over a flame with occasional swirling (hood). Maintain the temperature of the mixture just below the boiling point for 5 min. During this treatment a change can be noted as the sulfonyl chloride undergoes transformation to a more pasty suspension of the amide. Cool the suspension well in an ice bath, collect the p-acetaminobenzenesulfonamide by suction filtration, press the cake on the funnel, and allow it to drain thoroughly. Any excess water will unduly dilute the acid used in the next step.

Transfer the still moist amide to the well-drained reaction flask, add 5 ml of concd hydrochloric acid and 10 ml of water (for 5 g of aniline), boil the mixture gently until the solid has all dissolved (5–10 min), and then continue the heating at the boiling point for 10 min longer (do not evaporate to dryness). The solution when cooled to room temperature should deposit no solid amide, but, if it is deposited, heating should be continued for a further period. The cooled solution of sulfanilamide hydrochloride is shaken with decolorizing charcoal and filtered by suction. Place the solution in a beaker and cautiously add an aqueous solution of 5 g of sodium bicarbonate with stirring. After the foam has subsided, test the

Caution! Corrosive chemical, reacts violently with water. Withdraw with pipette and pipetter (Fig. 16.1)

Do not let the mixture stand before addition of ammonia

suspension with litmus, and, if it is still acidic, add more bicarbonate until the neutral point is reached. Cool thoroughly in ice and collect the granular, white precipitate of sulfanilamide. The crude product (mp 161–163°) on crystallization from alcohol or water affords pure sulfanilamide, mp 163–164°, with about 90% recovery.

38 p-*Di-t-butylbenzene* and 1,4-Di-t-butyl-2,5-dimethoxybenzene

---

KEYWORDS
Friedel-Crafts alkylation
Lewis acid catalyst, AlCl$_3$
t-Butyl chloride, benzene
Thiourea inclusion complex
Lewis acid catalyst, fuming H$_2$SO$_4$

t-Butyl alcohol,
1,4-dimethoxybenzene
Trimethylcarbonium ion
Strange crystallization
behavior

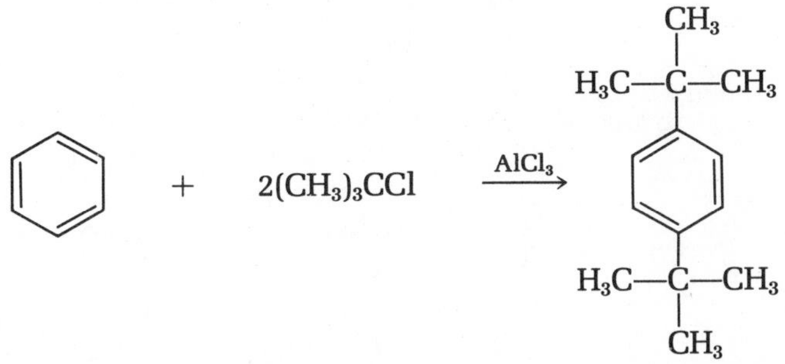

Benzene	*t*-Butyl chloride	*p*-Di-*t*-Butylbenzene
MW 78.11, den 0.88, bp 80°	MW 92.57, den 0.85, bp 51°	MW 190.32, mp 77–79°, bp 167°

The classical illustration of Friedel-Crafts alkylation, the reaction of benzyl chloride with benzene to form diphenylmethane, has the disadvantage that separation of the liquid reaction product from a complex mixture cannot be done satisfactorily with equipment ordinarily available. In this experiment *t*-butyl chloride is used instead of benzyl chloride. *t*-Butyl chloride reacts rapidly with benzene at 0° under the catalytic influence of aluminum chloride to give first *t*-butylbenzene, a liquid, and then *p*-di-*t*-butylbenzene, a beautifully crystalline solid (symmetrical structure). The crystalline reaction product is then isolable with ease in reasonable yield.

The chief factor limiting the yield appears to be the lability of the *t*-butyl group of the product. P. D. Bartlett (1954) reports the reaction of *p*-di-*t*-butylbenzene with *t*-butyl chloride and aluminum chloride (1.3 moles) at 0–5° to give *m*-di-*t*-butylbenzene, 1,3,5-tri-*t*-butylbenzene, and unchanged starting material. Thus, the mother liquor

By-products

from crystallization of p-di-t-butylbenzene probably contains t-butylbenzene, the desired p-di product, the m-di isomer, and 1,3,5-tri-t-butylbenzene.

Although the mother liquor probably contains a mixture of several components, the p-di-t-butylbenzene present can be isolated easily as an inclusion complex (compare Chapter 35, Oleic Acid). If you construct a molecular model of p-di-t-butylbenzene (Chapter 35) and try fitting it into cellulose acetate cylinders representing urea channels (dia 14.3 cm and 16.2 cm) and a thiourea channel (dia 26.4 cm), you will find that it fits as nicely in the thiourea channel as does adamantane. The adamantane complex is made up of 3.4 molecules of thiourea per molecule of hydrocarbon. Compare the length of the p-di-t-butylbenzene molecule with the length of n-alkanes (consult Table 2 of Chapter 35) and predict the host/guest ratio. You can then check your prediction experimentally.

Thiourea inclusion complex

Adamantane

EXPERIMENTS

1. *p*-Di-*t*-butylbenzene

Measure 20 ml of t-butyl chloride and 10 ml of benzene in a 125-ml filter flask (Erlenmeyer flask with side arm) and place the flask in an ice-water bath to cool. Rest a 10 × 75-cm test tube in an inverted No. 2 filter adapter, weigh 1 g of fresh aluminum chloride onto a creased paper, and scrape it with a small spatula into the tube; close the tube at once with a cork.[1] Connect the side arm of the flask to the aspirator (preferably one made of plastic) and operate it at a rate sufficient to carry away hydrogen chloride formed in the reaction. Cool the liquid to 0–3°, remove the thermometer, add about one quarter of the aluminum chloride and swirl vigorously in the ice bath. After an induction period of about 2 min a vigorous reaction sets in, with bubbling and liberation of hydrogen chloride. Add the remainder of the catalyst in three portions at intervals of about 2 min. Toward the end, the reaction product begins to separate as a white solid. When this occurs, remove the flask from the bath and let stand at room temperature for 5 min. Add ice and water to the reaction mixture and then ether for extraction of the product, stirring with a rod or spatula to help bring the solid into solution. Transfer the solution to a separatory funnel and shake; draw off the lower layer and wash the remaining layer with water and then with a saturated sodium chloride solution. Filter the solution through sodium sulfate into a tared Erlenmeyer flask, remove the ether by evaporation on the steam bath, and evacuate using the aspirator to remove traces of solvent until the weight is constant; yield of crude product, 15 g.

The oily product should solidify on cooling. For crystallization,

Convenient test tube holder

Reaction time about 15 min

Aspirator tube to remove ether vapor

---

[1] Alternative scheme: Put a wax pencil mark on the test tube 37 mm from the bottom and fill the tube with aluminum chloride to this mark.

dissolve the product in 20 ml of methanol and let the solution come to room temperature without disturbance. If you are in a hurry, lift the flask without swirling; place it in an ice-water bath and observe the result. After thorough cooling at 0°, collect the product and rinse the flask with a little ice-cold methanol. The yield of *p*-di-*t*-butylbenzene from the first crop is 8.2–8.6 g of satisfactory product. Save the product for the next step as well as the mother liquor, in case you later wish to work it up for a second crop.

Spontaneous crystallization gives beautiful needles or plates

In a 125-ml Erlenmeyer flask dissolve 5 g of thiourea and 3 g of *p*-di-*t*-butylbenzene in 50 ml of methanol (break up lumps with a flattened stirring rod) and let the solution stand for crystallization of the complex, which occurs with ice cooling. Collect the crystals and rinse with a little methanol and dry to constant weight; yield 5.8 g. Bottle a small sample, determine carefully the weight of the remaining complex, and place the material in a separatory funnel along with about 25 ml each of water and ether. Shake until the crystals disappear, draw off the aqueous layer containing thiourea, wash the ether layer with saturated sodium chloride, and filter through sodium sulfate into a tared 125-ml Erlenmeyer. Evaporate and evacuate as before, making sure the weight of hydrocarbon is constant before you record it. Calculate the number of molecules of thiourea per molecule of hydrocarbon (probably *not* an integral number).

Inclusion complex starts to crystallize in 10 min

$$\begin{matrix} & S \\ & \parallel \\ H_2NCNH_2 \end{matrix}$$
Thiourea
MW 76.12

To work up the mother liquor from the crystallization from methanol, first evaporate the solvent. Note that the residual oil does not solidify on ice cooling. Next, dissolve the oil, together with 5 g of thiourea, in 50 ml of methanol, collect the inclusion complex that crystallizes (3.2 g), and recover *p*-di-*t*-butylbenzene from the complex as before (0.8 g before crystallization).

Work-up of mother liquor

The infrared spectrum of *p*-di-*t*-butylbenzene is presented in Fig. 38.1 and the nmr spectrum in Fig. 38.2. Note the extreme simplicity of the nmr spectrum, a result of the high degree of symmetry of the molecule.

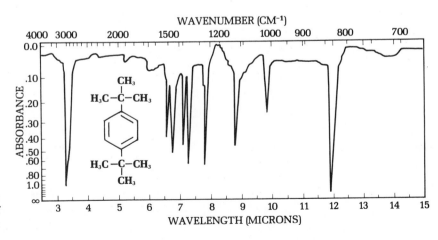

Figure 38.1
Infrared spectrum of
***p*-di-*t*-butylbenzene.**

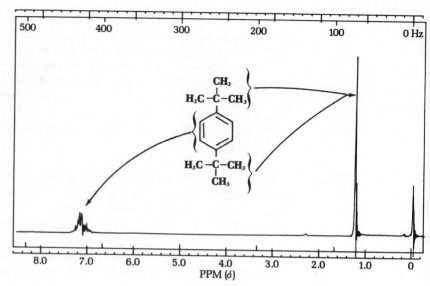

Figure 38.2
Nmr spectrum of
***p*-di-*t*-butylbenzene.**

2. 1,4-Di-*t*-butyl-2,5-dimethoxybenzene

Hydroquinone
dimethyl ether
MW 138.16, mp 57°

+ 2(CH₃)₃COH →[H₂SO₄]

***t*-Butyl alcohol**
MW 74.12, den 0.79,
mp 25.5°, bp 82.8°

1,4-Di-*t*-butyl-2,5-dimethoxybenzene
MW 250.37, mp 104–105°

Trimethylcarbonium ion

This experiment illustrates the Friedel-Crafts alkylation of an activated benzene molecule with a tertiary alcohol in the presence of sulfuric acid as the Lewis acid catalyst. As in the reaction of benzene and *t*-butyl chloride the substitution involves attack by the electrophilic trimethylcarbonium ion.

Place 6 g of hydroquinone dimethyl ether in a 125-ml Erlenmeyer flask, add 10 ml of *t*-butyl alcohol and 20 ml of acetic acid, and put the flask in an ice-water bath to cool. Measure 10 ml of concd sulfuric acid into a 50-ml Erlenmeyer, add 10 ml of 30% fuming sulfuric acid, and put the flask in the ice bath to cool. Put a thermometer in the large flask and swirl in the ice bath until the temperature is in the range 0–3°, and remove the thermometer (solid, if present, will dissolve later). Clamp a small separatory funnel in a position to deliver into the 125-ml Erlenmeyer so that the flask can remain in

Reaction time about 12 min

the ice-water bath, wipe the smaller flask dry, and pour the chilled sulfuric acid solution into the funnel. While constantly swirling the 125-ml flask in the ice bath, run in the chilled sulfuric acid by rapid drops during the course of 4–7 min. By this time considerable solid reaction product should have separated, and insertion of a thermometer should show that the temperature is in the range 20–25°. Swirl the mixture at about 25° for 5 minutes more and then cool in ice. Add enough ice to the mixture to decompose the sulfuric acid, then add water to nearly fill the flask, cool, and collect the product on a Büchner funnel with suction. Apply only very gentle suction at first to avoid breaking the filter paper, which is weakened by the strong sulfuric acid solution. Wash liberally with water and then turn on the suction to full force. Press down the filter cake with a spatula and let drain well. Meanwhile, cool a 30-ml portion of methanol for washing to remove a little oil and a yellow impurity. Release the suction, cover the filter cake with a third of the chilled methanol, and then apply suction. Repeat the washing a second and a third time.

Since air-drying of the crude reaction product takes time, the following short procedure is suggested: Place the moist material in a 125-ml Erlenmeyer, add a little methylene chloride to dissolve the organic material, and note the appearance of nonlipid droplets. Add enough calcium chloride to bind the nonlipid material and filter the supernatant into another 125-ml Erlenmeyer; notice that this process eliminates considerable extraneous purple pigment. As the last of the solution is passing through the paper, add 30 ml of methanol (bp 65°) to the filtrate and start evaporation (on the steam bath) to eliminate the methylene chloride (bp 41°). When the volume is estimated to be about 30 ml, let the solution stand for crystallization. When crystallization is complete cool in ice and collect. The yield of large plates of pure 1,4-di-*t*-butyl-2,5-dimethoxybenzene is 6–7 g.

Antics of growing crystals

R. D. Stolow of Tufts University reported[2] that growing crystals of the di-*t*-butyldimethoxy compound change shape in a dramatic manner: thin plates curl and roll up and then uncurl so suddenly that they propel themselves for a distance of several centimeters. If you do not observe this phenomenon during crystallization of a small sample, you may be interested in consulting the papers cited and pooling your sample with others for trial on a large scale. The solvent mixture recommended by the Tufts workers for observation of the phenomenon is 9.7 ml of acetic acid and 1.4 ml of water per gram of product.

Figure 38.3 and Fig. 38.4 present the infrared and nmr spectra of the starting hydroquinone dimethyl ether. Can you predict the appearance of the nmr spectrum of the product?

[2] R. D. Stolow and J. W. Larsen, *Chemistry and Industry*, 449 (1963). See also J. M. Blatchly and N. H. Hartshorne, *Transactions of the Faraday Society*, **62**, 512 (1966).

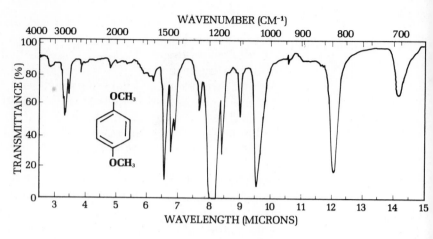

**Figure 38.3
Infrared spectrum of
p-dimethoxybenzene.**

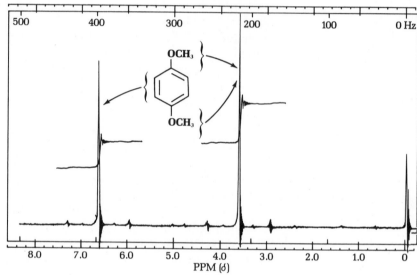

**Figure 38.4
Nmr spectrum of
p-dimethoxybenzene.**

39 Azoxybenzene, Azobenzene, Hydrazobenzene, and Benzidine

KEYWORDS   Reducing agents

Zinc dust, triethylene glycol, diimide, stannous chloride

Hydrogen peroxide + hydrazine = diimide

Benzidine rearrangement

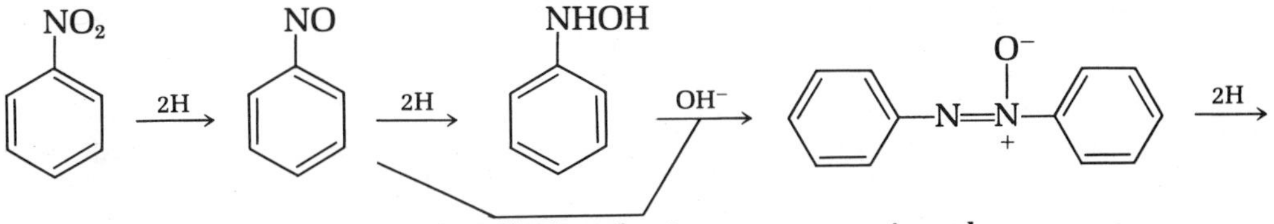

Nitrobenzene
MW 123.11,
den 1.20, bp 210.9°

Nitrosobenzene
MW 107.11,
mp 68°

Phenylhydroxylamine
MW 109.13

Azoxybenzene
MW 198.22,
mp 36°

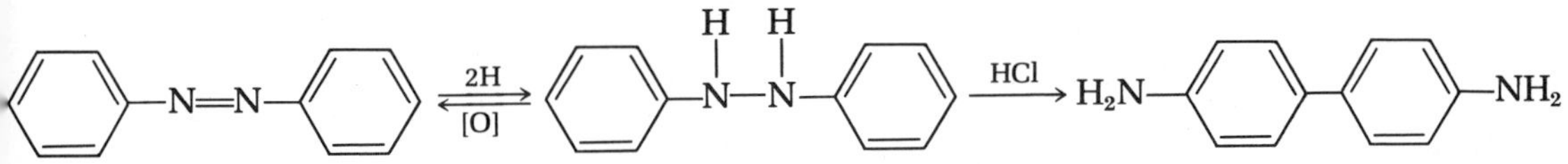

Azobenzene
MW 188.22,
mp 67°

Hydrazobenzene
MW 184.24,
mp 126°

Benzidine
(4,4'-Diaminobiphenyl)
MW 184.23, mp 126°

The formation of *trans*-azobenzene by reduction of nitrobenzene in an alkaline medium involves reduction to nitrosobenzene and phenylhydroxylamine, aldol-like condensation of these reactants to azoxybenzene, and reductive removal of the oxygen function.

The classical procedure calls for heating a mixture of nitrobenzene, zinc dust, and aqueous-methanolic sodium hydroxide under reflux with mechanical stirring for 10 hr. However, the reaction period can be reduced to 35 min by use of a higher boiling alcohol with greater solvent power and a more effective alkali.

Gentle heating of a mixture of nitrobenzene, zinc dust (2 equivalents), and potassium hydroxide in triethylene glycol initiates a strongly exothermic reaction which, if checked first at 80–85°, can be controlled and completed in 15 min at 135–140°. Surprisingly, very little zinc dust is consumed. The reducing agent is triethylene glycol, which is transformed into the corresponding aldehyde, isolable as the 2,4-dinitrophenylhydrazone.

Reduction by
$RCH_2OH \rightarrow RCHO$

$$2\ C_6H_5NO_2 + 4\ HOCH_2CH_2OCH_2CH_2OCH_2CH_2OH \rightarrow$$
$$C_6H_5N{=}NC_6H_5 + 4\ HOCH_2CH_2OCH_2CH_2OCH_2CHO$$

Azobenzene can be obtained in high yield from nitrobenzene, potassium hydroxide, and triethylene glycol without any zinc, but the period of heating at 135–140° must be extended from 15 min to 1 hr. Zinc is beneficial in accelerating this reaction if used in adequate quantity, but it is not effective in truly catalytic amount.

The intermediate, azoxybenzene, can be prepared by refluxing nitrobenzene with methanolic sodium hydroxide for three hr.

$$CH_3OH + NaOH \rightleftharpoons CH_3O^-Na^+ + H_2O$$

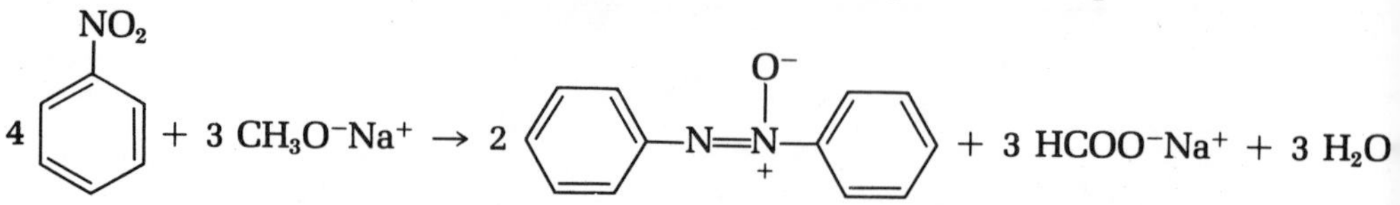

The reduction of orange colored azobenzene to colorless hydrazobenzene without reductive cleavage to aniline is accomplished efficiently by *in situ* generation of the highly reactive diimide by cupric ion-catalyzed oxidation of hydrazine with hydrogen peroxide. Diimide combines with azobenzene (or with an olefin) to form a cyclic transition state which collapses to nitrogen and hydrazobenzene.

$$H_2NNH_2 + H_2O_2 \xrightarrow{Cu^{2+}} H{-}N{=}N{-}H + 2\ H_2O$$

Hydrazine   **Diimide**

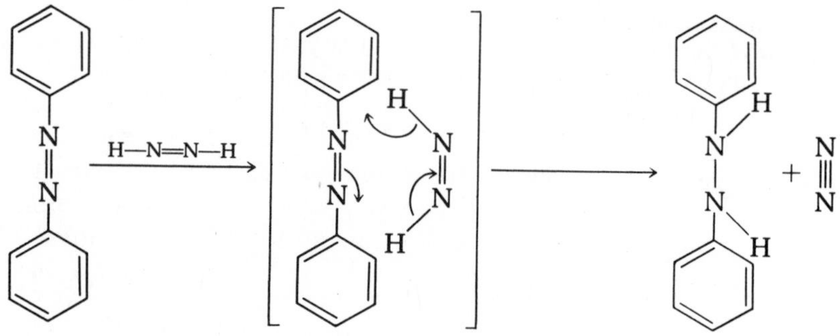

The procedure which follows affords hydrazobenzene as colorless crystals of high purity but, since this substance is very susceptible to air oxidation and darkens rapidly in storage, it is transformed by acid-catalyzed rearrangement into the more stable benzidine by reaction with hydrochloric acid in methanol-water at 0°. A side reaction which limits the yield of benzidine to about 70% is rearrangement to 2,4'-diaminobiphenyl, which is eliminated in the crystallization of benzidine hydrochloride. Another side reaction, acid-catalyzed air oxidation to azobenzene, can be counteracted by addition of stannous chloride to reduce the azo compound as formed.

EXPERIMENTS

1. Azobenzene

Weigh or measure the following reagents into a 250-ml Erlenmeyer flask: 100 ml of triethylene glycol, 16 g of zinc dust, 22 g of potassium hydroxide, and 10 ml of nitrobenzene. Insert a thermometer and, with vigorous swirling, heat the mixture on a 70-watt hot plate or over a steam bath in the course of about 5 min until the temperature rises to 80°. Remove the flask from the source of heat, continue to swirl, observe the temperature carefully, and avoid a rise above 85° by brief chilling under the tap. By appropriate heating or cooling, keep the greenish suspension in the range 80–85° for 10 min. Then, by heating and swirling, bring the temperature to 135°

Reaction time about 35 min

in the course of 3–5 min. Remove the flask from the hot plate, swirl, and prevent a rise above 140° by brief cooling. Then maintain a temperature of 135–140° for 15 min by appropriate heating or cooling; during this period alternately swirl the flask and stir the mixture using the thermometer. Note that the mixture soon reddens and gradually acquires a red color of maximal intensity.

Cool the mixture to 100° and add 50 ml of 95% ethanol, using it to rinse the thermometer. Then remove the zinc dust by filtering the hot solution by suction through an 80-mm Büchner funnel; use ethanol or methanol to rinse the flask and wash the filter cake. Chill the filtrate in an ice bath, add an equal volume of water, and collect the orange colored azobenzene by filtration, if it crystallizes, or by extraction with ether if it separates as an oil. Recrystallize the crude material from the least volume of methanol; collect the product after cooling the mixture in ice. Product collected by filtration need not be dried prior to this purification by crystallization. The yield of the crude azobenzene, mp 65–67°, is 7.5–8.0 g; the first crop of orange plates, mp 65–66°, weighs 5.5 g.

2. Azoxybenzene

Add 10 g of sodium hydroxide pellets to 40 ml of methanol in a 100-ml round-bottomed flask equipped with a reflux condenser. Warm the flask and swirl to hasten solution of the hydroxide, then add 5 ml of nitrobenzene and reflux the solution (boiling chip) on the steam bath for 3 hr. Pour the reaction mixture onto a mixture of 20 g of ice and 20 ml of water in a 250-ml beaker. Stir until the azoxybenzene solidifies, then filter the product on a Büchner funnel

Reaction time: 3 hr

with suction, wash well with water, and recrystallize the damp product from alcohol. Recrystallization is accomplished by dissolving the product in a small quantity of alcohol at room temperature and cooling the resulting solution well in ice with seeding and scratching, if necessary. A second crop of crystals can be collected by evaporating methanol from the mother liquor and repeating the crystallization process. Azoxybenzene forms pale yellow needles mp 36°. Yield 4 g.

3. Hydrazobenzene

Reaction time: About 15 min

In a 125-ml Erlenmeyer flask dissolve 3.0 g of the azobenzene prepared in the earlier experiment in 50 ml of 95% ethanol with heating on a steam bath and then cool the orange solution to room temperature. Add 6 ml of 95% hydrazine and 1 ml of a 1% solution of cupric sulfate pentahydrate and swirl the mixture vigorously in an ice-water bath until a large crop of azobenzene crystallizes and the temperature is close to 0° (a low temperature minimizes air oxidation). Measure 3.0 ml of 30% hydrogen peroxide in a graduate using a capillary dropping tube to obtain exactly 3.0 ml. The peroxide is to be added gradually to the azobenzene mixture in the course of 12–15 min while the flask is swirled vigorously in the ice bath, for example at the rate of one microdrop every 10 sec. The azobenzene dissolves as the reaction proceeds. Toward the end of the addition, the orange or yellow color should be almost completely discharged and colorless hydrazobenzene should begin to separate. If not, add 0.5 ml more hydrogen peroxide, gradually, as before. Heat the mixture on the hot plate to redissolve the product and filter the hot solution by gravity to remove a trace of copper oxide (even if the solution is yellow, do not use decolorizing charcoal as this catalyzes air oxidation). Reheat the filtrate, add water until crystallization starts (10–15 ml), and after a large crop of product has separated, cool in ice. Collect the colorless plates of hydrazobenzene and use the moist material in the next step. The dry weight of product, mp 125–126°, is 2.7 g.

CAUTION. H_2O_2 blisters the skin

4. Benzidine

Warning! Carcinogen[2]

Reaction time: 15 min

Place 2.7 g of hydrazobenzene (or the unweighed moist product) in a 125-ml Erlenmeyer flask and in another flask prepare a mixture of 20 ml of water, 20 ml of methanol, 4 ml of concd hydrochloric acid, and 1 ml of a stock solution of stannous chloride[1] and swirl it in an ice-water bath to bring the temperature to 0–2°. Pour the mix-

---

[1] Dissolve 3 g of stannous chloride dihydrate (freshly opened bottle) in 3 ml of concd hydrochloric acid and 10 ml of water.

[2] Benzidine has recently been found to be a potent carcinogen (cancer causing compound). If this experiment is performed, precautions should be taken to avoid contact with the skin (polyethylene gloves). The procedure has been included here to illustrate the method and to make the point that all organic compounds should be handled with care. The authors do not recommend routine assignment of this experiment.

ture onto the hydrazobenzene, note the time, and swirl the flask in the ice bath for 15 min. During this period plates of hydrazobenzene can be seen to give way to a white powder of benzidine dihydrochloride. Heat the mixture on the hot plate to dissolve the dihydrochloride and if the solution is yellow add a few drops of the stannous chloride solution and heat to discharge the color. Measure 4 ml of concd hydrochloric acid in a graduate and add it by microdrops while heating the solution on the hot plate. Soon each drop produces an amorphous precipitate of the dihydrochloride, which dissolves on heating and swirling. Finally (after about 3.5 ml of acid have been added) crystals begin to appear, and the flask is then let stand for crystallization, first at room temperature, then after cooling in ice. Collect the salt, rinse the flask with dil hydrochloric acid, and wash the benzidine dihydrochloride with methanol to promote rapid drying. The yield of colorless product is 3.7 g; mp (evacuated capillary) 385° dec.[3]

Isolation of minor product

[3] The minor product can be isolated from the mother liquor by neutralization with sodium hydroxide, extraction with ether, adsorption on acid-washed alumina and elution with benzene. Early fractions give an oil which when dissolved in dilute hydrochloric acid and then crystallized at 0° from very concd acid affords 2,4'-diaminodiphenyl dihydrochloride as fine silken needles, mp 303°, dec. (very soluble in methanol). The yield is about 7%. The more strongly adsorbed benzidine can be eluted with ether. The free diamine is very slow to crystallize.

40 Anthraquinone and Anthracene

KEYWORDS   Friedel-Crafts acylation   Reduction, stannous chloride
                o-Benzoylbenzoic acid   Sodium hydroxide,
                HCl trap   sodium hydrosulfite
                $AlCl_3$, two equivalents   Anthrone
                Cyclodehydration, H_2SO_4

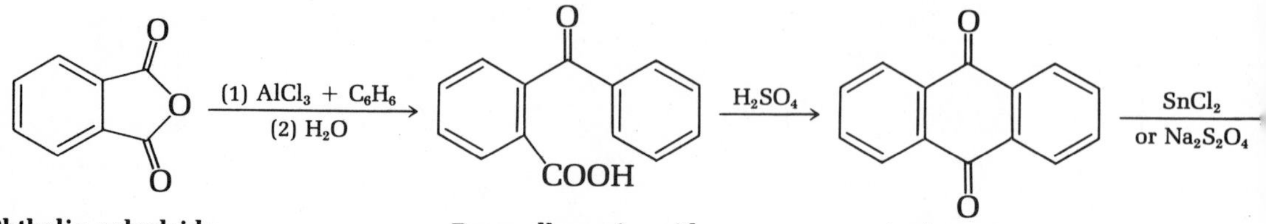

Phthalic anhydride
MW 148.11, mp 132°

(1) $AlCl_3$ + C_6H_6
(2) H_2O

o-Benzoylbenzoic acid
MW 226.22, mp 127°

H_2SO_4

Anthraquinone
MW 208.20, mp 286°

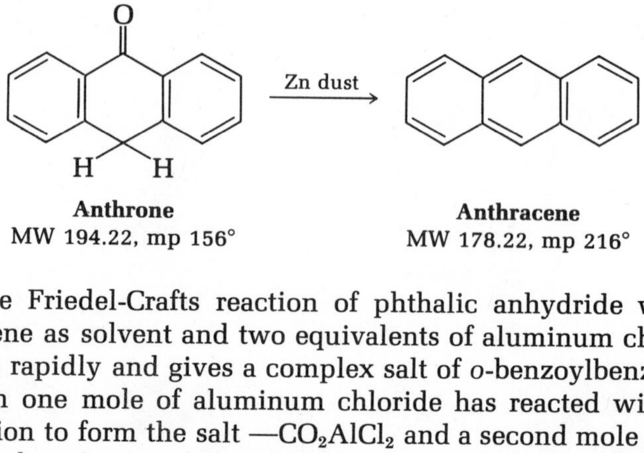

$SnCl_2$
or $Na_2S_2O_4$

Anthrone
MW 194.22, mp 156°

Zn dust

Anthracene
MW 178.22, mp 216°

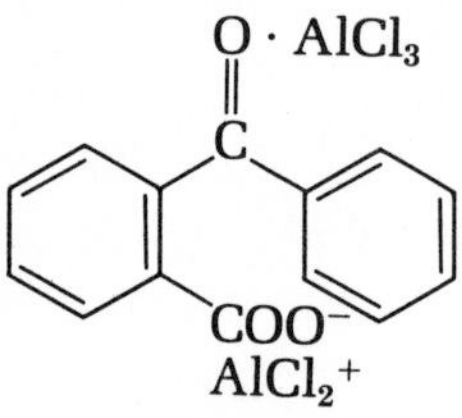

$O \cdot AlCl_3$

COO^-
$AlCl_2{}^+$

Complex salt

The Friedel-Crafts reaction of phthalic anhydride with excess benzene as solvent and two equivalents of aluminum chloride proceeds rapidly and gives a complex salt of o-benzoylbenzoic acid in which one mole of aluminum chloride has reacted with the acid function to form the salt —CO_2AlCl_2 and a second mole is bound to the carbonyl group. On addition of ice and hydrochloric acid the complex is decomposed and basic aluminum salts are brought into solution.

Treatment of o-benzoylbenzoic acid with concd sulfuric acid effects cyclodehydration to anthraquinone, a pale-yellow high-melting compound of great stability. Since anthraquinone can be sulfonated only under forcing conditions, a high temperature can be used to shorten the reaction time without loss in yield of product; the conditions are so adjusted that anthraquinone separates from the hot solution in crystalline form favoring rapid drying.

Reduction of anthraquinone to anthrone can be accomplished rapidly on a small scale with stannous chloride in acetic acid solution. A second method, which involves refluxing anthraquinone with an aqueous solution of sodium hydroxide and sodium hydrosulfite, is interesting to observe because of the sequence of color changes: anthraquinone is reduced first to a deep red vat containing anthrahydroquinone dianion; the red color then gives place to a yellow color characteristic of anthranol anion; as the alkali is neutralized by the conversion of $Na_2S_2O_4$ to $2\ NaHSO_3$, anthranol ketonizes to the more stable anthrone. The second method is preferred in the industry, because sodium hydrosulfite costs less than half as much as stannous chloride and because water is cheaper than acetic acid and no solvent recovery problem is involved.

Reduction of anthrone to anthracene is accomplished by refluxing in aqueous sodium hydroxide solution with activated zinc dust. The method has the merit of affording pure, beautifully fluorescent anthracene.

EXPERIMENTS   ### 1. o-Benzoylbenzoic Acid

This Friedel-Crafts reaction is carried out in a 500-ml round-bottomed flask equipped with a short condenser. A trap for collecting hydrogen chloride liberated is connected to the top of the condenser by rubber tubing of sufficient length to make it possible either to heat the flask on the steam bath or to plunge it into an ice bath. The trap is a suction flask half filled with water and fitted with a delivery tube inserted to within 1 cm of the surface of the water.

Fifteen grams of phthalic anhydride and 75 ml of thiophene-free benzene are placed in the flask and this is cooled in an ice bath until the benzene begins to crystallize. Ice cooling serves to moderate the vigorous reaction which otherwise might be difficult to control. Thirty grams of anhydrous aluminum chloride[1] is added, the condenser and trap are connected, and the flask is shaken well and warmed for a few minutes by the heat of the hand. If the reaction does not start, the flask is warmed *very gently* by holding it for a few seconds over the steam bath. At the first sign of vigorous boiling, or evolution of hydrogen chloride, the flask is held over the ice bath in readiness to cool it if the reaction becomes too vigorous. This gentle, cautious heating is continued until the reaction is proceeding smoothly enough to be refluxed on the steam bath. This point is reached in about 5 min. Continue the heating on the steam bath, swirl the mixture, and watch it carefully for sudden separation of the addition compound, since the heat of crystallization is such that it may be necessary to plunge the flask into the ice bath to moderate the process. Once the addition compound has separated as a thick

Short reaction period, high yield

---

[1] This is best weighed in a stoppered test tube. The chloride should be from a freshly opened bottle.

paste, heat the mixture for 10 min more on the steam bath, remove the condenser, and swirl the flask in an ice bath until cold. (Should no complex separate, heat for 10 min more and then proceed as directed.) Take the flask and ice bath to the hood, weigh out 100 g of ice, add a few small pieces of ice to the mixture, swirl and cool as necessary, and wait until the ice has reacted before adding more. After the 100 g of ice have been added and the reaction of decomposition has subsided, add 20 ml of concd hydrochloric acid, 100 ml of water, swirl vigorously, and make sure that the mixture is at room temperature. Then add 50 ml of water, swirl vigorously, and again make sure the mixture is at room temperature. Add 50 ml of ether and, with a flattened stirring rod, dislodge solid from the neck and walls of the flask and break up lumps at the bottom. To further promote hydrolysis of the addition compound, extraction of the organic product, and solution of basic aluminum halides, stopper the flask with a cork and shake vigorously for several minutes. When most of the solid has disappeared, pour the mixture through a funnel into a separatory funnel until the separatory funnel is nearly filled. Draw off and discard the lower aqueous layer. Pour the rest of the mixture into the separatory funnel, rinse the reaction flask with fresh ether, and again drain off the aqueous layer. To reduce the fluffy, dirty precipitate that appears at the interface add 10 ml of concd hydrochloric acid and 25 ml of water, shake vigorously for 2–3 min, and drain off the aqueous layer. If some interfacial dirty emulsion still persists, decant the benzene-ether solution through the mouth of the funnel into a filter paper for gravity filtration and use fresh ether to rinse the funnel. Clean the funnel and pour in the filtered benzene-ether solution. Shake the solution with a portion of dilute hydrochloric acid, and then isolate the reaction product by either of the following procedures:

Alternative procedures

(a) Add 50 ml of 10% sodium hydroxide solution, shake thoroughly, and separate the aqueous layer.[2] Extract with a further 25-ml portion of aqueous alkali and combine the extracts. Wash with 10 ml of water, and add this aqueous solution to the 75 ml of aqueous extract already collected. Discard the benzene-ether solution. Acidify the 85 ml of combined alkaline extract with concd hydrochloric acid to pH 1–2 and, if the o-benzoylbenzoic acid separates as an oil, cool in ice and rub the walls of the flask with a stirring rod to induce crystallization of the hydrate; collect the product and wash it well with water. This material is the monohydrate: $C_6H_5COC_6H_4CO_2H \cdot H_2O$. To convert it into anhydrous o-benzoyl-

Drying time:
About 1 hr

benzoic acid, put it in a tared, 250-ml round-bottomed flask, evacuate the flask at the full force of the aspirator, and heat it in the open rings of a steam bath covering the flask with a towel. Check the weight of the flask and contents for constancy after 45 min, 1 hr, and 1.25 hr. Yield 19–21 g, mp 126–127°.

---

[2] The nature of a yellow pigment that appears in the first alkaline extract is unknown; the impurity is apparently transient, for the final product dissolves in alkali to give a colorless solution.

Extinguish flames

(b) Filter the benzene-ether solution through anhydrous sodium sulfate for superficial drying, put it into a 250-ml round-bottomed flask, and distil over the steam bath through a condenser into an ice-cooled receiver until the volume in the distilling flask is reduced to about 55 ml. Add ligroin slowly to slight turbidity and let the product crystallize at 25° and then at 5°. The yield of anhydrous, colorless, well-formed crystals, mp 127–128°, is 18–20 g.

The o-benzoylbenzoic acid just prepared is to be used in the next preparation, anthraquinone. Any acid not used should be saved for the experiment of Chapter 41.

2. Anthraquinone

Reaction time: 10 min

Place 5.0 g of o-benzoylbenzoic acid (anhydrous) in a 125-ml round-bottomed flask, add 25 ml of concd sulfuric acid, and heat on the steam bath with swirling until the solid is dissolved. Then clamp the flask over a microburner, insert a thermometer, raise the temperature to 150°, and heat to maintain a temperature of 150–155° for 5 min. Let the solution cool to 100°, remove the thermometer after letting it drain, and, with a capillary dropping tube, add 5 ml of water by drops with swirling to keep the precipitated material dissolved as long as possible so that it will separate as small, easily filtered crystals. Let the mixture cool further, dilute with water until the flask is full, again let cool, collect the product by suction filtration, and wash well with water. Then remove the filtrate and wash the filter flask, return the funnel to the filter flask but do not apply suction, and test the filter cake for unreacted starting material as follows: Dilute 10 ml of concd ammonia solution with 50 ml of water, pour the solution onto the filter and loosen the cake so that it is well leached. Then apply suction, wash the cake with water, and acidify a few ml of the filtrate. If there is no precipitate upon acidifying, the yield of anthraquinone should be close to the theory, since it is insoluble in water. Dry the product to constant weight but do not take the melting point since it is so high (mp 286°).

In the next section adjust the quantities to fit your yield of anthraquinone (which should be quantitative).

3. Anthrone

(a) Stannous Chloride Reduction

Procedure (a)

In a 125-ml round-bottomed flask provided with a reflux condenser put 5.0 g of anthraquinone, 40 ml of acetic acid, and a solution made by warming a mixture of 13 g of stannous chloride dihydrate with 13 ml of concd hydrochloric acid. Add a boiling stone to the reaction mixture, note the time, and reflux gently until crystals of anthraquinone have completely disappeared (8–10 min); then reflux 15 min longer and record the total time. Disconnect the flask, heat it on the steam bath, and add water (about 12 ml) in 1-ml portions until the solution is saturated. Let the solution stand for crys-

tallization. Collect and dry the product and take the melting poin
(156° given). The yield of pale yellow crystals is 4.3 g.

(b) Hydrosulfite Reduction

Procedure (b)

In a 500-ml round-bottomed flask, which can be heated unde
reflux, put 5.0 g of anthraquinone, 6 g of sodium hydroxide, 15 g o
sodium hydrosulfite, and 130 ml of water. Heat over a free flame
and swirl for a few minutes to convert the anthraquinone into the
deep red vat containing the anthrahydroquinone anion. Note that
particles of different appearance begin to separate even before the
anthraquinone has all dissolved. Arrange for refluxing and reflux
for 45 min; cool, filter the product, and wash it well and let dry
Note the weight of the product and melting point of the crude material and then crystallize it from 95% ethanol; the solution may have
to be filtered to remove insoluble impurities. Record the approximate volume of solvent used and, if the first crop of crystals recovered is not satisfactory, concentrate the mother liquor and secure
a second crop.

4. Comparison of Results

Compare your results with those obtained by neighbors using the
alternative procedure with respect to yield, quality of product, and
working time. Which is the better laboratory procedure? Then consider the cost of the three solvents concerned, the cost of the two
reducing agents (current prices to be posted), and relative ease of
recovery of the organic solvents, and decide which method would
be preferred as a manufacturing process.

5. Special Experiment: Fluorescent Anthracene

Reflux time: 0.5 hr

Put 10 g of zinc dust into a 500-ml round-bottomed flask and
activate the dust by adding 60 ml of water and 1 ml of copper sulfate
solution (Fehling solution I) and swirling for a minute or two.
Add 4.0 g of anthrone, 10 g of sodium hydroxide, and 100 ml of
water; attach a reflux condenser, heat to boiling, note the time, and
start refluxing the mixture. Anthrone at first dissolves as the yellow
anion of anthranol, but anthracene soon begins to separate as a
white precipitate. In about 15 min the yellow color initially observed on the walls disappears, but refluxing should be continued
for a full 30 min. Then remove the flame, use a water wash bottle
to rinse down anthracene that has lodged in the condenser, and filter
the still hot mixture on a large Büchner funnel. It usually is possible to decant from, and so remove, a mass of zinc. After liberal
washing with water, blow or shake out the gray cake into a 400-ml
beaker and rinse funnel and paper with water. To remove most of
the zinc metal and zinc oxide, add 20 ml of concd hydrochloric acid,
heat on the steam bath with stirring for 20–25 min when initial
frothing due to liberated hydrogen should have ceased. Collect the

Methanol removes water without dissolving much anthracene

now nearly white precipitate on a large Büchner funnel and, after liberal washing with water, release the suction, rinse down the walls of the funnel with methanol, use enough more methanol to cover the cake, and then apply suction. Wash again with enough methanol to cover the cake and then remove solvent thoroughly by suction.

The product need not be dried before crystallization from benzene. Transfer the methanol-moist material to a 125-ml Erlenmeyer and add 75 ml of benzene; a liberal excess of solvent is used to avoid crystallization in the funnel. Make sure there are no flames nearby and heat the mixture on the hot plate to bring the anthracene into solution but leaving a small residue of zinc. Filter by gravity by the usual technique. Anthracene is obtained as thin, colorless, beautifully fluorescent plates; yield 2.8 g. In washing the equipment with acetone, you should be able to observe the striking fluorescence of very dilute solutions. The fluorescence is quenched by a bare trace of impurity.

CAUTION! Highly flammable solvent

41 *Benzophenone and Benzopinacol*

KEYWORDS   Photochemistry,
             photochemical reduction
          Decarboxylation, Cu catalyst
          Labile, allotropic
          Electronically excited,
             triplet state

Radicals
Benzopinacol, alkaline cleavage
Benzhydrol, benzophenone
Pinacol-pinacolone rearrangemen
Carbonium ion

Of the several methods possible for the preparation of benzophenone, the one used here is decarboxylation of *o*-benzoylbenzoic acid in the presence of copper catalyst derived from copper carbonate.

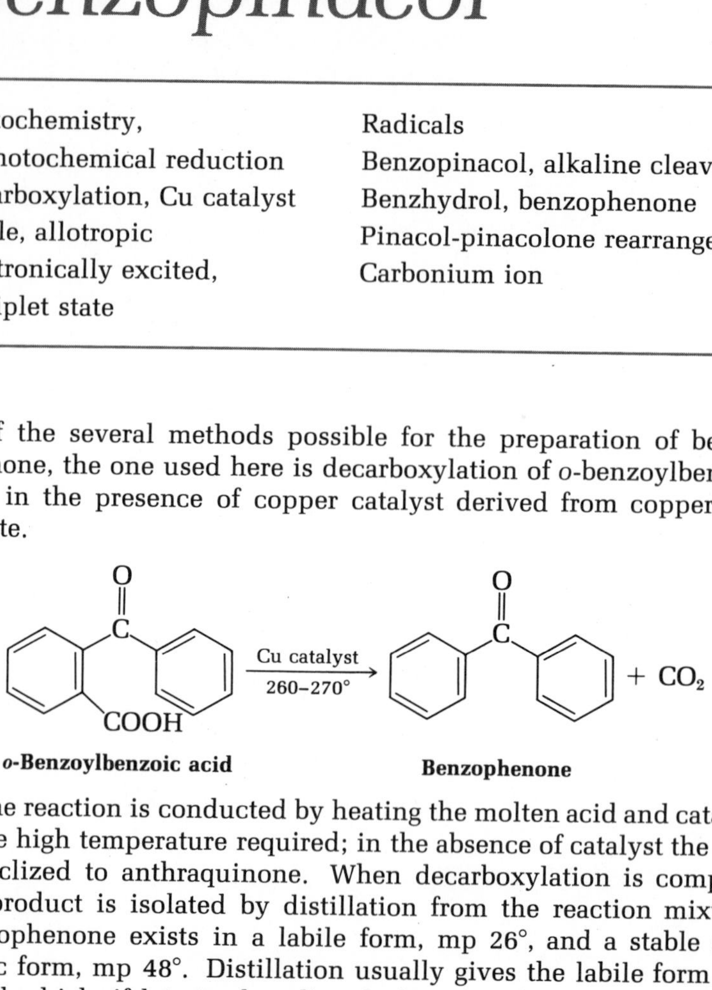

o-**Benzoylbenzoic acid**   **Benzophenone**

The reaction is conducted by heating the molten acid and catalyst to the high temperature required; in the absence of catalyst the acid is cyclized to anthraquinone. When decarboxylation is complete the product is isolated by distillation from the reaction mixture. Benzophenone exists in a labile form, mp 26°, and a stable allotropic form, mp 48°. Distillation usually gives the labile form as a liquid which, if let stand undisturbed in the absence of seed, very slowly solidifies to transparent crystals. However, if the liquid is scratched, is inoculated with the stable form, or is in an atmosphere carrying seed, it suddenly changes to the stable form and solidifies.

Benzophenone is colorless, and like aliphatic α,β-unsaturated ketones it absorbs ultraviolet light, resulting in an electronically excited molecule. This activated ketone, thought to be in the triplet state, abstracts a hydrogen atom from the solvent, isopropyl alcohol, producing two radicals, benzhydrol and hydroxyisopropyl. The hydroxyisopropyl radical transfers a hydrogen atom to neutral benzophenone, giving another benzhydrol radical and a molecule of acetone. Combination of the benzhydrol radicals produces the

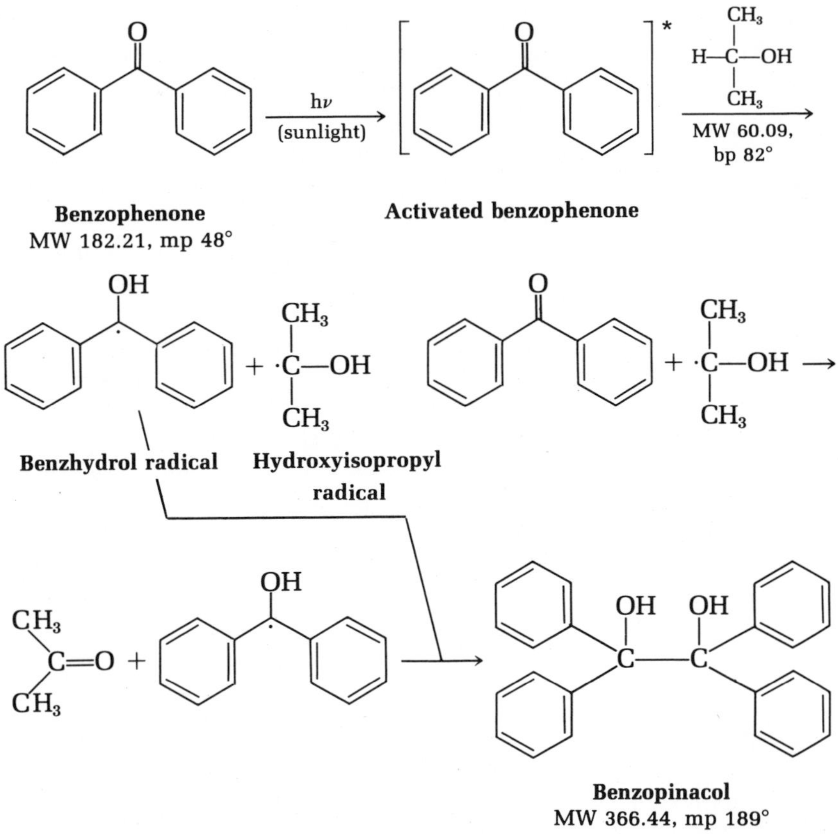

Benzophenone
MW 182.21, mp 48°

Activated benzophenone

Benzhydrol radical   **Hydroxyisopropyl radical**

Benzopinacol
MW 366.44, mp 189°

product benzopinacol, which crystallizes from the solution in dramatic fashion when the reactants are exposed to bright sunlight. The experiment should be done when there is good prospect for long hours of bright sunshine for several days. The benzopinacol is cleaved by alkali to benzhydrol and benzophenone (Experiment 3) and it is rearranged in acid to benzopinacolone (Experiment 4).

EXPERIMENTS

1. Benzophenone

Place 15 g of anhydrous o-benzoylbenzoic acid and 0.5 g of basic copper carbonate in a 50-ml round-bottomed flask and heat gently, with shaking, over a free flame to melt the acid and until the neutralization reaction is over. Attach a glass tube through a rubber stopper, evacuate the flask and heat it on the steam bath to eliminate the water formed. Then support the flask for heating with a free flame, insert a thermometer with the bulb submerged in the melt, heat with a small flame, and raise the temperature to 265°, at which point evolution of carbon dioxide starts. Note the time and, by intermittent heating, keep the temperature close to 265° until, in 20–25 min, metallic copper separates from the clear solution. Continue to heat at 265° for 5 min longer (when evolution of CO_2 should

Reaction time: 30 min

stop), remove the flame, and withdraw the thermometer. Fit the
flask with a stillhead-with-thermometer and an air condenser
(empty fractionating column); use a small tared Erlenmeyer as receiver and distil the benzophenone. The corrected boiling point is
306°; the uncorrected boiling point may be as low as 294°. Continue
the distillation until there is a marked rise in boiling point (4–5°) or
until the distillate becomes dark yellow.[1]

Record the weight of crude benzophenone (11–13 g) and see if it
will solidify on cooling or on being rubbed with a stirring rod. One
usually obtains at this point the labile form, which (particularly
since it is not quite pure) does not solidify easily. If this is the case
fix a barely visible particle of ordinary benzophenone on the end of
a stirring rod and rub it into the liquid against the side of the flask.
When crystallization sets in, note the warming against the palm of
the hand. Save a trace of seed. Benzophenone crystallizes well
from ligroin, bp 60–90° (alcohol leaves too much material in the
mother liquor). Without removing the material from the original receiving flask, cover it with ligroin (1.5 ml per gram) and dissolve by
heating on the steam bath. Without using charcoal or filtering, cool
the solution in an ice bath until it becomes cloudy and the product
oils out. Then remove the flask from the bath, allow the liquid to
come to rest, and add a seed crystal. As crystallization progresses,
return the flask to the ice bath and eventually stir the mixture and
cool it thoroughly. Collect the crystals on a suction funnel and use
30–40 ml of fresh, ice-cold solvent (ligroin) to wash any yellow
material into the mother liquor. Recrystallize the product if it is not
white; very large crystals can be obtained by allowing the solution
to cool slowly and without seeding; yield, 9.5–11.5 g, mp 47–48°.

2. Benzopinacol

In a 125-ml round-bottomed flask dissolve 10 g of benzophenone
in 60–70 ml of isopropyl alcohol by warming on the steam bath, fill
the flask to the neck with more of this alcohol, and add one drop of
glacial acetic acid. (If the acid is omitted enough alkali may be derived from the glass of the flask to destroy the reaction product by
the alkaline cleavage described in Experiment 3.) Stopper the flask
with a well-rolled, tight-fitting cork, which is then wired in place.
Invert the flask in a 100-ml beaker and place them where the mixture will be exposed to direct sunlight for the longest periods of
time. Since benzopinacol is but sparingly soluble in alcohol, its
formation can be followed by the separation from around the walls
of the flask of small, colorless crystals (benzophenone forms large,
thick prisms). If the reaction mixture is exposed to direct sunlight
throughout the daylight hours, the first crystals separate in about

---

Note to Student

[1] The dark residue in the distillation flask can be loosened and the mass of copper dislodged by adding benzene and heating the flask on the steam bath for one half hour. The dark solution is decanted and if necessary the process is repeated.

Reaction time: 4 days to two weeks

5 hr and the reaction is practically complete (95% yield) in four days. In winter the reaction may take as long as two weeks, and any benzophenone which crystallizes must be brought into solution by warming on the steam bath. When the reaction appears to be over, chill the flask if necessary and collect the product. The material should be pure, mp 188–189°. If the yield is poor, more material can be obtained by further exposure of the mother liquor to sunlight.

3. Alkaline Cleavage

Suspend a small test sample of benzopinacol in alcohol and heat to boiling, making sure that the amount of solvent is not sufficient to dissolve the solid. Add one drop of sodium hydroxide solution, heat for a minute or two, and observe the result. The solution contains equal parts of benzhydrol and benzophenone, formed by the following reaction:

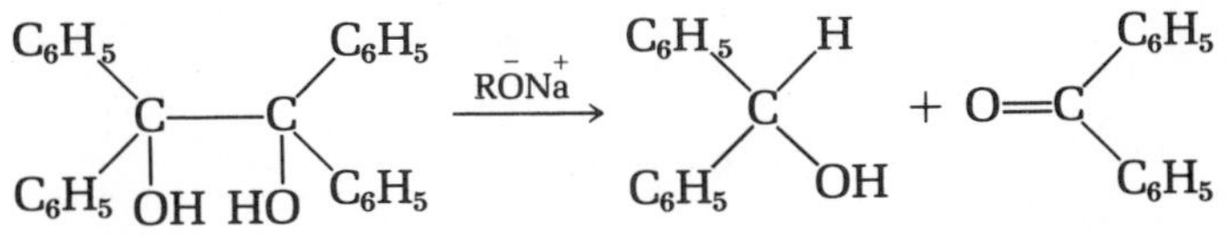

| Benzopinacol | Benzhydrol | Benzophenone |
| Mp 189° | Mp 68° | Mp 48° |

The low-melting products resulting from the cleavage are much more soluble than the starting material.

Benzophenone can be converted into benzhydrol in nearly quantitative yield by following the procedure outlined above for the preparation of benzopinacol, modified by addition of a very small piece of sodium (0.05 g) instead of the acetic acid. The reaction is complete when, after exposure to sunlight, the greenish-blue color disappears. To obtain the benzhydrol the solution is diluted with water, acidified, and evaporated. Benzopinacol is produced as before by photochemical reduction, but it is at once cleaved by the sodium alcoholate; the benzophenone formed by cleavage is converted into more benzopinacol, cleaved, and eventually consumed.

4. Pinacolone Rearrangement

This acid-catalyzed, carbonium ion rearrangement is characterized by its speed and by the high yield (see page 246).

In a 125-ml round-bottomed flask place 5 g of benzopinacol, 25 ml of acetic acid, and two or three very small crystals of iodine (0.05 g). Heat to the boiling point for a minute or two under a reflux condenser until the crystals are dissolved, and then reflux the red solution for 5 min. On cooling, the pinacolone separates as a stiff paste. Thin the paste with alcohol, collect the product, and wash it free from iodine with alcohol. The material should be pure; yield 95%.

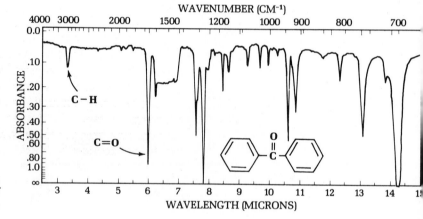

$$C_6H_5-\underset{\underset{C_6H_5}{|}}{\overset{\overset{OH}{|}}{C}}-\underset{\underset{C_6H_5}{|}}{\overset{\overset{OH}{|}}{C}}-C_6H_5 \;\;\overset{H^+}{\rightleftharpoons}\;\; \left[C_6H_5-\underset{\underset{C_6H_5}{|}}{\overset{\overset{+OH_2}{|}}{C}}-\underset{\underset{C_6H_5}{|}}{\overset{\overset{OH}{|}}{C}}-C_6H_5 \right] \;\;\overset{-H_2O}{\longrightarrow}\;\; \left[C_6H_5-\underset{\underset{C_6H_5}{|}}{\overset{+}{C}}-\underset{\underset{C_6H_5}{|}}{\overset{\overset{:OH}{|}}{C}}-C_6H_5 \right.$$

Benzopinacol   **(Carbonium ion)**

$$C_6H_5-\underset{\underset{C_6H_5}{|}}{\overset{\overset{C_6H_5}{|}}{C}}-\overset{\overset{O}{\|}}{C}-C_6H_5 \;\;\overset{-H^+}{\rightleftharpoons}\;\; \left[C_6H_5-\underset{\underset{C_6H_5}{|}}{\overset{\overset{C_6H_5}{|}}{C}}-\overset{\overset{+}{\overset{OH}{|}}}{C}-C_6H_5 \right.$$

Benzopinacolone
Mp 179–180°

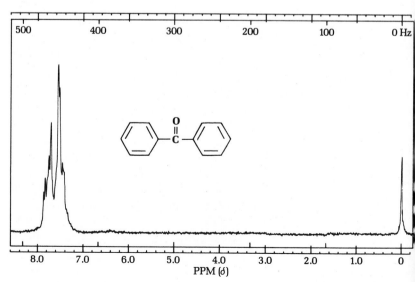

**Figure 41.1
Infrared spectrum of
benzophenone in CS$_2$.**

**Figure 41.2
Nmr spectrum of
benzophenone.**

42 Dichlorocarbene

KEYWORDS Carbene, bivalent carbon, Dichlorocarbene
electrophile 1,5-Cyclooctadiene
Cyclopropanes Diglyme
Sodium trichloroacetate Phase transfer catalysis
Trichloromethyl anion Benzyltriethylammonium chloride

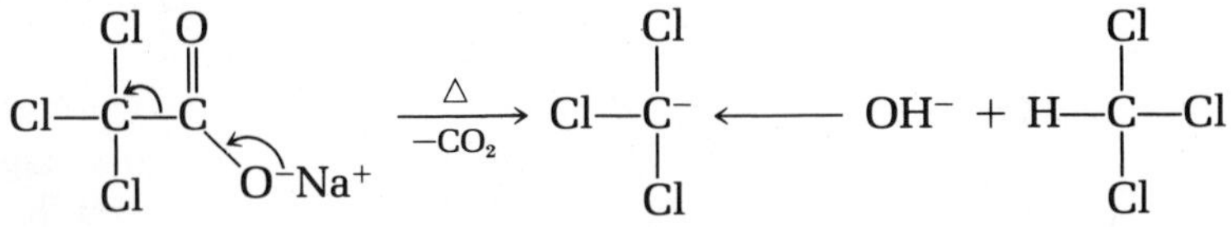

Sodium trichloroacetate **Trichloromethyl anion** **Chloroform**
MW 185.39 MW 119.39, bp 61°

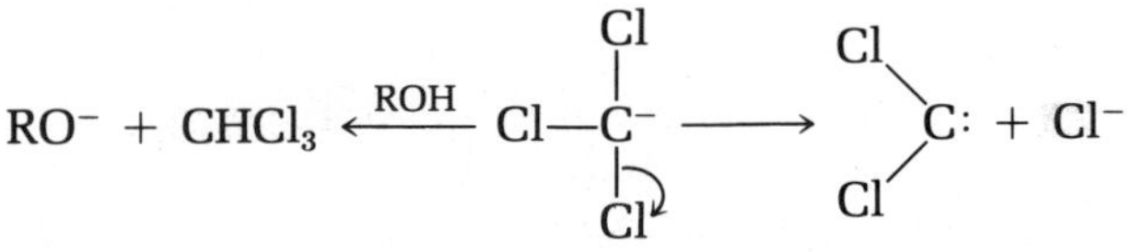

Eq. 1

Dichlorocarbene

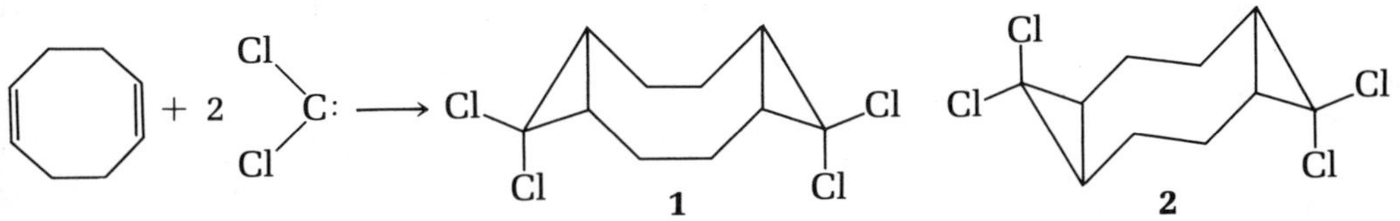

Eq. 2

s,cis-1,5-Cyclo- *cis*, mp 176° *trans*, mp 230° dec
octadiene
MW 108.14, **5,5,10,10-Tetrachlorotricyclo[7.1.0.0⁴·⁶]decane**
bp 149–150° MW 274.03

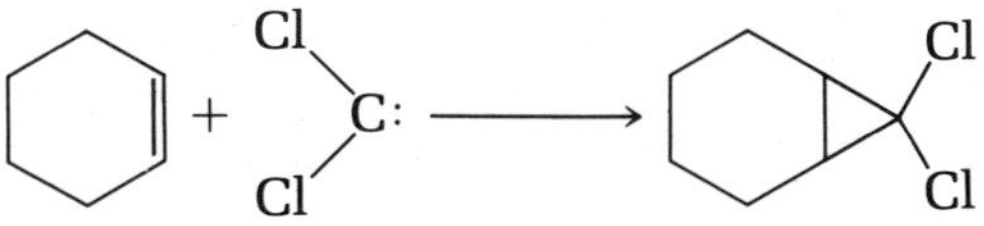

Eq. 3

Cyclohexene **7,7-Dichlorobicyclo[4.1.0]heptane**
MW 82.14, bp 83° MW 165.06, bp 198°

Eq. 4

Dichlorocarbene is a highly reactive intermediate of bivalent carbon with only six valence electrons around the carbon. It is electrically neutral and a powerful electrophile. As such it reacts with alkenes forming cyclopropane derivatives by *cis*-addition to the double bond.

Of the dozen or so ways by which dichlorocarbene may be gen-erated we shall investigate two. In the first experiment thermal de-composition of anhydrous sodium trichloroacetate in an aprotic solvent in the presence of cis, cis-1,5-cyclooctadiene gives **1**. In the second experiment, reaction of chloroform with aqueous hydroxide ion in the presence of a phase transfer catalyst and cyclohexene gives **3**.

EXPERIMENTS

1. Thermal Decomposition of Sodium Trichloroacetate and Reaction of Resulting Dichlorocarbene with 1,5-Cyclooctadiene

The thermal decomposition of sodium trichloroacetate initially gives the trichloromethyl anion (Eq. 1). In the presence of a proton-donating solvent (or moisture) this anion gives chloroform, in the absence of these reagents the anion decomposes by loss of chloride ion to give dichlorocarbene (Eq. 2).

The conventional method for carrying out the reaction is to add the salt portionwise, during 1–2 hr, to a magnetically stirred solution of the olefin in diethylene glycol dimethyl ether ("diglyme") at a temperature maintained in a bath at 120°. Under these conditions the reaction mixture becomes almost black, isolation of a pure prod-uct is tedious, and the yield is low.

$CH_3OCH_2CH_2OCH_2CH_2OCH_3$
Diglyme
MW 134.17, Bp 161°
Miscible with water

Tetrachloroethylene, a nonflammable solvent widely used in the dry cleaning industry, boils at 121° and is relatively inert toward electrophilic dichlorocarbene. On generation of dichlorocarbene from either chloroform or sodium trichloroacetate in the presence of tetrachloroethylene the yield of hexachlorocyclopropane (mp 104°) is only 0.2–10% (W. R. Moore, 1963; E. K. Fields, 1963). The first idea for simplifying the procedure[1] was to use tetrachloroethylene to control the temperature to the desired range, but sodium trichloro-acetate is insoluble in this solvent and no reaction occurs. Diglyme, or an equivalent, is required to provide some solubility. The re-action proceeds better in a 7:10 mixture of diglyme to tetrachloro-ethylene than in diglyme alone, but the salt dissolves rapidly in this mixture and has to be added in several small portions and the reac-tion mixture becomes very dark. The situation is vastly improved by the simple expedient of cutting down the amount of diglyme to a 2.5:10 ratio. The salt is so sparingly soluble in this mixture that all of it can be added at the start of the experiment and it dissolves slowly as the reaction proceeds. The boiling and evolution of carbon dioxide provide adequate stirring, the mixture can be left unat-tended, and what little color develops is eliminated by washing the crude product with methanol.

Rationale of the procedure

The main reaction product crystallizes from ethyl acetate in beautiful prismatic needles, mp 177–179°, and this was the sole product encountered in runs made in the solvent mixture recom-

---

[1] L. F. Fieser and David H. Sachs, *J. Org. Chem.* **29**, 1113 (1964).

mended. In earlier runs made in diglyme with manual control of temperature, the ethyl acetate mother liquor material on repeated crystallization from toluene afforded small amounts of a second isomer, mp 230°, dec. Analyses checked for a pair of *cis-trans* isomers and both gave negative permanganate tests. To distinguish

X-ray analysis

between them, the junior author of the paper cited[1] undertook X-ray analysis. Isomer I (lower melting) proved to be monoclinic and isomer II orthorhombic, a difference evident to an expert from patterns such as those shown in the drawings. Measurements and cal-

Crystal I   Crystal II

culations established in each case the precise space group and the dimensions of the unit cell. Measurement of the molecular dimensions on models showed that the *cis* isomer does not fit properly into the unit cell of the orthorhombic crystal, whereas the *trans* isomer

A striking reaction

does fit; hence the lower melting isomer I is *cis* and isomer II is *trans*.

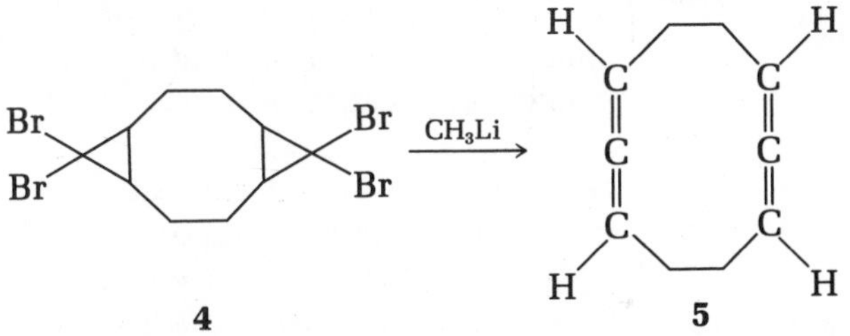

4
5,5,10,10-Tetrabromotricyclo-1
[7.1.0.0⁴·⁶]decane

5
Cyclodeca-1,2,6,7-tetraene

The bis adduct (**4**) of *cis,cis*-1,5-cyclooctadiene with dibromocarbene is described as melting at 174–180° and may be a *cis-trans* mixture (Lars Skattebøl, 1961). Treatment of the substance with methyl lithium at −40° gave a small amount of the ring-expanded bisallene (**5**).

Procedure

Place 12.8 g of trichloroacetic acid in a 125-ml Erlenmeyer with side tube, dissolve 3.2 g of sodium hydroxide pellets in 12 ml of water in a 50-ml Erlenmeyer, cool the solution thoroughly in an ice bath, swirl the flask containing the acid in the ice bath, while slowly pouring in about nine-tenths of the alkali solution. Then add a drop of 0.04% Bromocresol Green solution to produce a faint yellow color, visible when the flask is dried and placed on white paper. With a capillary dropping tube, titrate the solution to an endpoint where a single drop produces a change from yellow to blue. If the endpoint is overshot, add a few crystals of acid and titrate more carefully. Close the flask with a rubber stopper, support the suction tubing connected through a filter trap to the aspirator at the level of a steam bath by passing it over a clamp on a ringstand, connect it to the flask, and place the flask within the rings of the steam bath and wrap a towel around both for maximum heat. Turn on the water at full force for maximum suction. The evaporation requires no further attention and should be complete in 15–20 min. When you have an apparently dry white solid, scrape it out with a spatula and break up the large lumps. If you see any evidence of moisture, or in case the weight exceeds the theory, place the solid in a 25 × 150-mm test tube and rest this on its side on a drying tray mounted 5 cm above the base of a 70-watt hot plate and let the drier operate overnight.[2]

Make sure that the salt is completely dry

Place 18.1 g of dry sodium trichloroacetate in a 250-ml round-bottomed flask mounted over a microburner and add 20 ml of tetrachloroethylene, 5 ml of diglyme, and 5 ml (4.4 g) of *cis,cis*-1,5-cyclooctadiene. Attach a reflux condenser and in the top opening of the condenser insert a rubber stopper carrying a glass tube connected by rubber tubing to a short section of glass tube which can be inserted below the surface of 2–3 ml of tetrachloroethylene in a 20 × 150 mm test tube mounted at a suitable level. This bubbler will show when the evolution of carbon dioxide ceases; the solvent should be the same as that of the reaction mixture, should there be a suckback. Heat to boiling, note the time, and reflux gently until the reaction is complete. You will notice foaming, due to liberated carbon dioxide, and separation of finely divided sodium chloride. Inspection of the bottom of the flask will show lumps of undissolved sodium trichloroacetate which gradually disappear. A large flask is specified because it will serve later for removal of tetrachloroethylene by steam distillation. Make advance preparation for this operation.

Reflux time: About 75 min

When the reaction is complete add 75 ml of water to the hot mixture, heat with the small flame of a microburner, and steam distil until the tetrachloroethylene is eliminated and the product separates as an oil or semisolid. Cool the flask to room temperature, decant the supernatant liquid into a separatory funnel and extract with methylene chloride. Run the extract into the reaction flask and use enough more methylene chloride to dissolve the product; use a capillary dropping tube to rinse down material adhering to the adapter. Run the lower layer into an Erlenmeyer through a cone of anhydrous sodium sulfate and evaporate the solvent (bp 41°). The residue is a tan or brown solid (8 g). Cover it with methanol, break up the cake with a flattened stirring rod, and crush the lumps. Cool in ice, collect, and wash the product with methanol. The yield of

Easy workup

[2]See the electric dryer shown in Fig. 2.23.

colorless, or nearly colorless, 5,5,10,10-tetrachlorotricyclo[7.1.0.0^{4,6}]-decane is 3.3 g. This material, mp 174–175°, consists almost entirely of the *cis* isomer. Dissolve it in ethyl acetate (15–20 ml) and let the solution stand undisturbed for crystallization at room temperature. The pure *cis* isomer separates in large, prismatic needles, mp 175–176°.

2. Phase Transfer Catalysis—Reaction of Dichlorocarbene with Cyclohexene

Ordinarily water must be excluded from carbene generating reactions with scrupulous care. Both the intermediate trichloromethyl anion and dichlorocarbene react with water. But in the presence of a phase transfer catalyst[3] (a quaternary ammonium salt such as benzyltriethylammonium chloride) it becomes possible to carry out the reaction in aqueous medium. Chloroform, 50% aqueous sodium hydroxide, and the olefin, in the presence of a catalytic amount of the quaternary ammonium salt, are stirred for a few minutes to produce an emulsion, and after the exothermic reaction is complete (about 30 min) the product is isolated. Apparently the water soluble benzyltriethylammonium chloride reacts with hydroxide ion to give the water insoluble quaternary ammonium hydroxide ion (Eq. 5), which passes into the chloroform layer and generates the trichloromethyl anion and water (Eq. 6).

$$\overset{Cl^-}{\underset{\text{Sol in H}_2\text{O}}{(C_2H_5)_3\overset{+}{N}CH_2C_6H_5}} + OH^- \rightarrow \underset{\text{Insol in H}_2\text{O, sol in CHCl}_3}{\overset{OH^-}{(C_2H_5)_3\overset{+}{N}CH_2C_6H_5}} + Cl^- \qquad \textbf{Eq. 5}$$

$$\underset{\text{Sol in CHCl}_3}{\overset{OH^-}{(C_2H_5)_3\overset{+}{N}CH_2C_6H_5}} + CHCl_3 \rightarrow \overset{CCl_3^-}{(C_2H_5)_3\overset{+}{N}CH_2CH_2C_6H_5} + H_2O \qquad \textbf{Eq. 6}$$

$$\underset{\text{Sol in CHCl}_3}{\overset{CCl_3^-}{(C_2H_5)_3\overset{+}{N}CH_2CH_2C_6H_5}} \rightarrow \overset{Cl^-}{(C_2H_5)_3\overset{+}{N}CH_2C_6H_5} + :CCl_2 \qquad \textbf{Eq. 7}$$

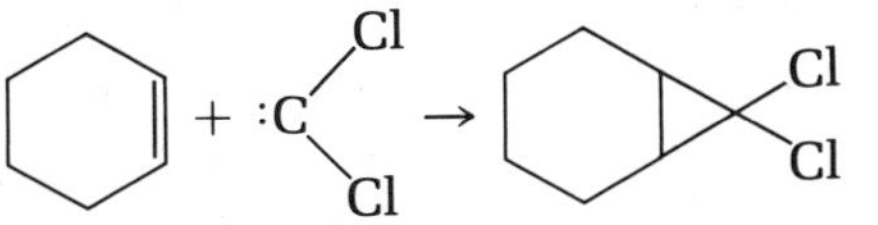

$$\textbf{Eq. 8}$$

The trichloromethyl anion decomposes in the organic layer to give dichlorocarbene and chloride ion (Eq. 7). The carbene reacts with the olefin to form the product (Eq. 8) and the chloride ion regenerates the quaternary ammonium ion which migrates back to the water layer to start the process anew.

[3] M. Makosza and M. Wawrzyniewicz, *Tetrahedron Lett.*, 4659 (1969).

Procedure

To a mixture of 8.2 g of cyclohexene, 12.0 g of chloroform, and 2[?] ml of 50% aqueous sodium hydroxide in a 125 ml Erlenmeyer flas[?] containing a thermometer add 0.2 g of benzyltriethylammoniu[?] chloride.[4] Swirl the mixture to produce a thick emulsion. The tem[?] perature of the reaction will rise gradually at first and then markedl[?] accelerate. As it approaches 60° prepare to immerse the flask in a[?] ice bath. With the flask both in and out of the ice bath, stir the thic[?] paste and maintain the temperature between 50 and 60°. After th[?] exothermic reaction is complete (about 10 min) allow the mixture t[?] cool spontaneously to 35° and then dilute with 50 ml of water[?] Separate the layers (test to determine which is the organic layer)[?] extract the aqueous layer once with 10 ml of ether, and wash th[?] combined organic layer and ether extract once with 20 ml of water[?] Dry the cloudy organic layer by shaking it with anhydrous sodiun[?] sulfate until the liquid is clear. Decant into a 50-ml, round-bottome[?] flask and distil the mixture (boiling chip), first from a steam bath t[?] remove ether and some unreacted cyclohexene and chloroform, ther[?] over a free flame. Collect 7,7-dichlorobicyclo[4.1.0]heptane ove[?] the range 195–200°. Yield 9 g.

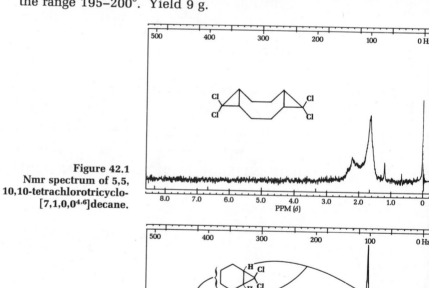

Figure 42.1
Nmr spectrum of 5,5,
10,10-tetrachlorotricyclo-
[7,1,0,0⁴·⁶]decane.

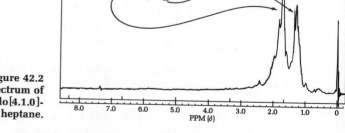

Figure 42.2
Nmr spectrum of
7,7-dichlorobicyclo[4.1.0]-
heptane.

[4] Prepared by adding 20.2 g (0.2 mole) of triethylamine to 25.3 g (0.2 mole) of benzyl chloride (α-chlorotoluene) in 50 ml of benzene with stirring. The benzene is removed at the water aspirator and the dry product pulverized in a mortar and stored in a desiccator over anhydrous calcium chloride and paraffin (to remove traces of benzene).

43 p-Chlorotoluene

KEYWORDS   Sandmeyer reaction   Diazonium salt
                Cuprous chloride   Toluidine
                Nitrous acid

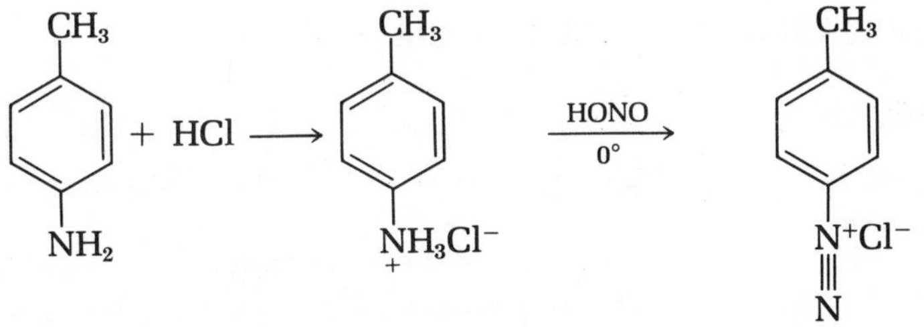

p-Toluidine
MW 107.15, mp 45°

**p-Toluidine
hydrochloride**

**p-Methylbenzenediazonium
chloride**

$$NaNO_2 + HCl \xrightarrow{0°} HONO + NaCl$$

Sodium nitrite
MW 69.01

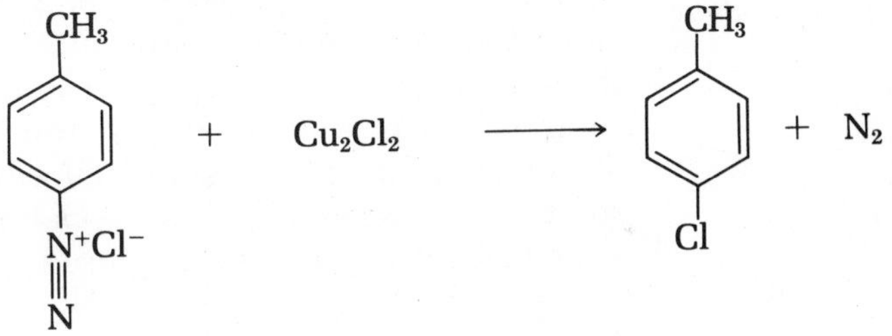

Cuprous chloride
MW 198.05

p-Chlorotoluene
MW 126.58, bp 162°

The Sandmeyer reaction is a versatile means of replacing an aromatic amine group with a chlorine or bromine atom, through reaction of the cuprous halide with a diazonium salt.

p-Toluidine is dissolved in the required amount of hydrochloric acid, two more equivalents of acid are added, and the mixture cooled in ice to produce a paste of the crystalline amine hydrochloride. When this is treated at 0–5° with one equivalent of sodium nitrite, nitrous acid is liberated and reacts to produce the diazonium salt. The hydrochloric acid beyond the two equivalents required to form the amine hydrochloride and react with sodium nitrite is to maintain acidity sufficient to prevent formation of the diazoamino compound and rearrangement of the diazonium salt.

Cuprous chloride is made by reduction of copper(II) sulfate with sodium sulfite (which is produced as required from the cheaper sodium bisulfite). The white solid is left covered with the reducing solution for protection against air oxidation until it is to be used and then dissolved in hydrochloric acid. On addition of the diazonium salt solution a complex forms and rapidly decomposes to give p-chlorotoluene and nitrogen. The mixture is badly discolored, but steam distillation leaves most of the impurities and all salts behind and gives material substantially pure except for the presence of a trace of yellow pigment which can be eliminated by distillation of the dried oil.

EXPERIMENTS   **1. Cuprous Chloride Solution**

$$2 \text{ CuSO}_4 \cdot 5\text{H}_2\text{O} + 4 \text{ NaCl} + \text{NaHSO}_3 + \text{NaOH} \rightarrow \text{Cu}_2\text{Cl}_2 + 3 \text{ Na}_2\text{SO}_4 + 2 \text{ HCl} + 10 \text{ H}_2\text{O}$$

MW 249.71   MW 58.45   MW 104.97   MW 40.01   MW 198.05

NaHSO$_3$ *and* Na$_2$S$_2$O$_4$

Stop here

In a 500-ml round-bottomed flask (to be used later for steam distillation) dissolve 30 g of copper(II) sulfate crystals (CuSO$_4$ · 5H$_2$O) in 100 ml of water by boiling and then add 10 g of sodium chloride, which may give a small precipitate of basic copper(II) chloride. Prepare a solution of sodium sulfite from 7 g of sodium bisulfite, 4.5 g of sodium hydroxide, and 50 ml of water and add this, not too rapidly, to the hot copper(II) sulfate solution (rinse flask and neck). Shake well and put the flask in a pan of cold water in a slanting position favorable for decantation and let the mixture stand to cool and settle while doing the diazotization. When you are ready to use the cuprous chloride, decant the supernatant liquid, wash the white solid once with water by decantation, and dissolve the solid in 45 ml of concd hydrochloric acid to form the double compound. The solution is susceptible to air oxidation and should not stand for an appreciable time before being used.

2. Diazotization

Put 11.0 g of p-toluidine and 15 ml of water in a 125-ml Erlenmeyer flask. Measure 25 ml of concd hydrochloric acid and add 10 ml of it to the flask. Heat over a free flame and swirl to dissolve the amine and hence insure that it is all converted into the hydrochloride. Add the rest of the acid and cool thoroughly in an ice bath and let the flask stand in the bath while preparing a solution of 7 g of sodium nitrite in 20 ml of water. To maintain a temperature of 0–5° during diazotization add a few pieces of ice to the amine hydrochloride suspension and add more later as the first ones melt. Pour in the nitrite solution in portions during 5 min with swirling in the ice bath. The solid should dissolve to a clear solution of the diazonium salt. After 3–4 min test for excess nitrous acid: dip a stirring rod in the solution, touch off the drop on the wall of the flask, put the rod in a small test tube, and add a few drops of water.

Then insert a strip of starch-iodide paper; an instantaneous deep blue color due to a starch-iodine complex indicates the presence of nitrous acid. (The sample tested is diluted with water because strong hydrochloric acid alone produces the same color on starch-iodide paper, after a slight induction.) Leave the solution in the ice bath.

3. Sandmeyer Reaction

Complete the preparation of cuprous chloride solution, cool it in the ice bath, pour in the solution of diazonium chloride through a long-stemmed funnel and rinse the flask. Swirl occasionally at room temperature for 10 min and observe initial separation of a complex of the two components and its decomposition with liberation of nitrogen and separation of an oil. Arrange for steam distillation (Fig. 9.2) but do not start until bubbling in the mixture has practically ceased and an oily layer has separated. Then steam distil and note that p-chlorotoluene, although lighter than the solution of inorganic salts in which it was produced, is heavier (den 1.07) than water. Extract the distillate with a little ether, wash the extract with 10% sodium hydroxide solution to remove any p-cresol present, then wash with saturated sodium chloride solution; filter through anhydrous sodium sulfate into a tared flask, evaporate the ether, and determine the yield and percentage yield of product (about 9 g).

Caution! Use aspirator tube

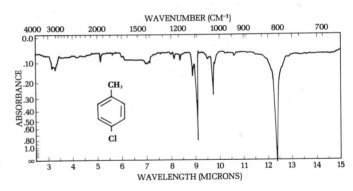

Figure 43.1
Infrared spectrum of
***p*-chlorotoluene in CS₂.**

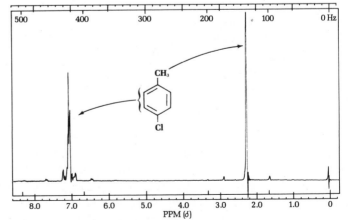

Figure 43.2
Nmr spectrum of
***p*-chlorotoluene.**

44 *Luminol*

KEYWORDS   Chemiluminescence, fluorescence, phosphorescence
Singlet, triplet, intersystem crossing
Excited state, ground state
Phthalic acid, hydrazine
Triethylene glycol, high bp solvent
Reduction, sodium hydrosulfite

The oxidation of 5-aminophthalhydrazide, **1**, commonly known as luminol, is attended with a striking emission of blue-green light. Most exothermic chemical reactions produce energy in the form of heat, but a few produce light and release little or no heat. This phenomenon, chemiluminescence, is usually an oxidation reaction. In the case of luminol an alkaline solution of the compound is allowed to react with a mixture of hydrogen peroxide and potassium ferricyanide. The dianion **2** is oxidized to the triplet excited state (two unpaired electrons of like spin) of the amino phthalate ion, **3**.

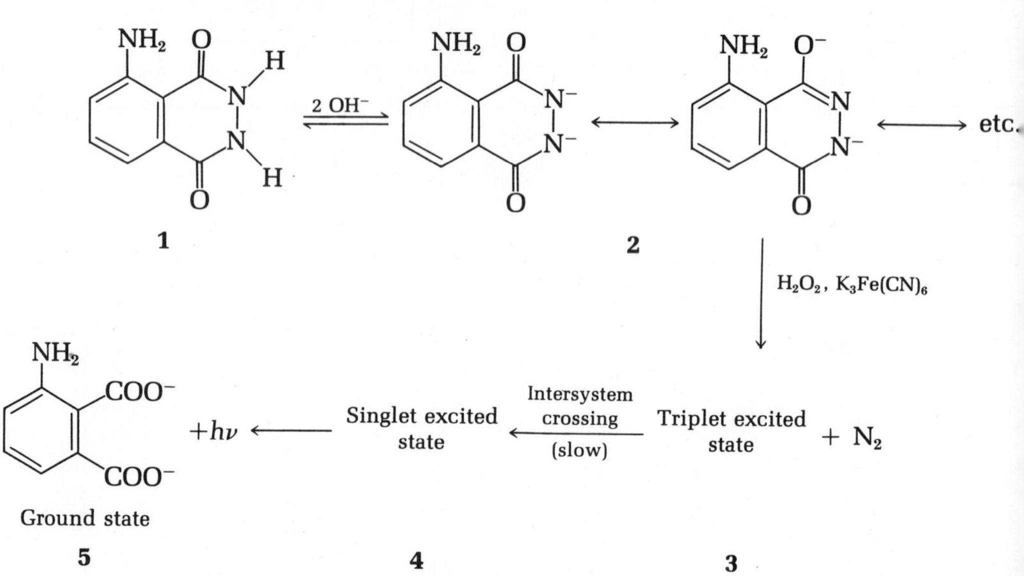

This slowly undergoes intersystem crossing to the singlet excited state (two unpaired electrons of opposite spin) **4**, which decays to the ground state ion, **5**, with the emission of one quantum of light (a photon) per molecule. Very few molecules are more efficient in chemiluminescence than luminol.

Luminol, **1**, is made by reduction of the nitro derivative, **8**, formed on thermal dehydration of a mixture of 3-nitrophthalic acid, **6**, and hydrazine, **7**.

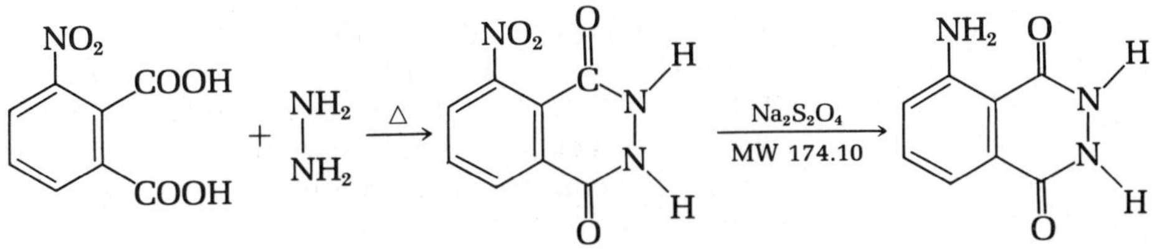

6	**7**	**8**	**1**
Mp 222°,			Mp 332°,
MW 211.13			MW 177.16

An earlier procedure for effecting the first step called for addition of hydrazine sulfate to an alkaline solution of the acid, evaporation to dryness, and baking the resulting mixture of the hydrazine salt and sodium sulfate at 165°, and required a total of 4.5 hr for completion. This working time can be drastically reduced by adding high-boiling (bp 290°) triethylene glycol to an aqueous solution of the hydrazine salt, distilling the excess water, and raising the temperature to a point where dehydration to **8** is complete within a few minutes. Nitrophthalhydrazide, **8**, is insoluble in dilute acid, but soluble in alkali, by virtue of enolization, and is conveniently reduced to luminol, **1**, by sodium hydrosulfite (sodium dithionite) in alkaline solution. In dilute, weakly acidic or neutral solution luminol exists largely as the dipolar ion **9** which exhibits beautiful blue fluorescence.[1]

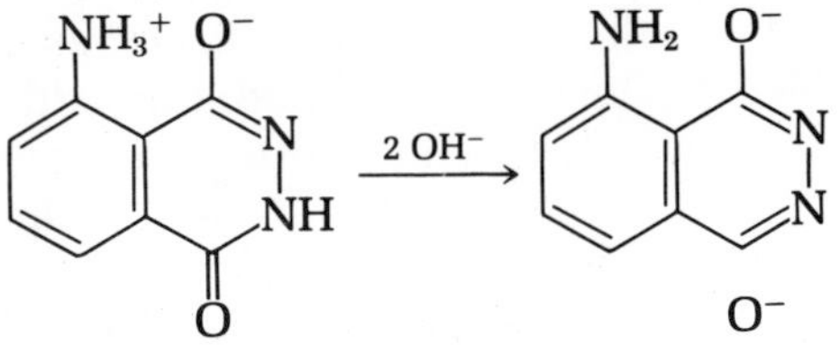

9	**10**

An alkaline solution contains the doubly enolized anion **10** and displays particularly marked chemiluminescence when oxidized with a combination of hydrogen peroxide and potassium ferricy-anide.

EXPERIMENT   First put a flask containing 15 ml of water on the steam bath to get hot. Then heat a mixture of 1 g of 3-nitrophthalic acid and 2 ml of

[1] Several methods of demonstrating the chemiluminescence of luminol are described by E. H. Huntress, L. N. Stanley, and A. S. Parker, *J. Chem. Ed.*, **11**, 142 (1934). The mechanism of the reaction has been investigated by Emil H. White and co-workers. (*J. Amer. Chem. Soc.*, **86**, 940 and 942 (1964)).

*The two-step
synthesis of a
chemiluminescent
substance can be
completed in 25 min*

an 8% aqueous solution of hydrazine[2] in a 20 × 150-mm test tube
over a free flame until the solid is dissolved, add 3 ml of triethylene
glycol, and clamp the tube in a vertical position above a microburner.
Insert a thermometer, a boiling chip, and an aspirator tube connected to a suction pump, and boil the solution vigorously to distil
the excess water (110–130°). Let the temperature rise rapidly until
(3–4 min) it reaches 215°. Remove the burner, note the time, and by
intermittent gentle heating maintain a temperature of 215–220° for
2 min. Remove the tube, cool to about 100° (crystals of the product
often appear), add the 15 ml of hot water, cool under the tap, and
collect the light yellow granular nitro compound (**8**). (Dry weight,
0.7 g).[3]

The nitro compound need not be dried and can be transferred at
once, for reduction, to the uncleaned test tube in which it was prepared. Add 5 ml of 10% sodium hydroxide solution, stir with a rod,
and to the resulting deep brown-red solution add 3 g of sodium
hydrosulfite dihydrate. Wash the solid down the walls with a little
water. Heat to the boiling point, stir, and keep the mixture hot for
5 min, during which time some of the reduction product may separate. Then add 2 ml of acetic acid, cool under the tap, and stir;
collect the resulting precipitate of light yellow luminol (**1**). The
filtrate on standing overnight usually deposits a further crop of
luminol (0.1–0.2 g).

$Na_2S_2O_4 \cdot 2 H_2O$
*Sodium hydrosulfite
dihydrate*
MW 210.15

*The light-producing
reaction*

Dissolve the first crop of moist luminol (dry weight, 0.2–0.3 g) in
10 ml of 10% sodium hydroxide solution and 90 ml of water; this
is stock solution A. Prepare a second stock solution B by mixing
20 ml of 3% aqueous potassium ferricyanide, 20 ml of 3% hydrogen
peroxide, and 160 ml of water. Now dilute 25 ml of solution A with
175 ml of water and, in a relatively dark corner or cupboard, pour
this solution and solution B simultaneously into a funnel resting in
the neck of a large Erlenmeyer flask. Swirl the flask and, to increase
the brilliance, gradually add further small quantities of alkali.

Ultrasonic sound can also be used to promote this reaction.
Prepare stock solutions A and B again but omit the hydrogen peroxide. Place the combined solutions in an ultrasonic cleaning bath
or immerse an ultrasonic probe into the reaction mixture. Spots of
light are seen where the ultrasonic vibrations produce hydroxyl
radicals.

---

[2] Dilute 31.2 g of the commercial 64% hydrazine solution to a volume of 250 ml.
[3] The reason for adding hot water and then cooling rather than adding cold water is that the solid
is then obtained in more easily filterable form.

45 Acetylsalicylic Acid (Aspirin)

KEYWORDS   Salicyclic acid
Acetylation catalysts
Sodium acetate, pyridine, boron trifluoride etherate, H₂SO₄
Commercial aspirin

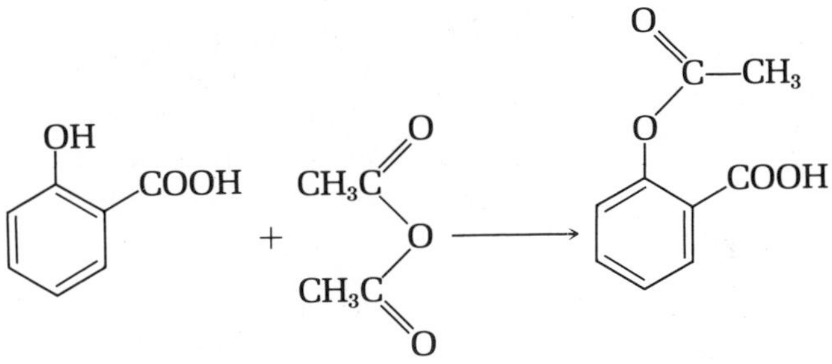

Salicyclic acid
MW 138.12, mp 159°   **Acetic anhydride**
MW 102.09, bp 140°   **Acetylsalicylic acid**
MW 180.15, mp 128–137°

This experiment demonstrates the action of four acetylation catalysts: two bases, sodium acetate and pyridine; a Lewis acid, boron trifluoride; and a mineral acid, sulfuric.

EXPERIMENT

Acetylation Catalysts

(1) CH₃COONa

(2)

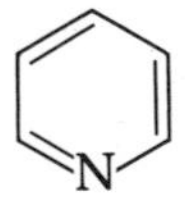

(3) BF₃·(C₂H₅)₂O

(4) H₂SO₄

Place 1 g of salicylic acid in each of four 13 × 100-mm test tubes and add to each tube 2 ml of acetic anhydride. To the first tube add 0.2 g of anhydrous sodium acetate, note the time, stir with a thermometer, and record the time required for a 4° rise in temperature and the estimated proportion of solid that has dissolved. Replace the thermometer and continue to stir occasionally while starting the next acetylation. Obtain a clean thermometer, put it in the second tube, add 5 micro drops of pyridine, observe as before, and compare with the first results. To the third and fourth tubes add 5 micro drops of boron trifluoride etherate[1] and 5 micro drops of concd sulfuric acid, respectively. What is the order of activity of the four catalysts as judged by the rates of the reactions?

Put all the tubes in a beaker of hot water for 5 min to dissolve solid and complete the reactions, and then pour all the solutions

Note for the instructor

[1] Commercial reagent if dark should be redistilled (bp 126°, water-white).

into a 125-ml Erlenmeyer containing 50 ml of water and rinse the
tubes with water. Swirl to aid hydrolysis of excess acetic anhydride
and then cool thoroughly in ice, scratch, and collect the crystalline
solid; yield 4 g.

Acetylsalicylic acid melts with decomposition at temperatures
reported from 128 to 137°. It can be crystallized by dissolving it in
ether, adding an equal volume of petroleum ether, and letting the
solution stand undisturbed in an ice bath.

Test the solubility of your sample in benzene and in hot water and
note the peculiar character of the aqueous solution when it is cooled
and when it is then rubbed against the tube with a stirring rod. Note
also that the substance dissolves in cold sodium bicarbonate solution
and is precipitated by addition of an acid. Compare a tablet of commercial aspirin with your sample. Test the solubility of the tablet in
water and in benzene and observe if it dissolves completely. Compare its behavior when heated in a melting point capillary with the
behavior of your sample. If an impurity is found present, it is probably some substance used as binder for the tablets. It is organic or
inorganic? What harmless, edible type of substance do you suppose
it is, judging from the various properties? Test with a drop of
iodine-potassium iodide solution.

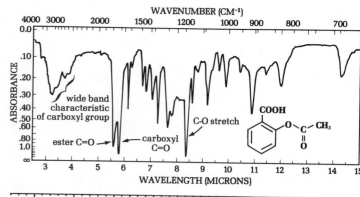

**Figure 45.1
Infrared spectrum of
acetylsalicylic acid
(aspirin) in CHCl₃.**

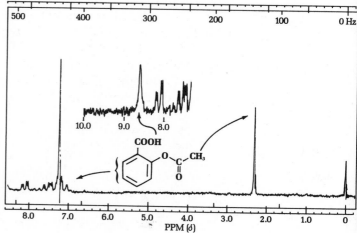

**Figure 45.2
Nmr spectrum of acetyl-
salicylic acid (aspirin).**

46 *Derivatives of 1,2-Diphenylethane*[1]

KEYWORDS   Benzoin condensation
Cyanide ion catalyst
Nitric acid oxidation
Quinoxaline
Sodium borohydride reduction
Reductive acetylation
Thionyl chloride
Dehydrohalogenation

Pyridinium hydro-
   bromide perbromide
Triethylene glycol
Perkin condensation
Decarboxylation
Copper chromite catalyst
Quinoline
Steric inhibition of resonance

Procedures are given in this chapter for rapid preparation of small samples of twelve related compounds starting with benzaldehyde and phenylacetic acid by transformations (see the following page). The quantities of reagents specified in the procedures are such as to provide somewhat more of each intermediate than is required for completion of all subsequent steps in the sequences of synthesis. If the experiments are dovetailed, the entire series of preparations can be completed in very short total working time. For example, one can start the preparation of benzoin (record the time of starting and do not rely on memory), and during the 30-min reflux period start the preparation of α-phenylcinnamic acid; this requires refluxing for 35 min, and while it is proceeding the benzoin preparation can be stopped when the time is up and the product let crystallize. The α-phenylcinnamic acid mixture can be let stand (and cool) until one is ready to work it up. Also, while a crystallization is proceeding one may want to observe the crystals occasionally but should utilize most of the time for other operations.

Points of interest concerning stereochemistry and reaction mechanisms are discussed in the introductions to the individual procedures. Since several of the compounds have characteristic ultraviolet or infrared absorption spectra, pertinent spectroscopic constants are recorded and brief interpretations of the data are presented (Figs. 46.1–46.11).

---

[1] If the work is well organized and proceeds without setbacks, the experiments can be completed in about four laboratory periods. The instructor may elect to name a certain number of periods in which the student is to make as many of the compounds as he can; he may also decide to require submission only of the end products in each series.

Note for the instructor

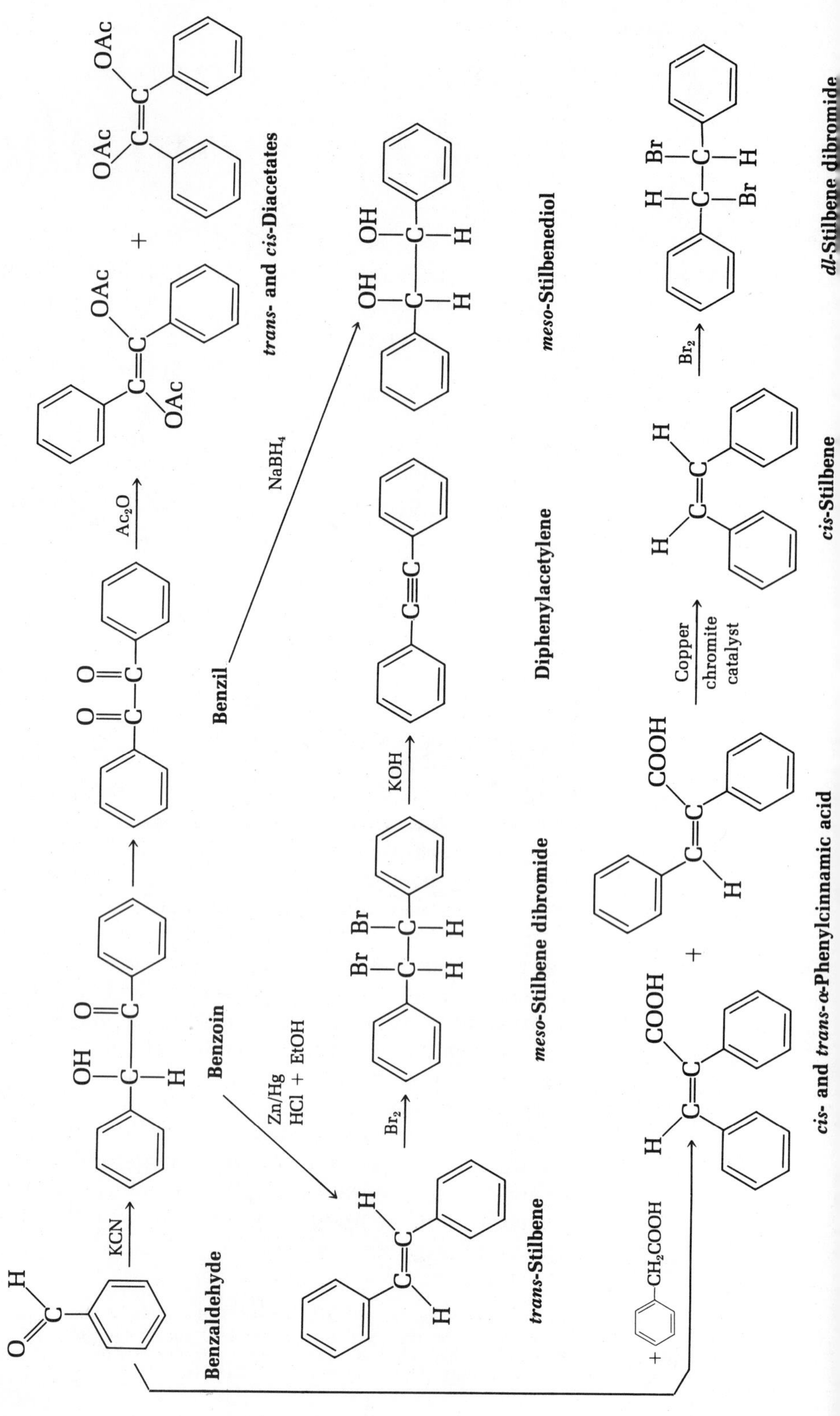
Benzaldehyde
KCN
Benzoin
Benzil
Ac_2O
trans- and *cis*-Diacetates
$NaBH_4$
meso-Stilbenediol
Zn/Hg
HCl + EtOH
trans-Stilbene
Br_2
meso-Stilbene dibromide
KOH
Diphenylacetylene
CH_2COOH
cis- and *trans*-α-Phenylcinnamic acid
Copper chromite catalyst
cis-Stilbene
Br_2
dl-Stilbene dibromide

EXPERIMENTS   **1. Benzoin**

Condensation of two moles of benzaldehyde under the specific catalytic influence of cyanide ion affords *dl*-benzoin:

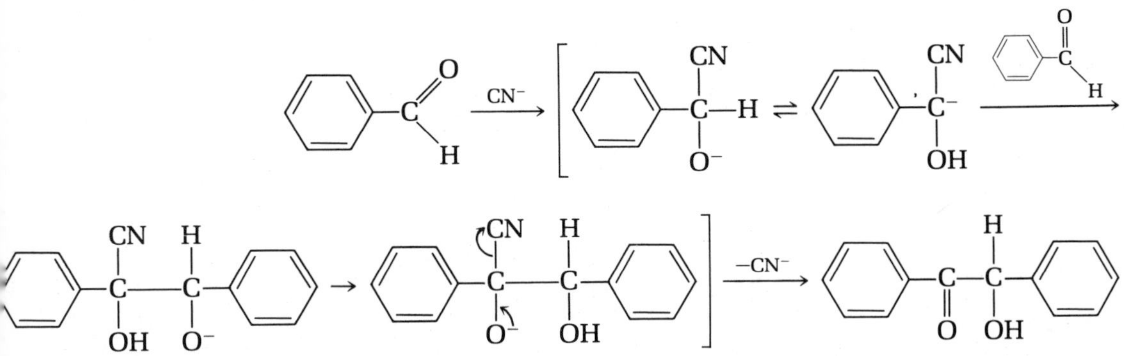

Benzoin
Mp 135°, MW 212.24

Being an α-ketol (compare D-fructose), benzoin reduces Fehling solution (in alcoholic solution) and forms an osazone (mp 225°); the 2,4-dinitrophenylhydrazone forms orange-yellow plates from ethanol, mp 239°.

Procedure

Place 1.5 g of potassium cyanide (**poison, do not spill!**) in a 125-ml round-bottomed flask, dissolve it in 15 ml of water, add 30 ml of 95% ethanol and 15 ml of pure benzaldehyde,[2] introduce a boiling stone, attach a short condenser, and reflux the solution gently on the steam bath for 30 min. Remove the flask and, if no crystals appear within a few minutes, withdraw a drop on a stirring rod and rub it against the neck of the flask to induce crystallization. When crystallization is complete, collect the product and wash it free of yellow mother liquor with a 1:1 mixture of 95% ethanol and water (wash the cyanide-containing mother liquor down the sink with plenty of water). Usually this first-crop material is colorless and of satisfactory melting point (134–135°); yield 10–12 g.[3]

Reaction time: 30 min

2. Preparation of Benzil and a Derivative*

Benzoin can be oxidized to the α-diketone, benzil, very efficiently by nitric acid or by copper sulfate in pyridine. On oxidation with

[2] Commercial benzaldehyde inhibited against autoxidation with 0.1% hydroquinone is usually satisfactory. If the material available is yellow or contains benzoic acid it should be shaken with equal volumes of 5% sodium carbonate solution until carbon dioxide is no longer evolved and the upper layer dried over calcium chloride and distilled (bp 178–180°), with avoidance of exposure of the hot liquid to air.

[3] Concentration of the mother liquor to a volume of 20 ml gives a second crop (1.8 g, mp 133–134.5°); best total yield 13.7 g (87%). Recrystallization can be accomplished with either methanol (11 ml/g) or 95% ethanol (7 ml/g) with 90% recovery in the first crop.

* Equation for the preparation of derivative on top page 264.

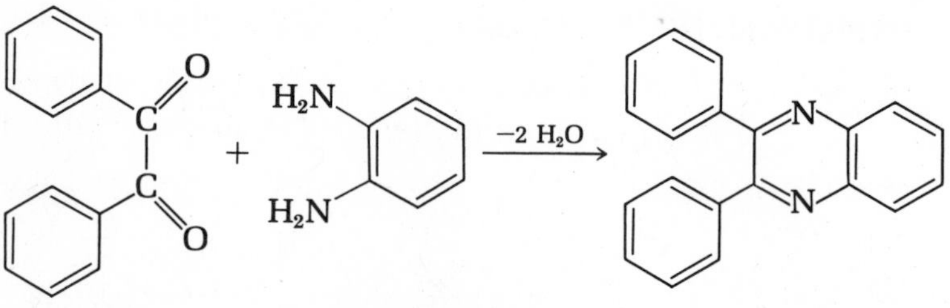

Benzil
Mp 96°, MW 210.22

o-**Phenylenediamine**
Mp 103°,
MW 108.14

Quinoxaline
derivative
Mp 126°, MW 282.33

sodium dichromate in acetic acid the yield is lower because some of the material is converted into benzaldehyde by cleavage of the bond between two oxidized carbon atoms and activated by both phenyl groups (a). Similarly, hydrobenzoin on oxidation with dichromate or permanganate yields chiefly benzaldehyde and only a trace of benzil (b).

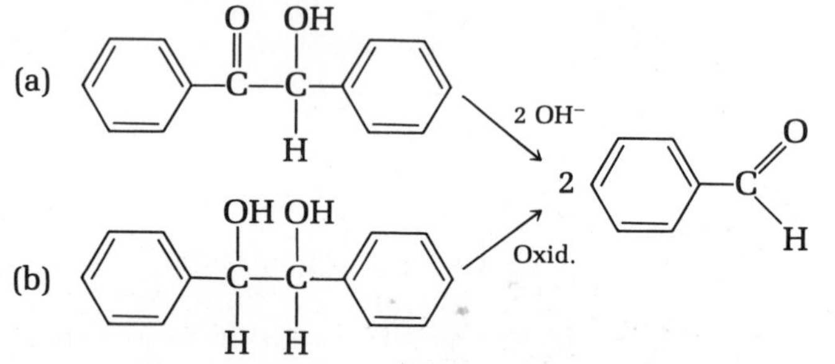

A reaction that characterizes benzil as an α-diketone is a condensation reaction with *o*-phenylenediamine to the quinoxaline derivative. The aromatic heterocyclic ring formed in the condensation is fused to a benzene ring to give a bicyclic system analogous to naphthalene.

Procedure (oxidation)

Reaction time: 11 min

Heat a mixture of 4 g of benzoin and 14 ml of concd nitric acid on the steam bath for 11 min. Use an aspirator tube near the top of the flask to remove nitrogen oxides. Add 75 ml of water to the reaction mixture, cool to room temperature, and swirl for a minute or two to coagulate the precipitated product; collect and wash the yellow solid on a Hirsch funnel, pressing the solid well on the filter to squeeze out the water. The crude product (dry weight 3.7–3.9 g) need not be dried but can be crystallized at once from ethanol. Dissolve the product in 10 ml of hot ethanol, add water dropwise to the cloud point, and set aside to crystallize. Record the yield, crystalline form, color, and mp of the purified benzil.

Figure 46.1
The ultraviolet spectrum of benzoin. λ_{max}^{EtOH} 247 nm ($\epsilon = 13{,}200$). Concentration: 12.56 g/l = 5.92 × 10^{-5} mole/l. See Chapter 15 (Ultraviolet Spectroscopy) for the relationship between the extinction coefficient, ϵ, absorbance, A, and concentration, C. The absorption band at 247 nm is attributable to the presence of the phenyl ketone group, C_6H_5—$\overset{\overset{O}{\|}}{C}$—, in which the carbonyl group is conjugated with the benzene ring. Aliphatic α,β-unsaturated ketones, R—CH=CH—C=O, show selective absorption of ultraviolet light of comparable wavelength.

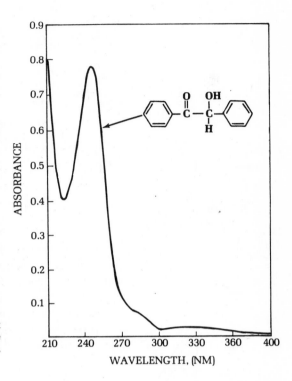

ABSORBANCE

WAVELENGTH, (NM)

Test for the Presence of Unoxidized Benzoin. Dissolve about 0.5 mg of crude or purified benzil in 0.5 ml of 95% ethanol or methanol and add one drop of 10% sodium hydroxide. If benzoin is present the solution soon acquires a purplish color owing to a complex of benzil with a product of autoxidation of benzoin. If no color develops in 2–3 min, an indication that the sample is free from benzoin, add a small amount of benzoin, observe the color that develops, and note that if the test tube is stoppered and shaken vigorously, end-to-end, the color momentarily disappears; when the solution is then let stand, the color reappears.

Benzil Quinoxaline Preparation

Commercial *o*-phenylenediamine is usually badly discolored (air oxidation) and gives a poor result unless purified. Before using in this procedure, purify as follows. Place 200 mg of material in the bottom of a 20 × 150-mm test tube, evacuate the tube at the maximum suction of the aspirator, clamp it in a horizontal position, and heat the bottom part of the tube with a free flame to distil or sublime colorless *o*-phenylenediamine away from the dark residue into the upper half of the tube. Let the tube cool in position until the melt has solidified, and scrape out the white solid.

Weigh 0.2 g of benzil (theory = 210 mg) and 0.1 g of your purified *o*-phenylenediamine (theory = 108 mg) into a 20 × 150-mm test tube and heat in the rings of a steam bath for 10 min, by which time the

Reaction time: 10 min

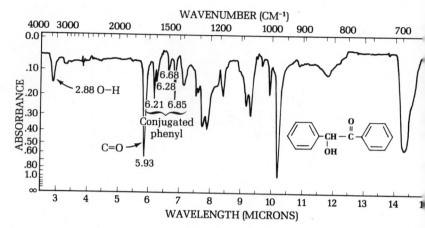

Figure 46.2
Infrared spectrum of benzoin in CHCl₃.

initially molten mixture should have changed to a light tan solid
Dissolve the solid in hot methanol (about 5 ml) and let the solution
stand undisturbed. If crystallization does not occur within 10 min
reheat the solution and dilute it with a little water to the point of
saturation. The crystals should be filtered as soon as formed, for
brown oxidation products accumulate on standing. The quinoxaline
forms colorless needles, mp 125–126°; yield 185 mg.

Benzil
MW 210.22

***meso*-Hydrobenzoin**
Mp 137°, MW 214.25

***dl*-Hydrobenzoin**
Mp 120°

3. Reduction of Benzil

Addition of two atoms of hydrogen to benzoin or of four atoms of
hydrogen to benzil gives a mixture of stereoisomeric diols, of which
the predominant isomer is the nonresolvable *meso*-hydrobenzoin.
Reduction is accomplished rapidly with sodium borohydride
($Na^+BH_4^-$) in ethanol. The high cost of the reagent is offset by its
low molecular weight and the fact that one mole of hydride reduces
four moles of a ketone.

$$4 R_2C{=}O + Na^+BH_4^- \rightarrow (R_2CHO)_4BNa$$
$$(R_2CHO)_4BNa + 2 H_2O \rightarrow 4 R_2CHOH + NaBO_2$$

The procedure that follows specifies use of benzil rather than benzoin because the progress of the reduction can then be followed by

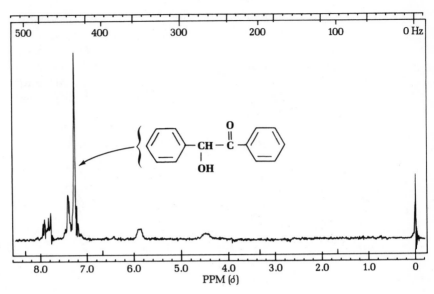

Figure 46.3
Nmr spectrum of
benzoin.

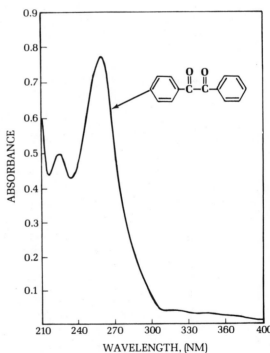

Figure 46.4
Ultraviolet spectrum of
benzil. λ_{max}^{EtOH} **260 nm**
($\epsilon = 19{,}800$). One-cm
cells and 95% ethanol
have been employed for
all the UV spectra in
this chapter.

the discharge of the yellow color. The amount of reagent required for reduction of 500 mg of benzil is calculated as follows:

$$500 \text{ mg} \times \frac{37.85}{210.22} \times \frac{2}{4} = 45 \text{ mg NaBH}_4$$

Procedure

In a 50-ml Erlenmeyer flask, dissolve 0.5 g of benzil in 5 ml of 95% ethanol and cool the solution under the tap to produce a fine sus-

Reaction time: 10 min

pension. Then add 0.1 g of sodium borohydride (large excess). The benzil dissolves, the mixture warms up, the yellow color disappears in 2–3 min. After a total of 10 min, add 5 ml of water, heat to the boiling point, filter in case the solution is not clear, dilute to the point of saturation with more water (10 ml), and set the solution aside to crystallize. meso-Hydrobenzoin separates in lustrous thin plates mp 136–137°; yield 0.35 g.

4. Reductive Acetylation of Benzil

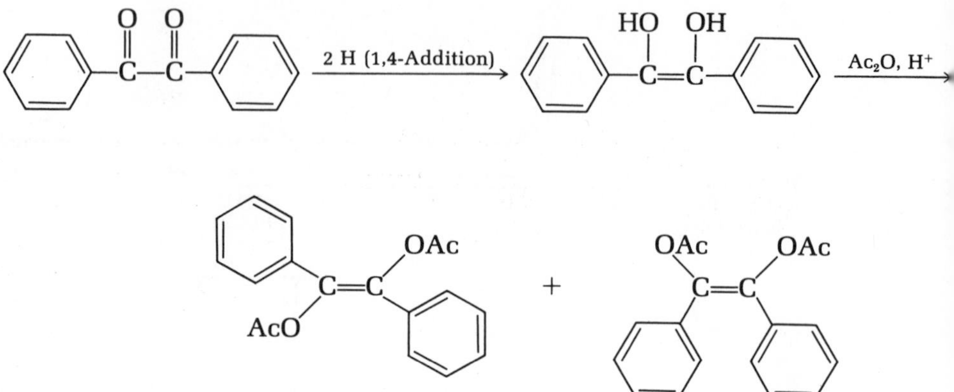

$$2 H (1,4\text{-Addition}) \longrightarrow$$

$$Ac_2O, H^+ \longrightarrow$$

+

1

(*trans*)
Mp 155°, λ_{max}^{EtOH} 271nm ($\epsilon = 23{,}400$),
MW 296.31

2

(*cis*)
Mp 119°, λ_{max}^{EtOH} 265nm ($\epsilon = 12{,}800$),
MW 296.31

In one of the first demonstrations of the phenomenon of 1,4-addition, Johannes Thiele (1899) established that reduction of benzil with zinc dust in a mixture of acetic anhydride-sulfuric acid involves 1,4-addition of hydrogen to the α-diketone grouping and acetylation of the resulting enediol before it can undergo ketonization to benzoin. The process of reductive acetylation results in a mixture of the cis and trans isomers **1** and **2**. Thiele and subsequent investigators isolated the more soluble, lower-melting cis-stilbenediol diacetate (**2**) in only impure form, mp 110°. Separation of the two isomers by chromatography is not feasible because they have the same degree of adsorbability on alumina. However, separation is possible by fractional crystallization (described in the following procedure) and both isomers are isolated in pure condition. In the method prescribed here for the preparation of the isomer mixture, hydrochloric acid is substituted for sulfuric acid because the latter acid gives rise to colored impurities and is reduced to sulfur and to hydrogen sulfide.[4]

The configurations of this pair of geometrical isomers remained unestablished for over 50 years, but the tentative inference that the

---

[4] If acetyl chloride (2 ml) is substituted for the hydrochloric acid-acetic anhydride mixture in the procedure, the cis-isomer is the sole product.

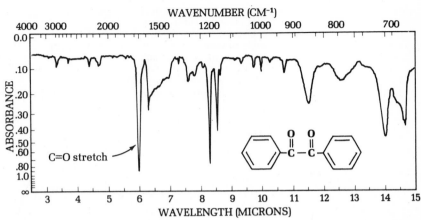

**Figure 46.5
Infrared spectrum of
benzil in CHCl₃.**

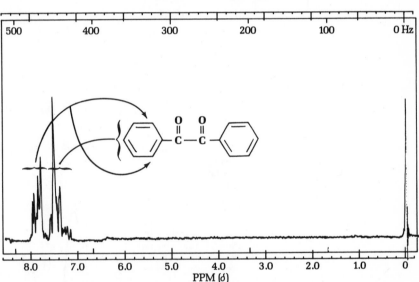

**Figure 46.6
Nmr spectrum of benzil.**

higher-melting isomer has the more symmetrical *trans* configuration
eventually was found to be correct. Evidence of infrared spectroscopy is of no avail; the spectra are nearly identical in the inter-
pretable region (2–8 μ) characterizing the acetoxyl groups, but differ
in the fingerprint region (8–12 μ). However, the isomers differ
markedly in ultraviolet absorption (Fig. 46.8) and, in analogy to
trans- and *cis*-stilbene (Section 5), the conclusion is justified that
the higher-melting isomer, since it has an absorption band at longer
wavelength and higher intensity than its isomer, does indeed have
the configuration **1**.

Procedure

Place one test tube (20 × 150-mm) containing 7 ml of acetic
anhydride and another (13 × 100-mm) containing 1 ml of concd
hydrochloric acid in an ice bath and, when both are thoroughly
chilled, transfer the acid to the anhydride dropwise in not less than

Reaction time: 10 min

Figure 46.7
Ultraviolet spectrum of
quinoxaline derivative.
λ_{max}^{EtOH} **244 nm ($\epsilon = 37,400$),**
345 nm ($\epsilon = 12,700$).
Spectrum recorded on
Cary Model 17
spectrometer.

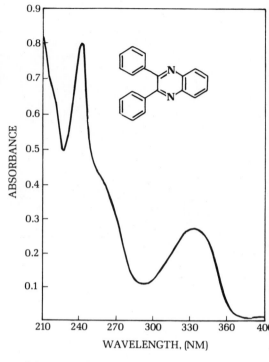

Figure 46.8
Ultraviolet spectra of
cis- **and** *trans-***stilbene**
diacetate. *cis:* λ_{max}^{EtOH} **223**
nm ($\epsilon = 20,500$), 269 nm
($\epsilon = 10,800$); *trans:*
λ_{max}^{EtOH} **272 nm ($\epsilon = 20,800$).**
In the *cis-***diacetate the**
two phenyl rings can not
be coplanar which pre-
vents overlap of the
*p-***orbitals of the phenyl**
groups with those of the
central double bond.
This steric inhibition of
resonance accounts for
the diminished intensity
of the *cis-***isomer relative**
to the *trans.* **Both spectra**
were run at the same
concentration.

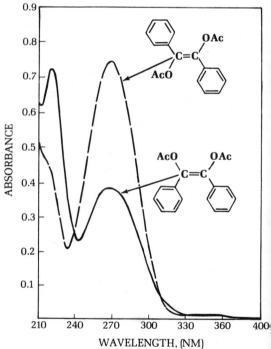

one minute by means of a capillary dropping tube. Wipe the test
tube dry, pour the chilled solution into a 50-ml Erlenmeyer flask
containing 1 g of pure benzil and 1 g of zinc dust, and swirl for 2–3

min in an ice bath. Remove the flask and hold it in the palm of the hand; if it begins to warm up, cool further in ice. When there is no further exothermic effect, let the mixture stand for 5 min and then add 25 ml of water. Swirl, break up any lumps of product, and allow a few minutes for hydrolysis of excess acetic anhydride. Then collect the mixture of product and zinc dust, wash with water, and press and apply suction to the cake until there is no further drip. Digest the solid (drying is not necessary) with 70 ml of ether to dissolve the organic material, add about 4 g of anhydrous sodium sulfate, and swirl briefly; filter the solution into a 125-ml Erlenmeyer flask, concentrate the filtrate (steam bath, boiling stone, water aspirator) to a volume of approximately 15 ml,[5] and let the flask stand, corked and undisturbed.

The *trans*-diacetate (**1**) soon begins to separate in prismatic needles, and after 20–25 min crystallization appears to have stopped. Removal and evaporation of the mother liquor can be done employing a rotary evaporator or as follows: Warm a 25-ml Erlenmeyer flask on the steam bath under an aspirator tube and, with use of a capillary dropping tube, suck up a portion of the mother liquor (ethereal solution) covering the crystals and transfer it, by drops, to the warm evaporation flask; if the ether is caused to evaporate at the same rate as the solution is added, the solvent can be eliminated rapidly without need for a boiling stone. When all the solution has been transferred, wash the remaining crystals with a little fresh ether and transfer the washings to the warm flask as previously. The crystals of *trans*-diacetate (**1**) can then be scraped out of the larger flask and the weight and mp determined (e.g., 294 mg, mp 154–156°).

Dissolve the white solid left in the evaporation flask in 10 ml of methanol, let the solution stand undisturbed for about 10 min, and drop in one tiny crystal of the *trans*-diacetate (**1**). This should give rise, in 20–30 min, to a second crop of the *trans*-diacetate (e.g., 58 mg, mp 153–156°). Then concentrate the mother liquor and washings to a volume of 7–8 ml, let cool to room temperature as before, and again seed with a crystal of *trans*-diacetate; this usually affords a third crop of the *trans*-diacetate (e.g., 62 mg, mp 153–155°).

At this point the mother liquor should be rich enough in the more soluble *cis*-diacetate (**2**) for its isolation. Concentrate the methanol mother liquor and washings from the third crop of **1** to a volume of 4–5 ml, stopper the flask, and let the solution stand undisturbed overnight. The *cis*-diacetate (**2**) sometimes separates spontaneously in large rectangular prisms of great beauty. If the solution remains supersaturated, addition of a seed crystal of **2** causes prompt separation of the *cis*-diacetate in a paste of small crystals (e.g., 215 mg, mp 118–119°; then: 70 mg, mp 116–117°).

---

[5] Measure 15 ml of a solvent into a second flask of the same size and compare the levels in the two flasks.

5. *trans*-Stilbene

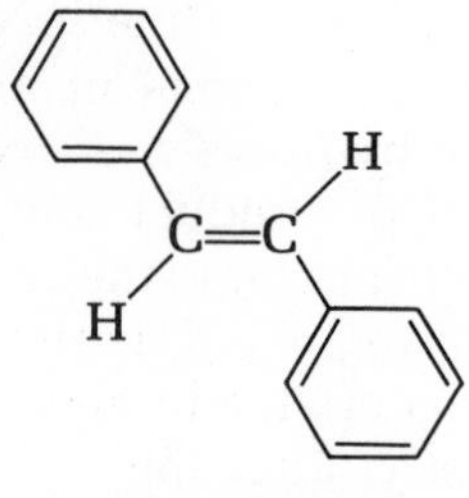

trans-Stilbene
Mp 125°, MW 180.24
λ_{max}^{EtOH}301nm ($\epsilon = 28,500$)
226nm ($\epsilon = 17,700$)
Heat of hydrogenation, −20.1 kcal/mole

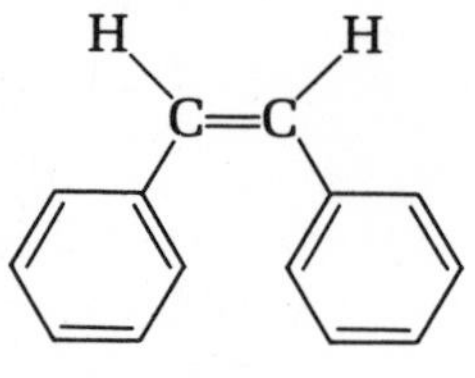

cis-Stilbene
Mp 6°, MW 180.24
λ_{max}^{EtOH}280nm ($\epsilon = 13,500$)
223nm ($\epsilon = 23,000$)
Heat of hydrogenation, −25.8 kcal/mole

One method of preparing *trans*-stilbene is reduction of benzoin with zinc amalgam in ethanol-hydrochloric acid, presumably through an intermediate:

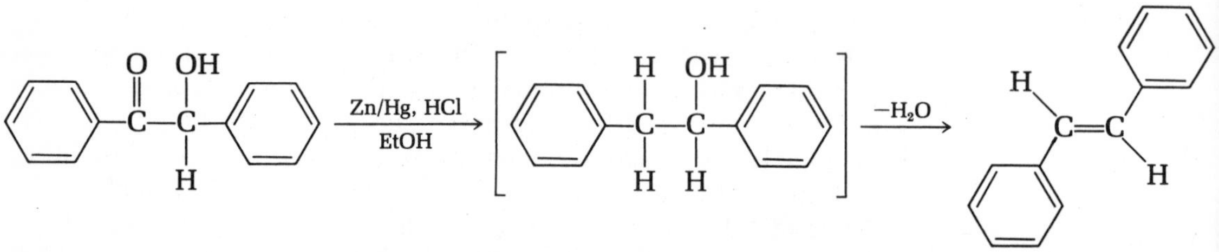

The procedure that follows is quick and affords hydrocarbon of high purity. It involves three steps: (1) replacement of the hydroxyl

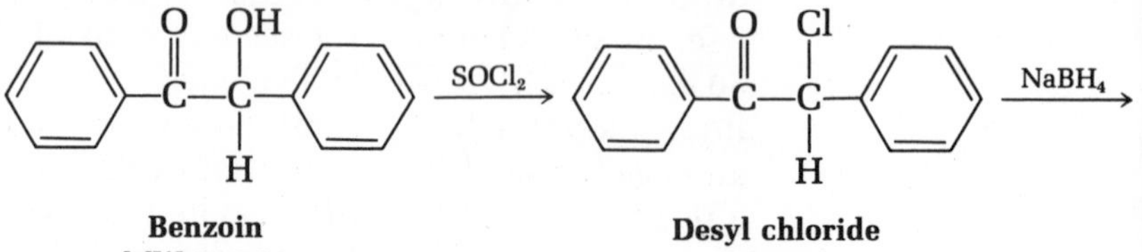

Benzoin
MW 212.24

Desyl chloride
Mp 68°

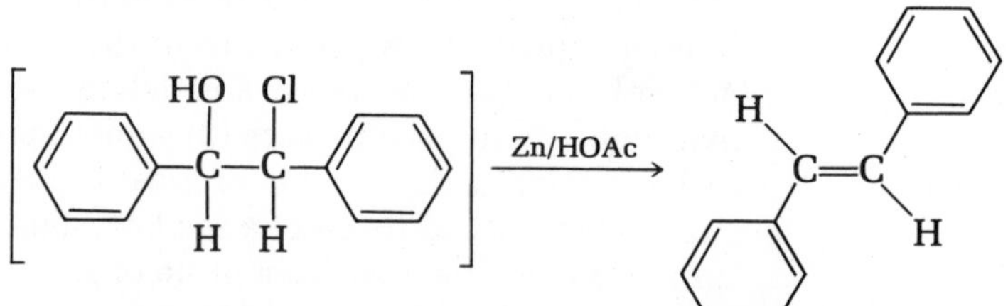

Mixture of *d,l*- and *d′,l′*-isomers

trans-Stilbene

group of benzoin by chlorine to form desyl chloride (2), reduction of the keto group with sodium borohydride to give what appears to be a mixture of the two diastereoisomeric chlorohydrins, and (3) elimination of the elements of hypochlorous acid with zinc and acetic acid. The last step is analogous to the debromination of an olefin dibromide.

Procedure

Place 4 g of benzoin (crushed to a powder) in a 125-ml round-bottomed flask, cover it with 4 ml of thionyl chloride,[6] warm gently on the steam bath (hood) until the solid has all dissolved, and then more strongly for 5 min.

Caution: If the mixture of benzoin and thionyl chloride is let stand at room temperature for an appreciable time before being heated, an undesired reaction intervenes[7] and the synthesis of *trans*-stilbene is spoiled.

To remove excess thionyl chloride (bp 77°), evacuate at the aspirator for a few minutes, add 10 ml of petroleum ether (bp 30–60°), boil it off, and evacuate again. Desyl chloride is thus obtained as a viscous, pale yellow oil (it will solidify if let stand). Dissolve this in 40 ml of 95% ethanol, cool under the tap, and add 360 mg of sodium borohydride (an excess is harmful). Stir, break up any lumps of the borohydride, and after 10 min add to the solution of chlorohydrins 2 g of zinc dust and 4 ml of acetic acid and reflux for 1 hr. Then cool under the tap. When white crystals separate add 50 ml of ether and decant the solution from the bulk of the zinc into a separatory funnel. Wash the solution twice with an equal volume of water containing 1–2 ml of concd hydrochloric acid (to dissolve basic zinc salts), then, in turn, with 5% sodium carbonate solution and saturated sodium chloride solution. Filter through a cone of anhydrous sodium sulfate (4 g), evaporate the filtrate to dryness, dissolve the residue in the minimum amount of hot 95% ethanol (30–40 ml), and let the product crystallize. *trans*-Stilbene separates in diamond-shaped iridescent plates, mp 124–125°; yield 1.8–2.2 g.

6. *meso*-Stilbene Dibromide

trans-Stilbene reacts with bromine predominantly by the usual process of *trans*-addition and affords the optically inactive, non-resolvable *meso*-dibromide (see page 274); the much lower-melting *dl*-dibromide (Section 9) is a very minor product of the reaction.

---

[6] The reagent can be dispensed from a burette or measured by pipette; in the latter case the liquid should be drawn into the pipette with a pipetter or by gentle suction at the aspirator, **not by mouth.**

[7]

$$\underset{\underset{\text{C}_6\text{H}_5\text{CHOH}}{|}}{\text{C}_6\text{H}_5\text{C}=\text{O}} \xrightarrow{\text{SOCl}_2} \underset{\underset{\text{C}_6\text{H}_5\text{C}-\text{O}}{\|}}{\underset{\text{C}_6\text{H}_5\text{C}-\text{O}}{}}\text{SO} \xrightarrow{\text{NaBH}_4} \underset{\underset{\text{C}_6\text{H}_5\text{CH}_2}{|}}{\text{C}_6\text{H}_5\text{CO}}$$

Desoxybenzoin

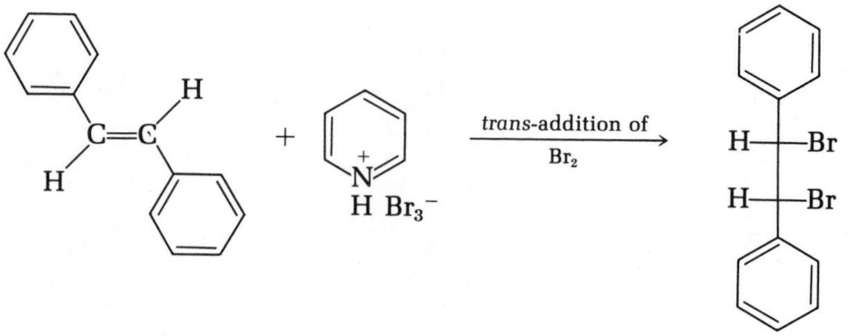

trans-Stilbene
MW 180.24

**Pyridinium hydro-
bromide perbromide**
MW 319.86

meso-Dibromide
Mp 238°, MW 340.0

Procedure

Total time required:
10 min

In a 125-ml Erlenmeyer flask dissolve 2 g of *trans*-stilbene in 40 ml of acetic acid, by heating on the steam bath, and then add 4 g of pyridinium hydrobromide perbromide. Mix by swirling, if necessary rinse crystals of reagent down the walls of the flask with a little acetic acid, and continue the heating for 1–2 min longer. The dibromide separates almost at once in small plates. Cool the mixture under the tap, collect the product, and wash it with methanol; yield of colorless crystals, mp 236–237°, 3.2 g. Use 0.5 g of this material for the preparation of diphenylacetylene and save the rest for a later experiment (Chapter 47).

7. Diphenylacetylene

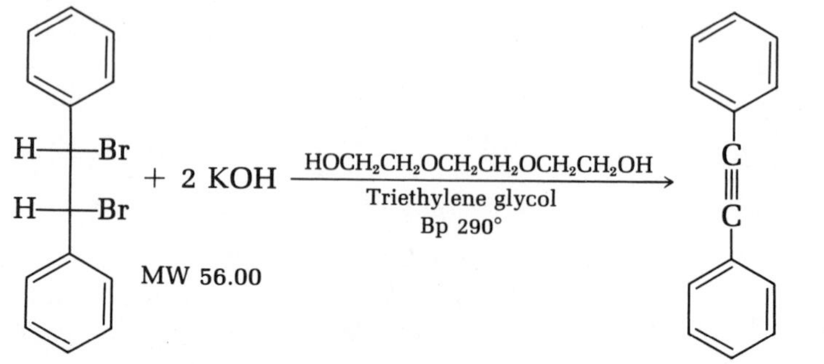

meso-Stilbene dibromide
MW 340.07

Diphenylacetylene
Mp 61°, MW 178.22

One method for the preparation of diphenylacetylene involves the oxidation of benzil dihydrazone with mercuric oxide; the intermediate diazide loses nitrogen under the conditions of its formation and affords the hydrocarbon:

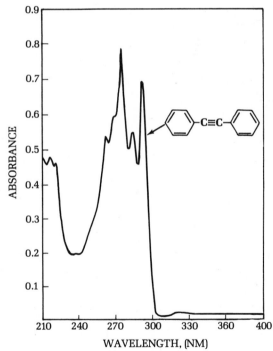

The method used in this procedure involves dehydrohalogenation of *meso*-stilbene dibromide. An earlier procedure called for refluxing the dibromide with 43% alcoholic potassium hydroxide in an oil bath at 140° for 24 hours. In the following procedure the reaction time is cut down to a few minutes by use of the high-boiling triethyleneglycol as solvent to permit operation at a higher reaction temperature.

Figure 46.9
Ultraviolet spectrum of diphenylacetylene. λ_{max}^{EtOH} **279 nm (ϵ = 31,400). This spectrum is characterized by considerable fine structure (multiplicity of bands) and a high extinction coefficient.**

Procedure

In a 20 × 150-mm test tube place 0.5 g of *meso*-stilbene dibromide, 3 pellets of potassium hydroxide[8] (250 mg), and 2 ml of triethylene

[8] Potassium hydroxide pellets are 85% KOH and 15% water.

Reaction time: 5 min

glycol. Insert a thermometer into a 10×75-mm test tube containing enough triethylene glycol to cover the bulb, and slip this assembly into the larger tube. Clamp the tube in a vertical position about two inches above a microburner, and heat the mixture with a very small flame to a temperature of 160°, when potassium bromide begins to separate. By intermittent heating, keep the mixture at 160–170 for 5 min more, then cool to room temperature, remove the thermometer and small tube, and add 10 ml of water. The diphenyl acetylene that separates as a nearly colorless, granular solid is collected by suction filtration. The crude product need not be dried but can be crystallized directly from 95% ethanol. Let the solution stand undisturbed in order to observe the formation of beautiful very large spars of colorless crystals. After a first crop has been collected, the mother liquor on concentration affords a second crop of pure product; total yield, 0.23 g; mp 60–61°.

8. α-Phenylcinnamic Acid (*cis* and *trans*)

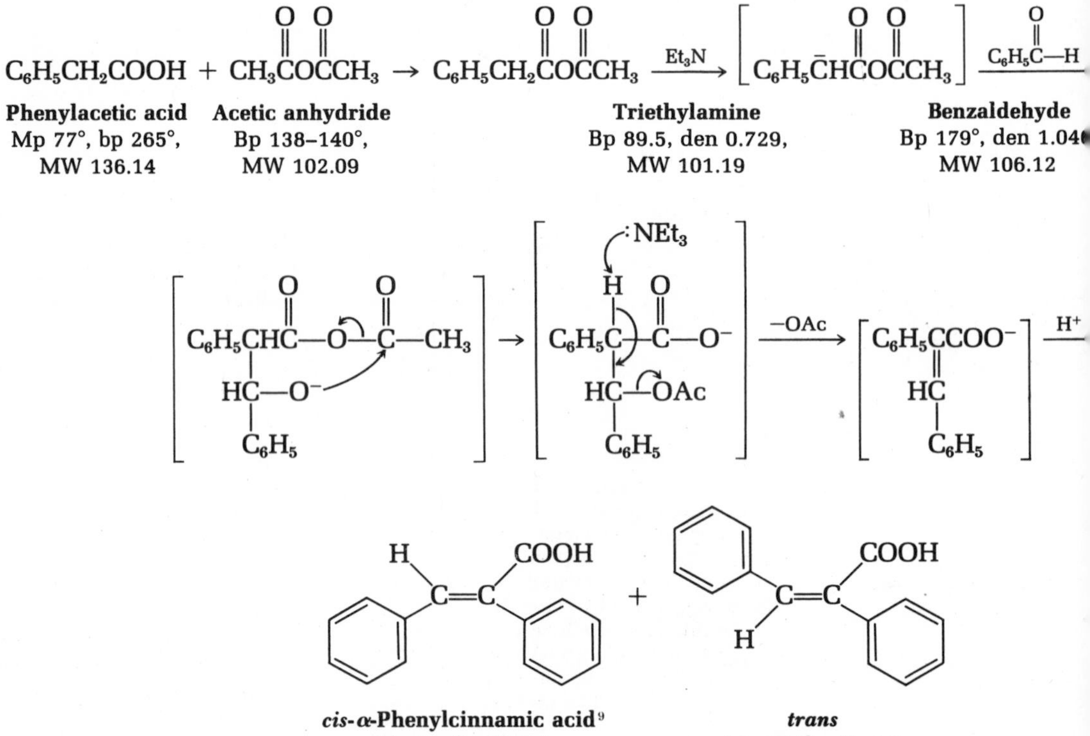

$$C_6H_5CH_2COOH + CH_3\overset{O}{\overset{\|}{C}}O\overset{O}{\overset{\|}{C}}CH_3 \rightarrow C_6H_5CH_2\overset{O}{\overset{\|}{C}}O\overset{O}{\overset{\|}{C}}CH_3 \xrightarrow{Et_3N} \left[C_6H_5\bar{C}H\overset{O}{\overset{\|}{C}}O\overset{O}{\overset{\|}{C}}CH_3 \right] \xrightarrow{C_6H_5\overset{O}{\overset{\|}{C}}-H}$$

Phenylacetic acid	**Acetic anhydride**	**Triethylamine**	**Benzaldehyde**
Mp 77°, bp 265°,	Bp 138–140°,	Bp 89.5, den 0.729,	Bp 179°, den 1.046
MW 136.14	MW 102.09	MW 101.19	MW 106.12

cis-α-**Phenylcinnamic acid**[9]
Mp 174°, pK_a 6.1

trans
Mp 138°, pK_a 4.8

MW 224.25

[9] pK_a measured in 60% ethanol.

The reaction of benzaldehyde with phenylacetic acid to produce a mixture of the α-carboxylic acid derivatives of *cis-* and *trans-*stilbene, a form of aldol condensation known as the Perkin reaction, is effected by heating a mixture of the components with acetic anhydride and triethylamine. In the course of the reaction the phenylacetic acid is probably present both as anion and as the mixed anhydride resulting from equilibration with acetic anhydride. A reflux period of 5 hrs specified in an early procedure has been shortened by a factor of 10 by restriction of the amount of the volatile acetic anhydride, use of an excess of the less expensive, high-boiling aldehyde component, and use of a condenser that permits some evaporation and consequent elevation of the reflux temperature.

*trans-*Stilbene is a by-product of the condensation, but experiment has shown that neither the *trans-* nor *cis-*acid undergoes decarboxylation under the conditions of the experiment.

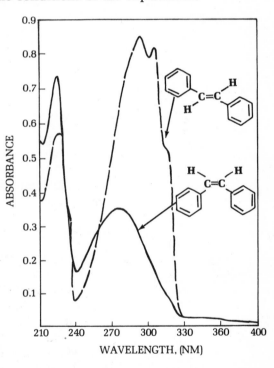

Figure 46.10
Ultraviolet spectra of
***cis-* and *trans-*stilbene.**
***cis:* λ_{max}^{EtOH} 224 nm ($\epsilon =$**
23,300), 279 nm ($\epsilon =$
11,100); *trans:* λ_{max}^{EtOH} 226
nm ($\epsilon = 18{,}300$), 295 nm
($\epsilon = 27{,}500$). Like the
diacetates, steric hin-
drance and lack of
coplanarity in these
hydrocarbons cause the
long wavelength absorp-
tion of the *cis-* isomer to
be of diminished inten-
sity relative to the
***trans-* isomer.**

ABSORBANCE

WAVELENGTH, (NM)

At the end of the reaction the α-phenylcinnamic acids are present in part as the neutral mixed anhydrides, but these can be hydrolyzed by addition of excess hydrochloric acid. The organic material is taken up in ether and the acids extracted with alkali. Neutralization with acetic acid (pK_a 4.76) then causes precipitation of only the less acidic *cis-*acid (see pK_a values under the formulas); the *trans-*acid separates on addition of hydrochloric acid.

Whereas *cis-*stilbene is less stable and lower melting than *trans-*stilbene, the reverse is true of the α-carboxylic acids, and in this preparation the more stable, higher-melting *cis-*acid is the pre-

Figure 46.11
Ultraviolet spectra of
cis- **and** *trans-α-phenyl-*
cinnamic acid, run at
identical concentrations.
cis: λ_{max}^{EtOH} **222 nm (ε =**
18,000), 282 nm (ε =
14,000); *trans:* λ_{max}^{EtOH} **222**
nm (ε = 15,500), 292 nm
(ε = 22,300). Note in
each pair of *cis-trans*
isomers the *trans* **isomer**
not only has the higher
extinction coefficient,
but also absorbs at
longer wavelength in the
long wavelength region
of the spectrum.

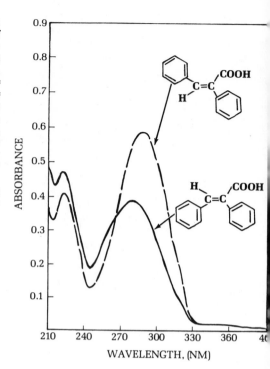

dominant product. Evidently the steric interference between the
carboxyl and phenyl groups in the *trans*-acid is greater than that
between the two phenyl groups in the *cis*-acid. Steric hindrance is
also evident from the fact that the *trans*-acid is not subject to Fischer
esterification whereas the *cis*-acid is.

Procedure

Reflux time: 35 min

　　Measure into a 25 × 150-mm test tube 2.5 g of phenylacetic acid,
3 ml of benzaldehyde, 2 ml of triethylamine, and 2 ml of acetic
anhydride. Insert a boiling stone, mount a cold-finger condenser as
in Fig. 1.3, and reflux the mixture for 35 min. Cool the yellow melt,
add 4 ml of concd hydrochloric acid, and swirl, where upon the
mixture sets to a stiff paste. Add ether, warm to dissolve the bulk
of the solid, and transfer to a separatory funnel with use of more
ether. Wash the ethereal solution twice with water and then extract
it with a mixture of 25 ml of water and 5 ml of 10% sodium hydroxide solution.[10] Repeat the extraction twice more and discard the
dark-colored ethereal solution.[11] Acidify the combined, colorless
alkaline extract to pH 6 by adding 5 ml of acetic acid, collect the
cis-acid that precipitates, and save the filtrate and washings. The
yield of *cis*-acid, mp 163–166°, is 2.9 g. Crystallize 0.3 g of material

---

[10] If stronger alkali is used the sodium salt may separate.
[11] For isolation of stilbene, wash this ethereal solution with saturated sodium bisulfite solution
for removal of benzaldehyde, dry, evaporate, and crystallize the residue from a little methanol.
Large, slightly yellow spars, mp 122–124°, separate (90 mg).

by dissolving it in 8 ml of ether, adding 8 ml of petroleum ether (bp 30–60°), heating briefly to the boiling point, and letting the solution stand. Silken needles form, mp 173–174°.

Addition of 5 ml of concd hydrochloric acid to the aqueous filtrate from precipitation of the cis-acid produces a cloudy emulsion which on standing for about one half hour coagulates to crystals of the trans-acid: 0.3 g, mp 136–137°.[12]

9. cis-Stilbene and dl-Stilbene Dibromide

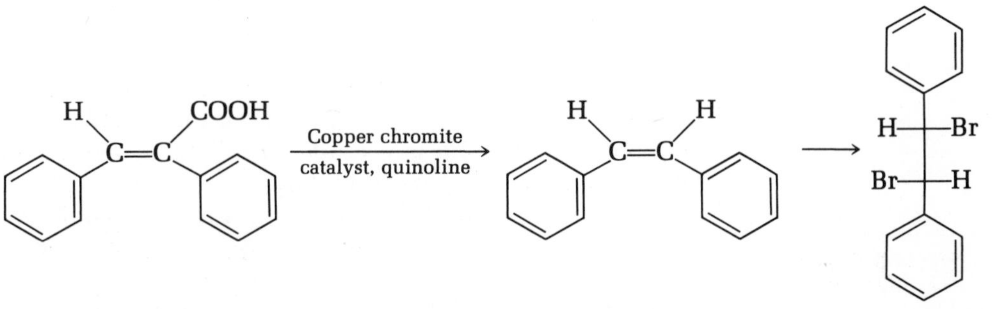

cis-α-Phenylcinnamic acid	cis-Stilbene	dl-Dibromide
MW 224.25	Mp 4°, bp 82–84°/0.4 mm, MW 180.24	Mp 114°, MW 340.07

Decarboxylation of cis-α-phenylcinnamic acid is effected by refluxing the acid in quinoline in the presence of a trace of copper chromite catalyst; both the basic properties and boiling point (237°) of quinoline make it a particularly favorable solvent. cis-Stilbene, a liquid at room temperature, can be characterized by trans addition of bromine to give the crystalline dl-dibromide. A little meso-dibromide derived from trans-stilbene in the crude hydrocarbon starting material is easily separated by virtue of its sparing solubility.

Although free rotation is possible around the single bond connecting the asymmetric carbon atoms of the stilbene dibromides and hydrobenzoins, evidence from dipole-moment measurements indicates that the molecules tend to exist predominantly in the specific shape or conformation in which the two phenyl groups repel each other and occupy positions as far apart as possible. The optimal conformations of the d- or l-dibromide and the meso-dibromide are represented in Fig. 46.12 by Newman projection formulas, in which the molecules are viewed along the axis of the bond connecting the two asymmetric carbon atoms. The carbon atom nearest to the eye is numbered 1; the carbon atom to the rear is 2. In the meso-dibromide the two repelling phenyl groups are on opposite sides of the molecule, and so are the two large bromine atoms. Hence, the structure is much more symmetrical than that of the d- (or l-) dibromide.

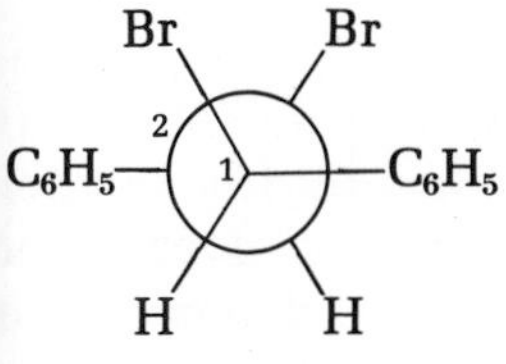

d-(or l)-Dibromide

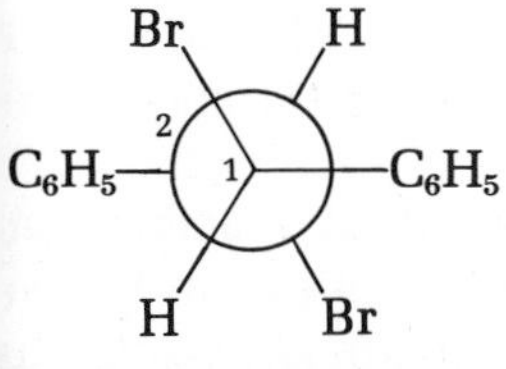

meso-Dibromide

Figure 46.12
Favored conformations
of stilbene dibromide.

[12] The trans-acid can be recrystallized by dissolving 0.3 g in 5 ml of ether, filtering if necessary from a trace of sodium chloride, adding 10 ml of petroleum ether (bp 30–60°), and evaporating to a volume of 5 ml; the acid separates as a hard crust of prisms, mp 138–139°.

X-ray diffraction measurements of the dibromides in the solid state confirm the conformations indicated in Fig. 46.12. The Br-Br distances found are: meso-dibromide, 4.50Å; dl-dibromide, 3.85Å. The difference in symmetry of the two optically inactive isomers accounts for the marked contrast in properties:

	Mp	Solubility in ether (18°)
dl-Dibromide	114°	1 part in 3.7 parts
meso-Dibromide	237°	1 part in 1025 parts

Procedure

Since a trace of moisture causes troublesome spattering, a drying operation is performed on the reactants and catalyst prior to decarboxylation. Stuff 2.5 g of crude, "dry" cis-α-phenylcinnamic acid and 0.2 g of copper chromite catalyst[13] into a 20 × 150-mm test tube add 3 ml of quinoline[14] (bp 237°), and let it wash down the solids Make connection with a rubber stopper to the aspirator and turn i on full force. Make sure that you have a good vacuum (pressure gauge) and heat the tube strongly on the steam bath with most of the rings removed. Heat and evacuate for 5–10 min to remove all traces of moisture. Then wipe the outside walls of the test tube dry insert a thermometer, clamp the tube over a microburner, raise the temperature to 230° and note the time. Then maintain a temperature close to 230° for 10 minutes. Cool the yellow solution containing suspended catalyst to 25°, add 30 ml of ether, and filter the solution by gravity (use more ether for rinsing). Transfer the solution to a separatory funnel and remove the quinoline by extraction twice with about 15 ml of water containing 3–4 ml of concd hydrochloric acid. Then shake the ethereal solution well with water containing a little sodium hydroxide solution, draw off the alkaline liquor, and acidify it. A substantial precipitate will show that decarboxylation was incomplete, in which case the starting material can be recovered and the reaction repeated. If there is only a trace of precipitate shake the ethereal solution with saturated sodium chloride solution for preliminary drying, filter the ethereal solution through sodium sulfate, and evaporate the ether. The residual brownish oil (1.3–1.8) is crude cis-stilbene containing a little trans-isomer formed by rearrangement during heating.

Dissolve the crude cis-stilbene (e.g., 1.5 g) in 10 ml of acetic acid and, in subdued light, add double the weight of pyridinium hydrobromide perbromide (e.g., 3.0 g). Warm on the steam bath until the reagent is dissolved, and then cool under the tap and scratch to effect separation of a small crop of plates of the meso-dibromide (10–20 mg). Filter the solution by suction, dilute extensively with

Reaction time in first step: 10 min

Quinoline, Bp 237°

Second step requires about one half hour

---

[13] The preparations described in J. Am. Chem. Soc., **54**, 1138 (1932) and ibid., **72**, 2626 (1950) are both satisfactory.
[14] Material that has darkened in storage should be redistilled over a little zinc dust.

water, and extract with ether. Wash the solution twice with water and then with 5% sodium bicarbonate solution until neutral; shake with saturated sodium chloride solution, filter through sodium sulfate, and evaporate to a volume of about 10 ml. If a little more of the sparingly soluble *meso*-dibromide separates, remove it by gravity filtration and then evaporate the remainder of the solvent. The residual *dl*-dibromide is obtained as a dark oil that readily solidifies when rubbed with a rod. Dissolve it in a small amount of methanol and let the solution stand to crystallize. The *dl*-dibromide separates as colorless prismatic plates, mp 113–114°; yield 0.6 g.

47 dl-Hydrobenzoin

KEYWORDS Spontaneous resolution, crystallization
dl, meso
AgOAc in wet HOAc, retention of configuration
Hemihedral, optically active
Dextro- and levorotatory
Seed crystals

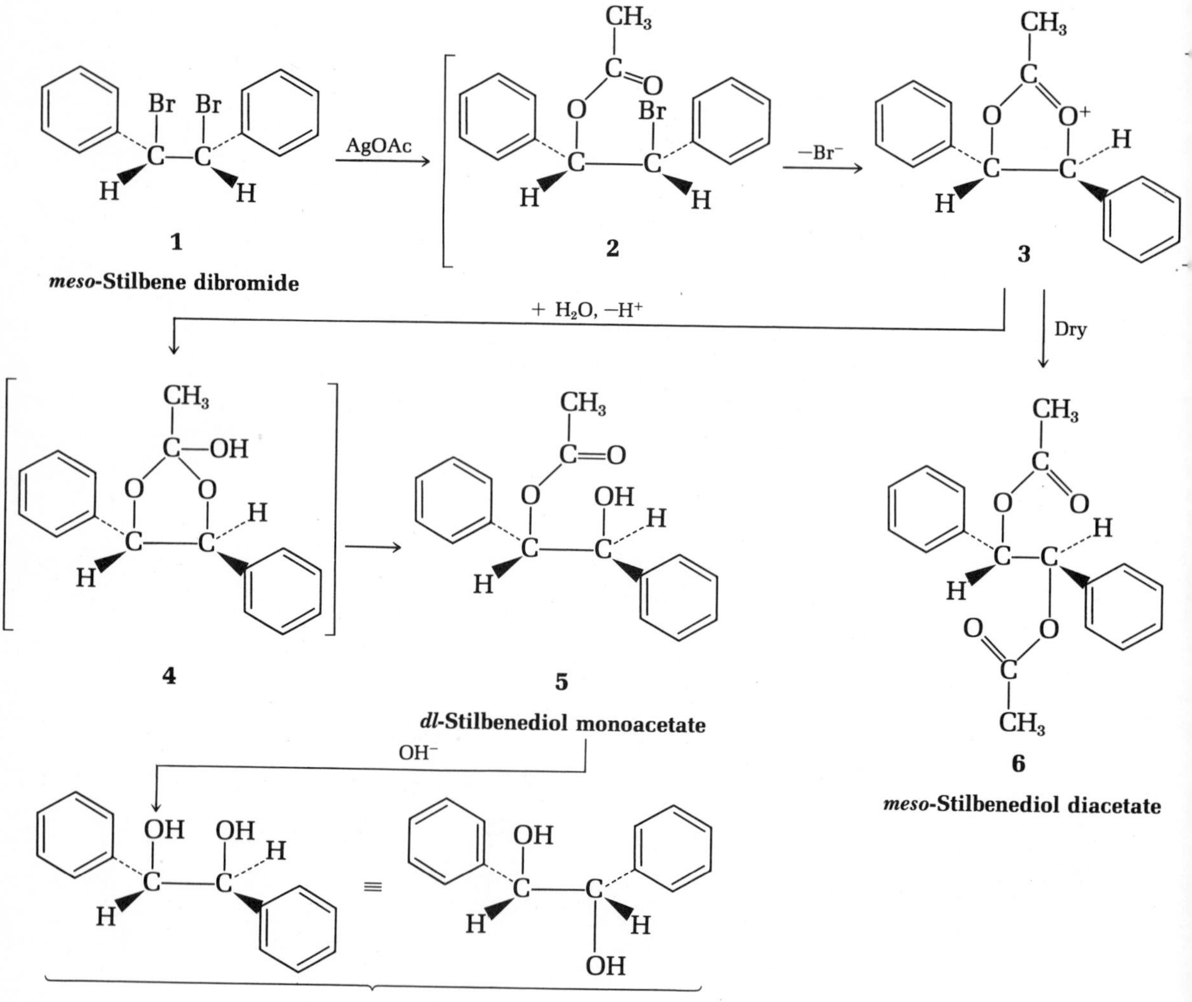

1
meso-**Stilbene dibromide**

2

3

4

5
dl-**Stilbenediol monoacetate**

6
meso-**Stilbenediol diacetate**

7
dl-**Hydrobenzoin**

dl-Hydrobenzoin (**7**) is one of a small number of *dl*-compounds that can be resolved into the optically active components by crystallization. Most methods of preparing these compounds give mixtures of *dl*- and *meso*-diols in which the more soluble *dl*-diol is the minor component. Such mixtures are very difficult to separate. However, silver acetate reacts with *meso*-stilbene dibromide in acetic acid containing 4% of water to give a mixture which contains considerable *dl*-diol monoacetate (**5**) and none of the isomeric *meso*-diol monoacetate.[1] Although the mixture contains both diacetates and both diols, the sole monoacetate present can be separated from these substances by chromatography.

EXPERIMENT

0.5 hr of work prior to chromatography

Place 2 g of *meso*-stilbene dibromide (Chapter 46, Section 6), 2 g of silver acetate, 25 ml of acetic acid, and 1 ml of water in a 50-ml Erlenmeyer flask. Heat the mixture on the steam bath for 10 min with frequent swirling. The initially uniform white suspension changes to a curdy, dense precipitate of silver bromide, a transient pink color disappears, and the solution clears and becomes colorless. Cool to room temperature, filter by suction, and wash the silver bromide with a little alcohol or acetone. Pour the filtrate into a separatory funnel, add water (60 ml) to produce a milky suspension, and extract with ether (40 ml). Wash the ethereal extract twice with water and once with 10% sodium hydroxide (25 ml), dry, filter, and evaporate the ether (use aspirator tube).

Dissolve the residual oil in 3–4 ml of benzene, pour the solution onto a column of 25 g of alumina, and rinse the flask with a little more benzene. Chromatograph in 25-ml fractions with the following sequence of solvents: 50 ml of petroleum ether, 50 ml of 1:1 petroleum ether-benzene, 50 ml of benzene, 50 ml of 1:1 benzene-ether, 50 ml of ether. If early fractions afford oily products (diacetates) and are then followed by negative fractions, the oils can be discarded and the flasks cleaned and reused. The *dl*-diol monoacetate (mp 87°) should appear in intermediate fractions as an oil that slowly solidifies on standing (e.g., overnight). If you obtain a consecutive set of solid fractions, scrape out a little of the first and the last and take the melting points. If the two end fractions correspond, then these and the intermediate ones can be combined. Dissolve each one in a little ether, transfer with a capillary dropper to a 50-ml Erlenmeyer flask heated on the steam bath (aspirator tube), and determine the weight of product (0.6 g). Dissolve the monoacetate in 10 ml of 95% ethanol, add 5 ml of 10% sodium hydroxide, and heat for 10 min on the steam bath. Dilute with 30 ml of water, cool and scratch to cause separation of *dl*-hydrobenzoin. Since the diol often separates from the aqueous medium as a low-melting hydrate, it is best collected by ether extraction. The dried and filtered ethereal solution is evaporated to a volume of about

---

[1] If anhydrous acetic acid is used the chief product is *meso*-hydrobenzoin diacetate (**6**). According to the mechanism postulated in the formulation, Walden inversions occur in the reactions **2 → 3** and **3 → 6**. The net result is retention of configuration at the carbon bearing the bromine atom.

10 ml, 15 ml of petroleum ether is added, and the solution is evaporated slowly until crystals begin to separate. Mp 120°, yield 0.35 g

Formation of separate hemihedral prisms of the optically active *d*- and *l*-forms requires slow recrystallization from ether. It is best t pool several preparations, in order to have sufficient material t work with.[2]

---

[2] If crystals of opposite hemihedrism are not at first identifiable, individual crystals will be found to give solutions that are strongly dextro- or levorotatory. Crystals obtained by evaporation of *d*- and *l*-solutions can then be introduced as seed to a concentrated ethereal solution of *dl*-diol, one on each side of the flask. If the flask is plugged with absorbent cotton and let stand in an explosion-proof refrigerator, two masses of crystals develop in which opposite hemihedrism is discernible.

48 Martius Yellow

KEYWORDS   Rapid laboratory work, 3–4 hr
Small samples, seven crystalline
   compounds
A laboratory contest
Martius Yellow, dye
Sulfonation, nitration

Sodium hydrosulfite,
   reduction
Stable imine
Quinone
Acetylation

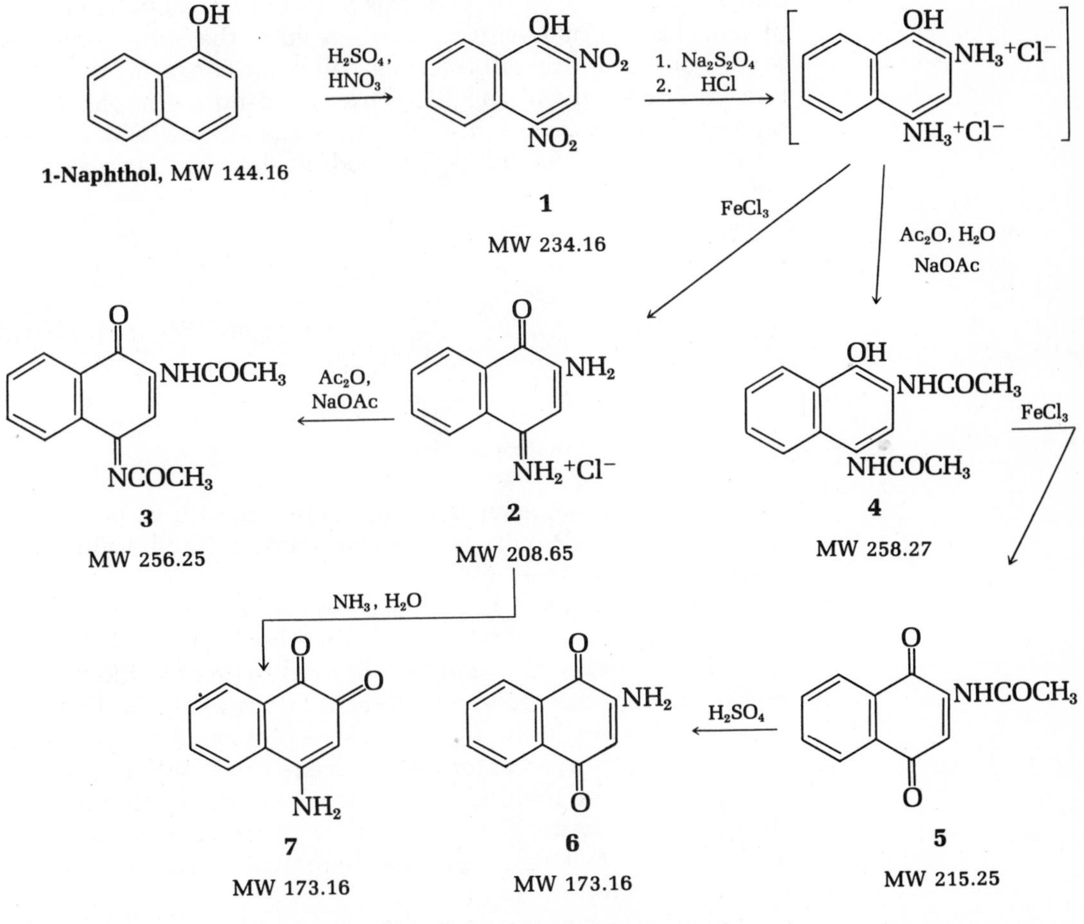

1-Naphthol, MW 144.16

1   MW 234.16

3   MW 256.25

2   MW 208.65

4   MW 258.27

7   MW 173.16

6   MW 173.16

5   MW 215.25

The series of experiments described here should afford the student an opportunity of gaining some experience in the rapid handling of small quantities of materials. Starting with 5 g of 1-naphthol, a skilled operator familiar with the procedures can prepare pure samples of the seven compounds in 3–4 hours. In a first trial of the experiment, a particularly competent student, who plans his work in advance, can complete the program in two laboratory periods (6 hr).

The first compound of the series, Martius Yellow, a mothproofing dye for wool (1 g of Martius Yellow dyes 200 g of wool) discovered in 1868 by Karl Alexander von Martius, is the ammonium salt of 2,4-dinitro-1-naphthol (1). Compound 1 in the series of equations is obtained by sulfonation of 1-naphthol with sulfuric acid and treatment of the resulting disulfonic acid with nitric acid in aqueous medium. The exchange of groups occurs with remarkable ease, and it is not necessary to isolate the disulfonic acid. The advantage of introducing the nitro groups in this indirect way is that 1-naphthol is very sensitive to oxidation and would be partially destroyed on direct nitration. Martius Yellow is prepared by reaction of the acidic phenolic group of 1 with ammonia to form the ammonium salt. A small portion of this salt (Martius Yellow) is converted by acidification and crystallization into pure 2,4-dinitro-1-naphthol (1), a sample of which is saved. The rest is suspended in water and reduced to diaminonaphthol with sodium hydrosulfite according to the equation:

$$+ \; 6 \; Na_2S_2O_4 \; + \; 8 \; H_2O \; \rightarrow \qquad\qquad\qquad + \; 11 \; NaHSO_3 \; + \; Na(NH_4)SO_4$$

Martius Yellow   **Diaminonaphthol**

The diaminonaphthol separates in the free condition, rather than as an ammonium salt, because the diamine, unlike the dinitro compound, is a very weak acidic substance.

Since 2,4-diamino-1-naphthol is exceedingly sensitive to air oxidation as the free base, it is at once dissolved in dilute hydrochloric acid. The solution of diaminonaphthol dihydrochloride is clarified with decolorizing charcoal and divided into equal parts. One part on oxidation with ferric chloride affords the fiery red 2-amino-1,4-naphthoquinonimine hydrochloride (2). Since this substance, like many other salts, has no melting point, it is converted for identification to the yellow diacetate, 3. Compound 2 is remarkable in that it is stable enough to be isolated. On hydrolysis it affords the orange 4-amino-1,2-naphthoquinone (7).

The other part of the diaminonaphthol dihydrochloride solution is treated with acetic anhydride and then sodium acetate; the reaction in aqueous solution effects selective acetylation of the amino groups and affords 2,4-diacetylamino-1-naphthol (4). Oxidation of

4 is attended with cleavage of the acetylamino group at the 4-position and the product is 2-acetylamino-1,4-naphthoquinone (**5**). This yellow substance is hydrolyzed by sulfuric acid to the red 2-amino-1,4-naphthoquinone (**6**), the last member of the series. The reaction periods are brief and the yields high.[1]

EXPERIMENTS

1. Preparation of 2,4-Dinitro-1-naphthol

1

Place 5 g of pure 1-naphthol[2] in a 125-ml Erlenmeyer flask, add 10 ml of concd sulfuric acid, and heat the mixture with swirling on the steam bath for 5 min, when the solid should have dissolved and an initial red color should be discharged. Cool in an ice bath, add 25 ml of water, and cool the solution rapidly to 15°. Measure 6 ml of concd nitric acid into a test tube and transfer it with a capillary dropping tube in small portions (0.5 ml) to the chilled aqueous solution while keeping the temperature in the range 15–20° by swirling the flask vigorously in the ice bath. When the addition is complete and the exothermic reaction has subsided (1–2 min), warm the mixture gently to 50° (1 min), when the nitration product should separate as a stiff yellow paste. Apply the full heat of the steam bath for 1 min more, fill the flask with water, break up the lumps and stir to an even paste, collect the product (**1**) on a Büchner funnel, wash it well with water, and then wash it into a 600-ml beaker with water (100 ml). Add 150 ml of hot water and 5 ml of concd ammonia solution (den 0.90), heat to the boiling point, and stir to dissolve the solid. Filter the hot solution by suction if it is dirty, add 10 g of ammonium chloride to the filtrate to salt out the ammonium salt (Martius Yellow), cool in an ice bath, collect the orange salt, and wash it with water containing 1–2% of ammonium chloride. The salt does not have to be dried (dry weight 7.7 g, 88.5%).

Avoid contact of the yellow product and its orange NH$_4^+$ salt with the skin

Martius Yellow

Set aside an estimated 0.3 g of the moist ammonium salt. This sample is to be dissolved in hot water, the solution acidified (HCl), and the free 2,4-dinitro-1-naphthol (**1**) crystallized from methanol or ethanol (decolorizing charcoal); it forms yellow needles, mp 138°.

2. Preparation of 2,4-Diamino-1-naphthol

Wash the rest of the ammonium salt into a beaker with a total of about 200 ml of water, add 40 g of sodium hydrosulfite, stir until the

---

[1] This series of reactions lends itself to a laboratory competition, the rules for which might be as follows: (1) No practice or advance preparation is allowable except collection of reagents not available at the contestant's bench (ammonium chloride, sodium hydrosulfite, ferric chloride solution, acetic anhydride). (2) The time scored is the actual working time, including that required for bottling the samples and cleaning the apparatus and bench; labels can be prepared out of the working period. (3) Time is not charged during an interim period (overnight) when solutions are let stand to crystallize or solids are let dry, on condition that during this period no adjustments are made and no cleaning or other work is done. (4) Melting point and color test characterizations are omitted. (5) Successful completion of the contest requires preparation of authentic and macroscopically crystalline samples of all seven compounds. (6) Judgment of the winners among the successful contestants is based upon quality and quantity of samples, technique and neatness, and working time. (Superior performance: 3–4 hr.)

[2] If the 1-naphthol is dark it can be purified by distillation at atmospheric pressure. The colorless distillate is most easily pulverized before it has completely cooled and hardened.

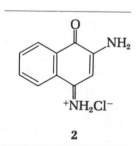

**2,4-Diamino-1-naphthol
dihydrochloride**

original orange color has disappeared and a crystalline tan precipitate has formed (5–10 min), and cool in ice. Make ready a solution
of 1–2 g of sodium hydrosulfite in 100 ml of water for use in washing
and a 400-ml beaker containing 6 ml of concd hydrochloric acid and
25 ml of water. In collecting the precipitate by suction filtration, use
the hydrosulfite solution for rinsing and washing, avoid even briefly
sucking air through the cake after the reducing agent has been
drained away, and wash the solid at once into the beaker containing
the dilute hydrochloric acid and stir to convert all the diamine to
the dihydrochloride.

The acid solution, often containing suspended sulfur and filter
paper, is clarified by filtration by suction through a moist charcoal
bed made by shaking 2 g of the decolorizing carbon with 25 ml of
water in a stoppered flask to produce a slurry and pouring this on
the paper of an 85-mm Büchner funnel. Pour the water out of the
filter flask and then filter the solution of dihydrochloride. Divide
the pink or colorless filtrate into approximately two equal parts and
at once add the reagents for conversion of one part to **2** and the other
part to **4**.

3. Preparation of 2-Amino-1,4-naphthoquinonimine Hydrochloride (2)

To one half of the diamine dihydrochloride solution add 25 ml of
1.3 M ferric chloride solution,[3] cool in ice, and, if necessary, initiate
crystallization by scratching. Rub the liquid film with a glass stirring
rod at a single spot slightly above the surface of the liquid. If efforts
to induce crystallizations are unsuccessful, add more hydrochloric
acid. Collect the red product and wash with dilute HCl. Dry weight
2.4–2.7 g.

Divide the moist product into three equal parts and spread out one
part to dry for conversion to **3**. The other two parts can be used
while still moist for conversion to **7** and for recrystallization. Dissolve the other two parts by gentle warming in a little water containing 2–3 drops of hydrochloric acid, shake for a minute or two with
decolorizing charcoal, filter, and add concd hydrochloric acid to
decrease the solubility.

4. Preparation of 2-Amino-1,4-naphthoquinonimine Diacetate (3)

A mixture of 0.5 g of the dry quinonimine hydrochloride (**2**), 0.5 g
of sodium acetate (anhydrous), and 3 ml of acetic anhydride is
stirred in a test tube and warmed gently on a hot plate or steam bath.
With thorough stirring the red salt should soon change into yellow
crystals of the diacetate. The solution may appear red, but as soon
as particles of red solid have disappeared the mixture can be poured
into about 10 ml of water. Stir until the excess acetic anhydride has
either dissolved or become hydrolyzed, collect and wash the prod-

2

3

Note for the instructor

[3] Dissolve 90 g of FeCl₃·6H₂O (MW 270.32) in 100 ml of water and 100 ml of concd hydrochloric
acid by warming, cool and filter (248 ml of solution).

uct (dry weight 0.5 g), and (drying is unnecessary) crystallize it from ethanol or methanol; yellow needles, mp 189°.

5. Preparation of 2,4-Diacetylamino-1-naphthol (4)

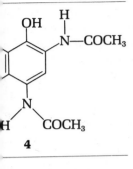

4

To one half of the diaminonaphthol dihydrochloride solution saved from Section 2 add 3 ml of acetic anhydride, stir vigorously, and add a solution of 3 g of sodium acetate (anhydrous) and about 100 mg of sodium hydrosulfite in 20–30 ml of water. The diacetate may precipitate as a white powder or it may separate as an oil that solidifies when chilled in ice and rubbed with a rod. Collect the product and, to hydrolyze any triacetate present, dissolve it in 5 ml of 10% sodium hydroxide and 50 ml of water by stirring at room temperature. If the solution is colored, a pinch of sodium hydrosulfite may bleach it. Filter by suction and acidify by gradual addition of well-diluted hydrochloric acid (2 ml of concd acid). The diacetate tends to remain in supersaturated solution and hence, either to initiate crystallization or to insure maximum separation, it is advisable to stir well, rub the walls with a rod, and cool in ice. Collect the product, wash it with water, and divide it into thirds (dry weight 2.1–2.6 g).

Two thirds of the material can be converted without drying into **5** and the other third used for preparation of a crystalline sample. Dissolve the third reserved for crystallization (moist or dry) in enough hot acetic acid to bring about solution, add a solution of a small crystal of stannous chloride in a few drops of dilute hydrochloric acid to inhibit oxidation, and dilute gradually with 5–6 volumes of water at the boiling point. Crystallization may be slow, and cooling and scratching may be necessary. The pure diacetate forms colorless prisms, mp 224°, dec.

6. Preparation of 2-Acetylamino-1,4-naphthoquinone (5)

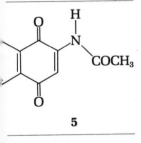

5

Dissolve 1.5 g of the moist diacetylaminonaphthol (**4**) saved from Section 5 in 10 ml of acetic acid (hot), dilute with 20 ml of hot water, and add 10 ml of 0.13 M ferric chloride solution. The product separates promptly in flat, yellow needles, which are collected (after cooling) and washed with a little alcohol; yield 1.2 g. Dry one-half of the product for conversion to **6** and crystallize the rest from 95% ethanol; mp 204°.

7. Preparation of 2-Amino-1,4-naphthoquinone (6)

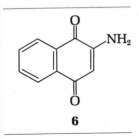

6

To 0.5 g of the 2-acetylamino-1,4-naphthoquinone (**5**) dried in Section 6 and contained in a 25-ml Erlenmeyer flask add 2 ml of concd sulfuric acid and heat the mixture on the steam bath with swirling to promote rapid solution (1–2 min). After 5 min cool the deep red solution, dilute extensively with water, and collect the precipitated product; wash it with water and crystallize the moist sample (dry weight 0.37 g) from alcohol or alcohol-water; red needles, mp 206°.

8. Preparation of 4-Amino-1,2-naphthoquinone

Moist 2 is satisfactory

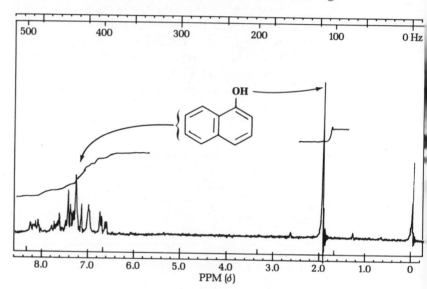

7

Dissolve 1 g of the aminonaphthoquinonimine hydrochloride (2 reserved in Section 3 in 25 ml of water, add 2 ml of concd ammonia solution (den 0.90), and boil the mixture for 5 min. The free quinonimine initially precipitated is hydrolyzed to a mixture of the aminoquinone **7** and the isomer **6**. Cool, collect the precipitate, and suspend it in about 50 ml of water, and add 25 ml of 10% sodium hydroxide. Stir well, filter out the small amount of residual 2-amino-1,4-naphthoquinone (**6**), and acidify the filtrate with acetic acid The orange precipitate of **7** is collected, washed, and crystallized while still wet from 500–600 ml of hot water (the separation is slow). The yield of orange needles, dec about 270°, is 0.4 g.

Figure 48.1
Nmr spectrum of
α-naphthol.

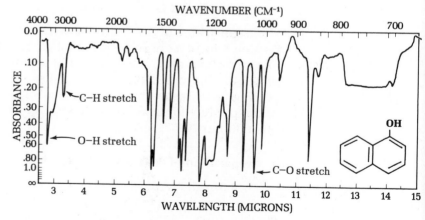

Figure 48.2
Infrared spectrum of
α-naphthol.

49 Tetraphenylcyclopentadienone

KEYWORDS Base-catalyzed condensation reaction
Triethylene glycol, high bp solvent
Benzyltrimethylammonium hydroxide, basic catalyst

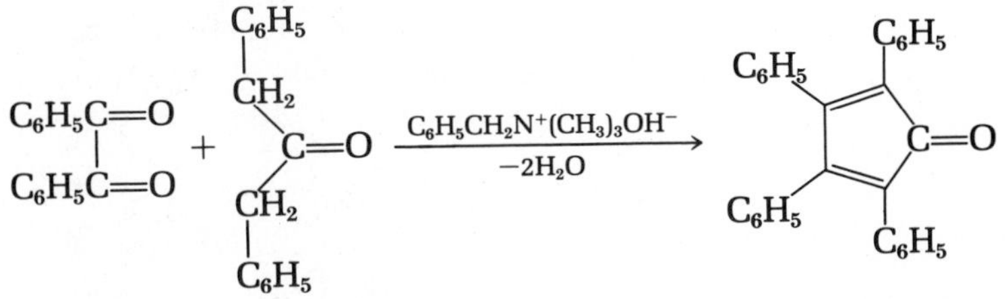

$$C_6H_5C{=}O \atop C_6H_5C{=}O \quad + \quad {C_6H_5 \atop CH_2} {C{=}O \atop CH_2} {\atop C_6H_5} \xrightarrow[-2H_2O]{C_6H_5CH_2N^+(CH_3)_3OH^-}$$

Benzil
Mp 96°,
MW 210.22

Dibenzyl ketone
Mp 35°,
MW 210.26

Tetraphenylcyclopentadienone
Mp 219°, MW 384.45

Cyclopentadienone is an elusive compound which has been sought for many years but with little success. Molecular orbital calculations predict that it should be highly reactive, and so it is; it exists only as the dimer. The tetraphenyl derivative of this compound is to be synthesized in this experiment. This derivative is stable, and reacts readily with dienophiles. It is used not only for the synthesis of highly aromatic, highly arylated compounds, but also for examination of the mechanism of the Diels-Alder reaction itself. Tetraphenylcyclopentadienone has been carefully studied by means of molecular orbital methods in attempts to understand its unusual reactivity, color, and dipole moment. In Chapter 51 this highly reactive molecule is used to trap the fleeting benzyne to form tetraphenylnaphthalene. Indeed this reaction constitutes evidence that benzyne does exist.

The literature procedure for condensation of benzil with dibenzyl ketone in ethanol with potassium hydroxide as basic catalyst suffers from the low boiling point of the alcohol and the limited solubility of both potassium hydroxide and the reaction product in this solvent. Triethylene glycol is a better solvent and permits operation at a higher temperature. In the procedure that follows, the glycol is used with benzyltrimethylammonium hydroxide, a strong base readily soluble in organic solvents, that serves as catalyst.

EXPERIMENT

Short reaction period

Measure into a 25 × 150-mm test tube 2.1 g of benzil, 2.1 g of di benzyl ketone, and 10 ml of triethylene glycol, using the solvent to wash down the walls. Support the test tube over a microburner, stir the mixture with a thermometer, and heat with a small flame until the benzil is dissolved and remove the flame. Measure 1 ml of a 40% solution of benzyltrimethylammonium hydroxide in methanol into a 10 × 75-mm test tube, adjust the temperature of the solution to exactly 100°, remove the flame, add the catalyst, and stir once to mix. Crystallization usually starts in 10–20 seconds. Let the temperature drop to about 80° and then cool under the tap, add 10 ml of methanol, stir to a thin crystal slurry, collect the product, and wash it with methanol until the filtrate is purple-pink, not brown. The yield of deep purple crystals is 3.3–3.7 g. If either the crystals are not well formed or if the melting point is low, place 1 g of material and 10 ml of triethylene glycol in a vertically supported test tube, stir with a thermometer, raise the temperature to 220° to bring the solid into solution, and let stand for crystallization (if initially pure material is recrystallized, the recovery is 92%).

QUESTION   *Write a detailed mechanism for this reaction.*

50 Derivatives of Tetraphenylcyclo-pentadienone

KEYWORDS Diels-Alder reaction
Acetylenic dienophile
o-Dichlorobenzene,
  high bp solvent
Carbon monoxide

Dienophile, diene
Diphenyl ether, high bp
  recrystallization solvent
Soxhlet extractor
Cold finger condenser

EXPERIMENT   **1. Dimethyl Tetraphenylphthalate**

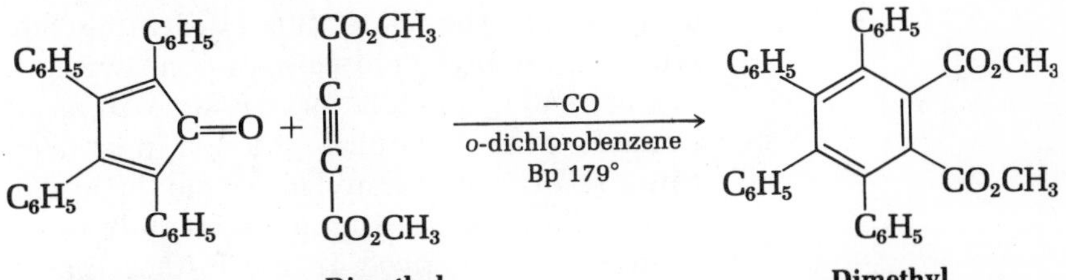

MW 384.45 — **Dimethyl acetylenedicarboxylate** Liq., MW 142.11 — **Dimethyl tetraphenylphthalate** Mp 258°, MW 474.53

Tetraphenylcyclopentadienone undergoes Diels-Alder reaction with dienophiles with expulsion of carbon monoxide. An acetylenic dienophile is selected for this experiment because it affords an adduct which is directly aromatic. The solvent specified, o-dichlorobenzene, has a boiling point such that the reaction is over in a minute or two.

Procedure

Caution! See footnote one

Measure into a 25 × 150-mm test tube 2 g of tetraphenylcyclopentadienone, 10 ml of o-dichlorobenzene, and 1 ml (1.1 g) of dimethyl acetylenedicarboxylate.[1] Clamp the test tube over a microburner, insert a thermometer and an aspirator tube (to remove carbon

---

[1] This ester is a powerful lachrymator and vesicant and should be dispensed from a bottle provided with a pipette and a pipetter. Even a trace of ester on the skin should be washed off with methanol, followed by soap and water.

Reaction time
about 5 min

monoxide), and raise the temperature to the boiling point (180–185°)
Boil gently until there is no further color change and let the rim of
condensate rise just high enough to wash down the walls. The pure
adduct is colorless, and if the starting ketone is of adequate purity
the color changes from purple to pale tan. A 5-min boiling period
should be adequate. Cool to 100°, slowly stir in 15 ml of 95% ethanol, and let crystallization proceed. Cool under the tap, collect the
product, and rinse out the tube with methanol. Yield of colorless
crystals 2.1–2.2 g.

2. Hexaphenylbenzene

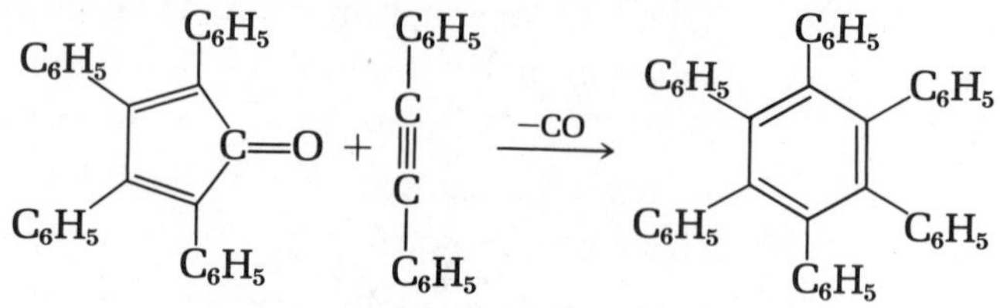

MW 384.45   MW 178.22   **Hexaphenylbenzene**
MW 534.66, mp 465°

Diphenylacetylene is a less reactive dienophile than dimethyl
acetylenedicarboxylate, but by heating the less active dienophile
with tetraphenylcyclopentadienone without solvent a temperature
(ca. 380–400°) suitable for reaction can be attained. In the following
procedure the dienophile is taken in large excess to serve as solvent.
Since refluxing diphenylacetylene (bp about 300°) keeps the temperature below the melting point of the product, the excess diphenylacetylene is removed in order that the reaction mixture can be
melted to ensure completion of the reaction.

Diphenyl ether (mp 27°, bp 259°) is selected for crystallization of
the sparingly soluble product because of its high boiling point and
its superior solvent power.

Procedure

Place 0.5 g each of tetraphenylcyclopentadienone and diphenylacetylene in a 25 × 150-mm test tube supported in a clamp and heat
the mixture strongly with the free flame of a microburner held in the
hand (do not insert a thermometer into the test tube; temperature is
too high). Soon after the reactants have melted with strong bubbling,
white masses of the product become visible. Let the diphenylacetylene reflux briefly on the walls of the tube and then remove some of
the diphenylacetylene by letting it condense for a minute or two

A hydrocarbon
of very high mp

on a cold finger filled with water, but without fresh water running
through it. Remove the flame, withdraw the cold finger, and wipe it
with a towel. Repeat the operation until you are able, by strong heating, to melt the mixture completely. Then let the melt cool and
solidify. Add 10 ml of diphenyl ether, using it to rinse down the

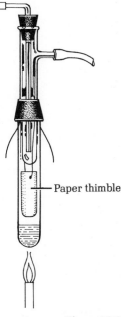

Figure 50.1
Soxhlet-type extractor.

walls. Heat *carefully* over a free flame to bring the solid into solution and let the product crystallize. When cold, add 10 ml of benzene to thin out the mixture, collect the product, and wash with
benzene. The yield of colorless plates, mp 465°, is 0.6–0.7 g.

Notes:

1. The melting point of the product can be determined with a
Mel-Temp apparatus and a 500°-thermometer. To avoid oxidation,
seal the sample in a capillary tube evacuated at the water pump by
the technique described in Chapter 6.

2. In case the hexaphenylbenzene is contaminated with insoluble
material, crystallization from a filtered solution can be accomplished
as follows: Place 10 ml of diphenyl ether in a 25 × 150-mm test tube
and pack the sample into a 10-mm extraction thimble and suspend
this in the test tube with two nichrome wires, as shown in Fig. 50.1.
Insert a cold finger condenser supported by a filter adapter and adjust
the condenser and the wires so that condensing liquid will drop into
the thimble. Let the diphenyl ether reflux until the hexaphenylbenzene in the thimble is in solution, and then let the product crystallize, add benzene, collect the product, and wash with benzene as
described previously.

QUESTIONS   1. *Draw the structures of the bicyclic compounds which must be
intermediates in these Diels-Alder reactions.*

2. *What is the driving force for loss of carbon monoxide from
these intermediates?*

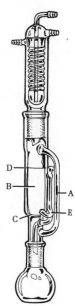

Figure 50.2
**Research quality Soxhlet
extractor. Solvent vapors
from the flask rise
through A and up into
the condenser. As they
condense the liquid just
formed returns to B,
where sample to be ex-
tracted is placed.
(Bottom of B is sealed
at C.) Liquid rises in B
to level D at which time
the automatic siphon, E,
starts. Extracted material
accumulates in the flask
as more pure vapor boils
off to repeat the process.**

51 1,2,3,4-Tetraphenyl-naphthalene via Benzyne

KEYWORDS Benzyne
Trapping agent
Diels-Alder reaction
Iodonium salts

Potassium persulfate
o-Iodobenzoic acid synthesis
Anthranilic acid, diazotization

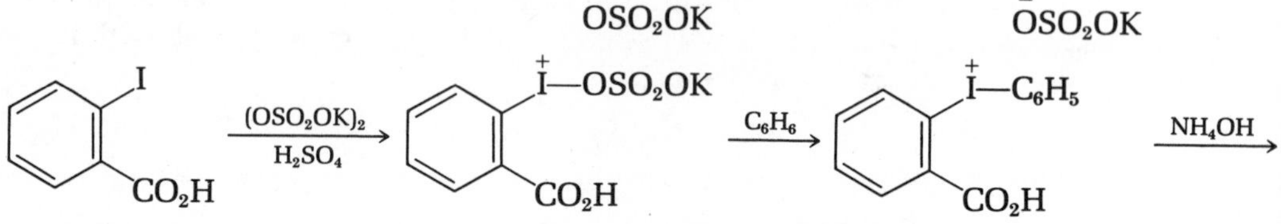

1
MW 247.01

2

3

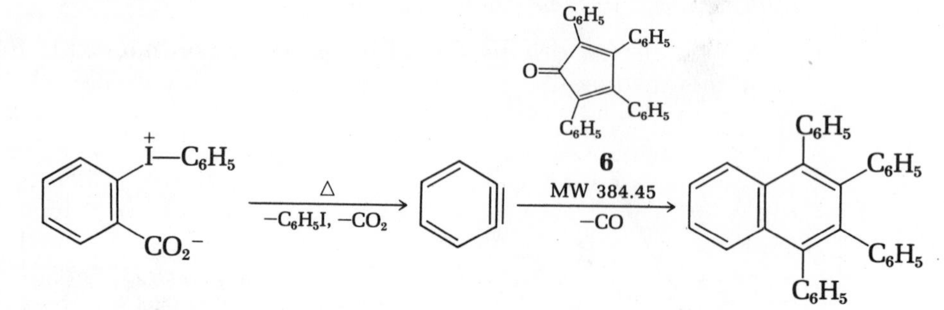

4
MW 341.13(monohydrate)

5

6
MW 384.45

7
MW 432.53

Synthetic use of an intermediate not known as such

This synthesis of 1,2,3,4-tetraphenylnaphthalene (**7**) demonstrates the transient existence of benzyne (**5**), a hydrocarbon which has not been isolated as such. The precursor, diphenyliodonium-2-carboxylate (**4**), is heated in an inert solvent to a temperature at which it decomposes to benzyne, iodobenzene, and carbon dioxide in the presence of tetraphenylcyclopentadienone (**6**) as trapping agent. The preparation of the precursor **4** illustrates oxidation of a derivative of iodobenzene to an iodonium salt (**2**) and the Friedel-Crafts-like reaction of the substance with benzene to form the diphenyl-iodonium salt **3**. Neutralization with ammonium hydroxide then liberates the precursor, inner salt, **4**, which when obtained by crystallization from water is the monohydrate.

The procedure affords about twice the amount of precursor (**4**) required for the synthesis of (**7**). Hence samples of both (**4**) and (**7**) should be submitted for evaluation.

EXPERIMENTS

1. Diphenyliodonium-2-carboxylate monohydrate (4)

Measure 8 ml of concd sulfuric acid into a 25-ml Erlenmeyer flask and place the flask in an ice bath to cool. Place 2.0 g of *o*-iodo-benzoic acid[1] and 2.6 g of potassium persulfate[2] in a 125-ml Erlenmeyer flask, swirl the flask containing sulfuric acid vigorously in the ice bath for 2–3 min and then remove it and wipe it dry. Place the larger flask in the ice bath and pour the chilled acid down the walls to dislodge any particles of solid. Swirl the flask in the ice bath for 4–5 min to produce an even suspension and then remove it and note the time. The reaction mixture foams somewhat and acquires a succession of colors. After it has stood at room temperature for 20 min, swirl the flask vigorously in an ice bath for 3–4 min, add 2 ml of benzene, and swirl and cool until the benzene freezes. Then remove and wipe the flask and note the time at which the benzene melts. Warm the flask in the palm of the hand and swirl frequently at room temperature for 20 min to promote completion of reaction in the two-phase mixture. During this period place three 50-ml Erlenmeyer flasks in an ice bath to chill: one containing 19 ml of distilled water, another 23 ml of 29% ammonium hydroxide solution, and another 40 ml of methylene chloride (bp 40.8°). At the end of the 20-min reaction period, chill the reaction mixture thoroughly in an ice bath, mount a separatory funnel to deliver into the flask containing the reaction mixture in benzene, and place in the funnel the chilled 19 ml of water. Swirl the reaction flask vigorously while running in the water slowly. The solid that separates is **3**, the potassium bisulfate salt of diphenyliodonium-2-carboxylic acid. Pour the chilled ammonia solution into the funnel and pour the chilled methylene chloride into the reaction flask so that it will be available for efficient extraction of the reaction product (**4**) as it is liberated from **3** on neutralization. While swirling the flask vigorously in the ice bath, run in the chilled ammonia solution during the course of about 10 min. The mixture must be alkaline (pH 9). If it isn't, add more ammonia solution. Pour the mixture into a separatory funnel and rinse the flask with a little fresh methylene chloride. Let the layers separate and run the lower layer into a tared 125-ml Erlenmeyer flask, through a funnel fitted with a paper containing anhy-

1st step, about 20 min

2nd step, about 30 min

Note for the instructor

---

[1] See note at end of this chapter.
[2] The fine granular material supplied by Fisher Scientific Co. is satisfactory. Persulfate in the form of large prisms should be finely ground prior to use.

drous sodium sulfate. Extract the aqueous solution with two 10-ml portions of methylene chloride and run the extracts through the drying agent into the tared flask. Evaporate the dried extracts to dryness on the steam bath and remove solvent from the residual cake of solid by connecting the flask, with a rubber stopper, to the suction pump and heating the flask on the steam bath until the weight is constant; yield 2.4 g.

For crystallization, dislodge the bulk of the solid with a spatula and transfer it onto weighing paper and then into a 50-ml Erlenmeyer flask. Measure 28 ml of distilled water into a flask and use part of the water to dissolve the residual material in the tared 125 ml flask by heating the mixture to the boiling point over a free flame. Pour this solution into the 50-ml flask. Add the remainder of the 28 ml of water to the 50 ml flask and bring the solid into solution at the boiling point. Add a small portion of charcoal for decolorization of the pale tan solution, swirl, and filter at the boiling point through a funnel, preheated on the steam bath and fitted with moistened filter paper. Diphenyliodonium-2-carboxylate monohydrate (4), the benzyne precursor, separates in colorless, rectangular prisms, mp 219–220°, dec.; yield 2.1 g.

2. Preparation of 1,2,3,4-Tetraphenylnaphthalene (7)

Diels-Alder reaction

Place 1.0 g of the diphenyliodonium-2-carboxylate monohydrate just prepared and 1.0 g of tetraphenylcyclopentadienone in a 25 × 150-ml test tube. Add 6 ml of triethylene glycol dimethyl ether (bp 222°) in a way that the solvent will rinse down the walls of the tube. Support the test tube vertically, insert a thermometer, and heat with a microburner. When the temperature reaches 200° remove the burner and note the time. Then keep the mixture at 200–205° by intermittent heating until the purple color is discharged, the evolution of gas (CO_2 + CO) subsides, and a pale yellow solution is obtained. In case a purple or red color persists after 3 min at 200–205°, add additional small amounts of the benzyne precursor and continue to heat until all the solid is dissolved. Let the yellow solution cool to 90° while heating 6 ml of 95% ethanol to the boiling point on the steam bath. Pour the yellow solution into a 25-ml Erlenmeyer flask and use portions of the hot ethanol, drawn into a capillary dropping tube, to rinse the test tube. Add the remainder of the ethanol to the yellow solution and heat at the boiling point. If shiny prisms do not separate at once, add a few drops of water by micro drops at the boiling point of the ethanol until prisms begin to separate. Let crystallization proceed at room temperature and then at 0°. Collect the product and wash it with methanol. The yield of colorless prisms is 0.8–0.9 g. The pure hydrocarbon (7) has a double melting point, the first of which is in the range 196–199°. Let the bath cool to about 195°, remove the thermometer and let the sample solidify. Then determine the second melting point which, for the pure hydrocarbon, is 203–204°.

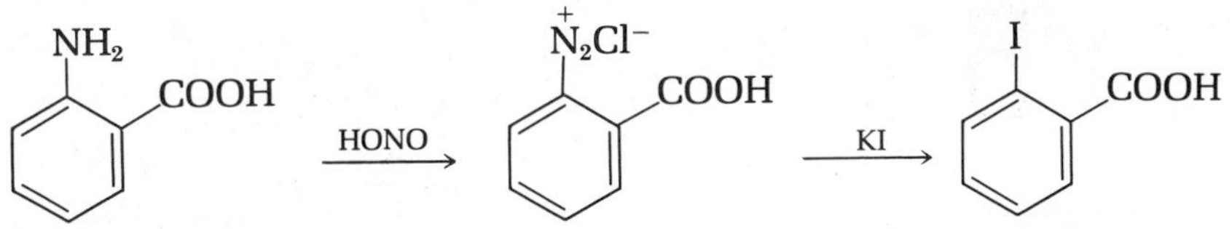

Note. Commercial o-iodobenzoic acid is expensive, but the re-agent can be prepared easily by diazotization of anthranilic acid followed by a Sandmeyer-like reaction with iodide ion. A 500-ml round-bottomed flask containing 13.7 g of anthranilic acid, 100 ml of water, and 25 ml of concd hydrochloric acid is heated until the solid is dissolved. The mixture is then cooled in ice while bubbling in nitrogen to displace the air. When the temperature reaches 0–5° a solution of 7.1 g of sodium nitrite is added slowly. After 5 min a solution of 17 g of potassium iodide in 25 ml of water is added, when a brown complex partially separates. The mixture is let stand with-out cooling for 5 min (under nitrogen) and then warmed to 40°, at which point a vigorous reaction ensues (gas evolution, separation of a tan solid). After reacting for 10 min the mixture is heated on the steam bath for 10 min and then cooled in ice. A pinch of sodium bi-sulfite is added to destroy any iodine present and the granular tan product collected and washed with water. The still moist product is dissolved in 70 ml of 95% ethanol, 35 ml of hot water is added, and the brown solution treated with decolorizing charcoal, filtered, diluted at the boiling point with 35–40 ml of water, and let stand. o-Iodobenzoic acid separates in large, slightly yellow needles of satisfactory purity (mp 164°) for the experiment; yield 17.6 g (71%).

52 Triptycene

KEYWORDS Benzyne, anthracene   Maleic anhydride adduct
              Diazotization, anthranilic acid   Triglyme,
              Isoamyl nitrite $(CH_3OCH_2CH_2OCH_2)_2$
              Aprotic solvent   Optional project

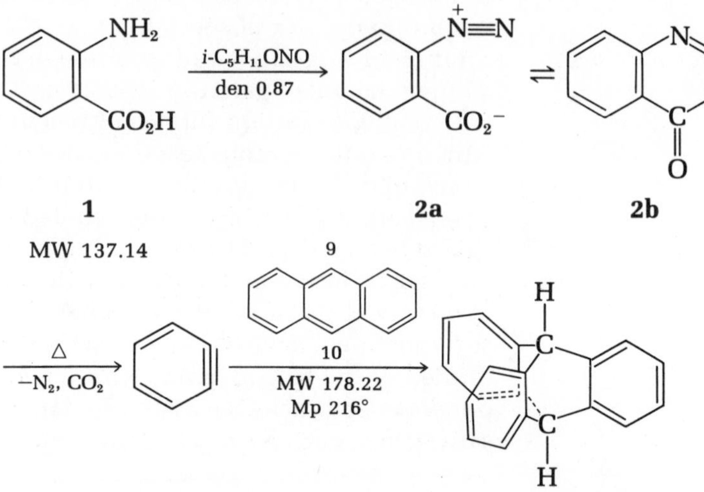

1
MW 137.14

2a

2b

9

10
MW 178.22
Mp 216°

3

4

Triptycene
MW 254.31, mp 255°

This interesting cage-ring hydrocarbon results from 9,10-addition of benzyne to anthracene. In one procedure presented in the literature benzyne is generated under nitrogen from o-fluorobromobenzene and magnesium in the presence of anthracene, but the work-up is tedious and the yield only 24%. Diazotization of anthranilic acid to benzenediazonium-2-carboxylate (**2a**) or to the covalent form (**2b**), gives another benzyne precursor (M. Stiles, 1963), but the isolated substance can be kept only at a low temperature and is sometimes explosive. However, L. Friedman found (1963) that isolation of the precursor is not necessary. On slow addition of anthranilic acid to a solution of anthracene and isoamyl nitrite in an aprotic solvent, the precursor (**2**) reacts with anthracene as fast as the precursor is formed. If the anthranilic acid is all present at the start, a side reaction of this substance with benzyne drastically reduces the yield. Friedman used a low-boiling solvent (CH_2Cl_2, bp 41°), in which the desired reaction goes slowly, and added a solution of

Aprotic means "no protons"; HOH and ROH are protic

anthranilic acid dropwise over a period of 4 hours. To bring the reaction time into the limits of a laboratory period, the higher boiling solvent 1,2-dimethoxyethane (bp 83°, water soluble) is specified in this procedure and use made of a large excess of anthranilic acid and isoamyl nitrite. Treatment of the dark reaction mixture with alkali removes acidic by-products and most of the color, but the crude product inevitably contains anthracene. However, brief heating with maleic anhydride at a suitable temperature leaves the triptycene untouched and converts the anthracene into its maleic anhydride adduct. Treatment of the reaction mixture with alkali converts the adduct into a water-soluble salt and affords colorless, pure triptycene.

Removal of anthracene

EXPERIMENT

Technique for slow addition of a solid

Reaction time: 1 hr

Place 2 g of anthracene,[1] 2 ml of isoamyl nitrite, and 20 ml of 1,2-dimethoxyethane in a 125-ml round-bottomed flask mounted over a microburner and fitted with a short reflux condenser. Insert a filter paper into a 55-mm short stem funnel, moisten the paper with 1,2-dimethoxyethane, and rest the funnel in the top of the condenser. Weigh 2.6 g of anthranilic acid on a folded paper, scrape the acid into the funnel with a spatula, and pack it down. Bring the mixture in the reaction flask to a gentle boil and note that the anthracene does not all dissolve. Measure 20 ml of 1,2-dimethoxyethane into a graduate and use a capillary dropping tube to add small portions of the solvent to the anthranilic acid in the funnel, to slowly leach the acid into the reaction flask. If you make sure that the condenser is exactly vertical and the top of the funnel is centered, it should be possible to arrange for each drop to fall free into the flask and not run down the condenser wall. Once dripping from the funnel has started, add fresh batches of solvent to the acid but only 2–3 drops at a time. Plan to complete leaching the first charge of anthranilic acid in a period of not less than 20 min, using about 10 ml of solvent. Then add a second 2.6-g portion of anthranilic acid to the funnel, remove the burner, and by lifting the funnel up a little run in 2 ml of isoamyl nitrite through the condenser. Replace funnel, resume heating, and leach the anthranilic acid as before in about 20 min time. Reflux for 10 min more and then add 10 ml of 95% ethanol and a solution of 3 g of sodium hydroxide in 40 ml of water to produce a suspension of solid in a brown alkaline liquor. Cool thoroughly in ice, and also cool a 4:1 methanol-water mixture for rinsing. Collect the solid on a small Büchner funnel and wash it with the chilled solvent to remove brown mother liquor. Transfer the moist, nearly colorless solid to a tared 125-ml round-bottomed flask and evacuate on the steam bath until the weight is constant; the anthracene-triptycene mixture (mp about 190–230°) weighs 2.1 g. Add 1 g of maleic anhydride and 20 ml of triethylene glycol dimethyl ether ("triglyme," bp 222°), heat the mixture to the bp under reflux, and

---

[1] Eastman Practical Grade anthracene and anthranilic acid is satisfactory.

reflux for 5 min. Cool to about 100°, add 10 ml of 95% ethanol and a solution of 3 g of sodium hydroxide in 40 ml of water; then cool in ice, along with 25 ml of 4:1 methanol-water for rinsing. Triptycene separates as nearly white crystals from the slightly brown alkaline liquor. The washed and dried product weighs 1.5 g and melts at 255°. It will appear to be colorless, but it contains a trace of black insoluble material. Dissolve the material in excess methylene chloride (10 ml/g), filter to remove the specks of black solid, and rinse the flask and filter paper with a liberal amount of methylene chloride. Then add a boiling stone, heat on the hot plate to boiling, slowly add about two volumes of methanol (bp 65°) for each volume of methylene chloride present, and boil on the hot plate to eliminate the solvent of lower bp (41°) and higher solvent power. Concentrate the solution until crystals just begin to separate, and let stand for crystallization. Better formed crystals can be obtained by recrystallization from methylcyclohexane (23 ml/g). Crystals thus produced are flat, rectangular, laminated prisms.

Optional Project

In Chapter 51 diphenyliodonium-o-carboxylate is utilized as a benzyne precursor in the synthesis of 1,2,3,4-tetraphenyl-naphthalene. For reasons unknown, the reaction of anthracene with benzyne generated in this way proceeds very poorly and gives only a trace of triptycene. What about the converse proposition? Would benzyne generated by diazotization of anthranilic acid in the presence of tetraphenylcyclopentadienone afford 1,2,3,4-tetraphenyl-naphthalene in satisfactory yield? In case you are interested in exploring this possibility, plan and execute a procedure and see what you can discover.

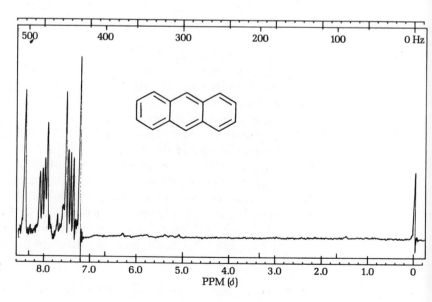

Figure 52.1
Nmr spectrum of anthracene.

53 Dyes and Dyeing[1]

KEYWORDS Acetate rayon, Dacron, Nylon, cotton, wool, Orlon
Azo dyes, diazotize, diazonium salt
Indicators, Methyl Orange, Orange II
Direct dyes, Martius Yellow, picric acid, eosin
Substantive dyes, Congo Red
Mordants, triphenylmethane dyes, Crystal Violet
Developed dyes, ingrain dyes
Vat dyes
Leuco derivatives, indigo
Disperse dyes, fiber reactive dyes
Brighteners-fluorescent dyes

Since prehistoric times man has been dyeing cloth. The "wearing of the purple" has long been synonymous with royalty, attesting to the cost and rarity of Tyrian purple, a dye derived from the sea snail *Murex brandaris*. The organic chemical industry originated with William Henry Perkin's discovery of the first synthetic dye, Perkin's Mauve, in 1856.

In this experiment several dyes will be synthesized and these and other representative dyes will be used to dye a representative group of natural and man-made fibers.

You will receive several 2½″ squares of Multifiber Fabric 10A[2] which have six strips of different fibers woven into it: spun acetate rayon, cotton, a spun polyamide (Nylon 6,6), silk, viscose rayon, and wool.

Acetate rayon is cellulose (from any source) in which about two of the hydroxyl groups in each unit have been acetylated. This renders the polymer soluble in acetone from which it can be spun into fiber. The smaller number of hydroxyl groups in acetate rayon compared to cotton makes direct dyeing of rayon more difficult than cotton.

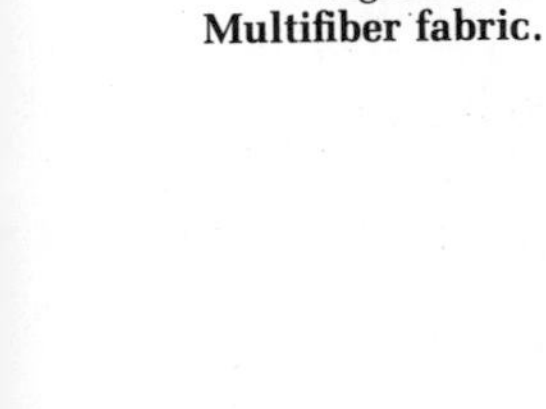

Cotton is pure cellulose. Nylon is a polyamide and made by polymerizing adipic acid and hexamethylenediamine. The nylon polymer chain can be prepared with one acid and one amine group at the termini, or with both acids or both amines. Except for these terminal groups, there are no polar centers in nylon and consequently it is difficult to dye. Similarly Dacron, a polyester made by polymerizing ethylene glycol and terephthalic acid, has few polar centers within the polymer, and consequently is difficult to dye. Even more diffi-

Acetate rayon
Cotton
Nylon
Silk
Viscose rayon
Wool

Figure 53.1
Multifiber fabric.

---

[1] For a detailed discussion of the chemistry of dyes and dyeing see *Topics in Organic Chemistry*, by Louis F. Fieser and Mary Fieser, Reinhold Publishing Corp., New York, 1963. pp. 357–417.
[2] Obtained from Testfabrics, Inc., P.O. Box 118, 200 Blackford Ave., Middlesex, N.J. 08846. Cut into 2½″ squares which include all six fibers.

cult to dye is Orlon, a polymer of acrylonitrile. Wool is a polypeptide cross-linked with disulfide bridges. The acidic and basic amino acids (e.g., glutamic acid and lysine) provide many polar groups in wool and silk to which a dye can bind, making these fabrics easy to dye. In this experiment note the marked differences in shade produced by the same dye on different fibers.

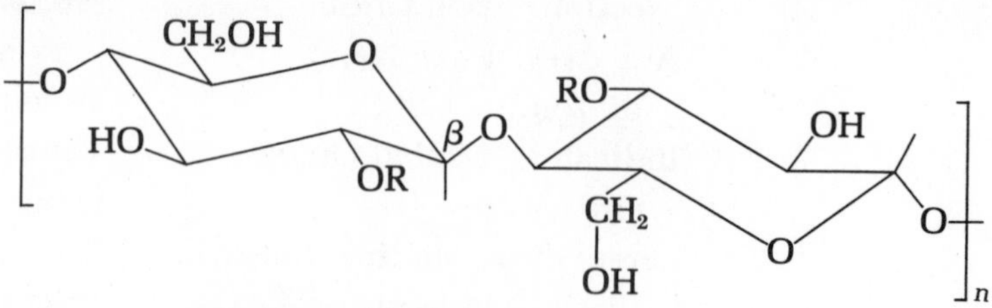

Cellulose (Cotton, R=H)
Acetylated Cellulose (Acetate rayon, R=OAc)

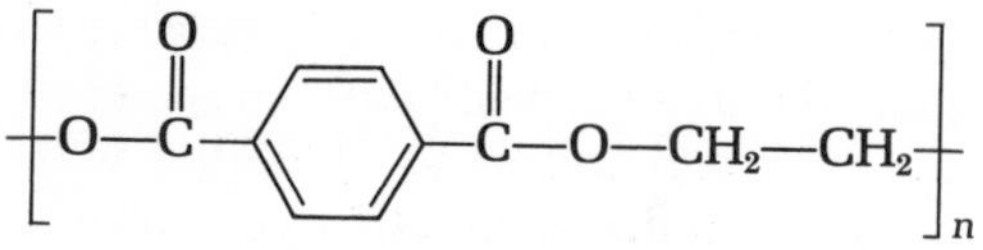

Polyethyleneglycol terephthalate (Dacron)

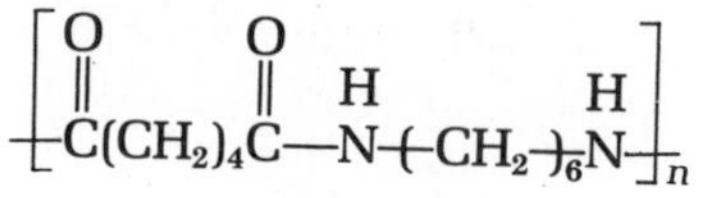

Nylon

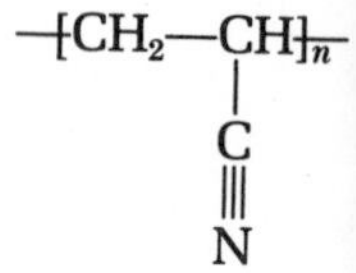

Polyacrylonitrile (Orlon)

PART I. Dyes

The most common dyes are the azo dyes, formed by coupling diazotized amines to phenols. The dye can be made in bulk, or, as we shall see, the dye molecule can be developed on and in the fiber by combining the reactants in the presence of the fiber.

One dye, Orange II, is made by coupling diazotized sulfanilic acid with 2-naphthol in alkaline solution; another, Methyl Orange, is prepared by coupling the same diazonium salt with N,N-dimethylaniline in a weakly acidic solution. Methyl Orange is used as an indicator as it changes color at pH 3.2–4.4. The change in color is due to transition from one chromophore (azo group) to another (quinonoid system).

You are to prepare one of these two dyes and then exchange samples with a neighbor and do the tests with both dyes. Both substances dye wool, silk, and skin, and you must work carefully to avoid getting them on your hands or clothes. The dye will eventually wear off your hands or they can be cleaned by soaking them in warm, slightly acidic (H_2SO_4) permanganate solution until heavily stained with manganese dioxide and then removing the stain in a bath of warm, dilute bisulfite solution.

EXPERIMENTS   **1. Diazotization of Sulfanilic Acid**

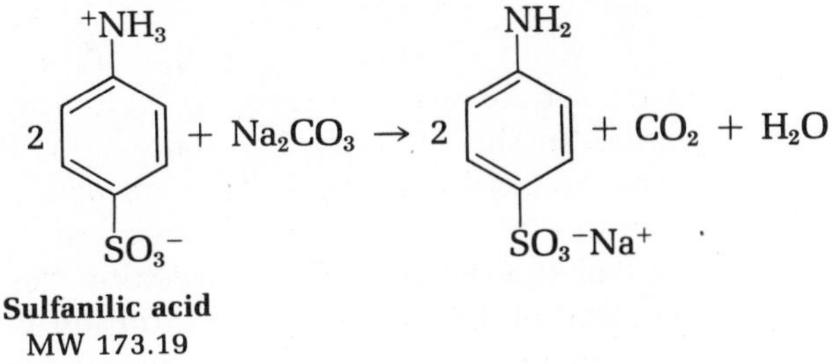

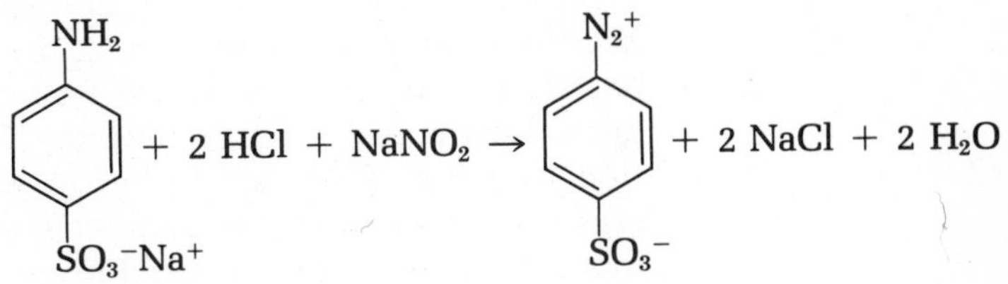

$$2 \quad \underset{\underset{\text{SO}_3^-}{\overset{+\text{NH}_3}{\bigcirc}}}{} + \text{Na}_2\text{CO}_3 \rightarrow 2 \quad \underset{\underset{\text{SO}_3^-\text{Na}^+}{\overset{\text{NH}_2}{\bigcirc}}}{} + \text{CO}_2 + \text{H}_2\text{O}$$

Sulfanilic acid
MW 173.19

$$\underset{\underset{\text{SO}_3^-\text{Na}^+}{\overset{\text{NH}_2}{\bigcirc}}}{} + 2 \text{ HCl} + \text{NaNO}_2 \rightarrow \underset{\underset{\text{SO}_3^-}{\overset{\text{N}_2^+}{\bigcirc}}}{} + 2 \text{ NaCl} + 2 \text{ H}_2\text{O}$$

In a 125-ml Erlenmeyer flask dissolve, by boiling, 4.8 g of sulfanilic acid crystals (monohydrate) in 50 ml of 2.5% sodium carbonate solution (or use 1.33 g of anhydrous sodium carbonate and 50 ml of water). Cool the solution under the tap, add 1.9 g of sodium nitrite, and stir until it is dissolved. Pour the solution into a flask containing about 25 g of ice and 5 ml of concd hydrochloric acid. In a minute or two a powdery white precipitate of the diazonium salt should separate and the material is then ready for use. The product is not collected but is used in the preparation of the dye Orange II while in suspension. It is more stable than most diazonium salts and will keep for a few hours.

2. Orange II (1-*p*-Sulfobenzeneazo-2-naphthol Sodium Salt)

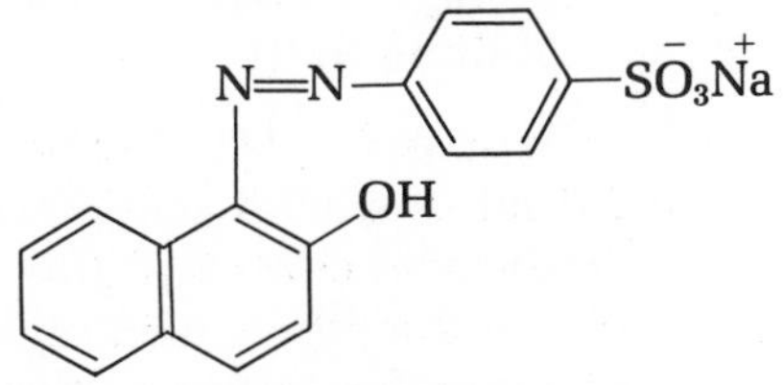

Orange II

In a 400-ml beaker dissolve 3.6 g of 2-naphthol in 20 ml of cold 10% sodium hydroxide solution and pour into this solution, with stirring, the suspension of diazotized sulfanilic acid from Section 1 (rinse). Coupling occurs very rapidly and the dye, being a sodium salt, separates easily from the solution on account of the presence of a considerable excess of sodium ion (from the carbonate, the nitrite,

Choice of 2 or 3

and the alkali added). Stir the crystalline paste thoroughly to effect good mixing and, after 5–10 min, heat the mixture until the solid dissolves. Add 10 g of sodium chloride to further decrease the solubility of the product, bring this all into solution by heating and stirring, set the beaker in a pan of ice and water, and let the solution cool undisturbed. When near room temperature, cool further by stirring and collect the product on a Büchner funnel. Use saturated sodium chloride solution rather than water for rinsing the material out of the beaker and for washing the filter cake free of the dark colored mother liquor. The filtration is somewhat slow.[3]

The product dries slowly and it contains about 20% of sodium chloride. The crude yield is thus not significant, and the material need not be dried before being purified. This particular azo dye is too soluble to be crystallized from water; it can be obtained in a fairly satisfactory form by adding saturated sodium chloride solution to a hot, filtered solution in water and cooling, but the best crystals are obtained from aqueous ethanol. Transfer the filter cake to a beaker, wash the material from the filter paper and funnel with water, and bring the cake into solution at the boiling point. Avoid a large excess of water, but use enough to prevent separation of solid during filtration (use about 50 ml). Filter by suction through a Büchner funnel that has been preheated on the steam bath. Pour the filtrate into an Erlenmeyer flask (wash), estimate the volume, and if greater than 60 ml, evaporate by boiling. Cool to 80°, add 100–125 ml of ethanol, and allow crystallization to proceed. Cool the solution well before collecting the product. Rinse the beaker with mother liquor and wash finally with a little ethanol. The yield of pure, crystalline material is 6.8 g. Orange II separates from aqueous alcohol with two molecules of water of crystallization and allowance for this should be made in calculating the yield. If the water of hydration is eliminated by drying at 120° the material becomes fiery red.

Extinguish flames!

3. Methyl Orange (*p*-Sulfobenzeneazo-4-dimethylaniline Sodium Salt)

In a test tube, thoroughly mix 3.2 ml of dimethylaniline and 2.5 ml of glacial acetic acid. To the suspension of diazotized sulfanilic acid contained in a 400-ml beaker add, with stirring, the solution of dimethylaniline acetate (rinse). Stir and mix thoroughly and within a few minutes the red, acid-stable form of the dye should separate. A stiff paste should result in 5–10 min and 35 ml of 10%

---

[3] If the filtration must be interrupted, close the rubber suction tubing (while the pump is still running) with a screw pinchclamp placed close to the filter flask and then disconnect the tubing from the pump. Fill the funnel and set the unit aside; thus, suction will be maintained, and filtration will continue.

sodium hydroxide solution is then added to produce the orange
sodium salt. Stir well and heat the mixture to the boiling point,
when a large part of the dye should dissolve. Place the beaker in a
pan of ice and water and allow the solution to cool undisturbed.

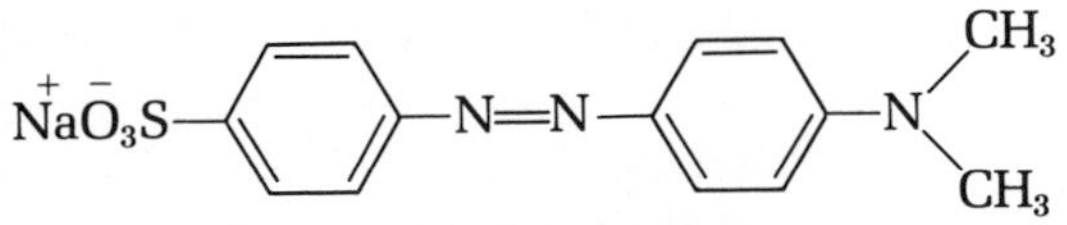

Methyl Orange
(alkali-stable form)

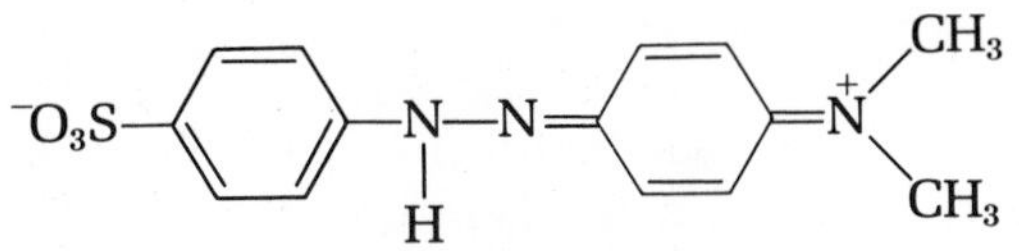

Methyl Orange
(acid-stable form)

When cooled thoroughly, collect the product on a Büchner funnel,
using saturated sodium chloride solution rather than water to rinse
the flask and to wash the dark mother liquor from the filter cake.

The crude product need not be dried but can be crystallized from
water after making preliminary solubility tests to determine the
proper conditions. The yield is 5–6 g.

Tests

Solubility and Color   Compare the solubility in water of Orange II
and Methyl Orange and account for the difference in terms of structure. Treat the first solution with alkali and note the change in shade
due to salt formation; to the other solution alternately add acid
and alkali.

Reduction   Characteristic of an azo compound is the ease with
which the molecule is cleaved at the double bond by reducing agents
to give two amines. Since amines are colorless, the reaction is easily
followed by the color change. The reaction is of use in preparation
of hydroxyamino and similar compounds, in analysis of azo dyes by
titration with a reducing agent, and in identification of azo compounds from an examination of the cleavage products.

Dissolve about 0.5 g of stannous chloride in 1 ml of concd hydrochloric acid, add a small quantity of the azo compound (0.1 g), and
heat. A colorless solution should result and no precipitate should
form on adding water. The aminophenol or the diamine derivative
is present as the soluble hydrochloride; the other product of cleavage, sulfanilic acid, is sufficiently soluble to remain in solution.

PART II.   Dyeing

EXPERIMENTS   ### 1. Direct Dyes

The sulfonate groups on the Methyl Orange and Orange II molecules are polar and thus enable these dyes to combine with polar sites in the fibers. Wool and silk have many polar sites on their polypeptide chains and hence bind strongly to a dye of this type. Martius Yellow, picric acid, and eosin are also highly polar dyes and thus dye directly to wool and silk.

Orange II or Methyl Orange

The dye bath is prepared from 0.5 g of Orange II or Methyl Orange, 5 ml of sodium sulfate solution, 300 ml of water, and 5 drops of concd sulfuric acid. Place a piece of test fabric in the bath for 5 min at a temperature near the boiling point. Remove fabric from the bath and let cool.

Replace one-half of the dyed fabric in the bath, make the solution alkaline with sodium carbonate, and add sodium hydrosulfite $(Na_2S_2O_4)$ until the color of the bath is discharged. Account for the result.

Picric Acid or Martius Yellow

Dissolve 0.5 g of the acid dye in a little hot water to which a few drops of dilute sulfuric acid have been added. Heat a piece of test fabric in this bath for one minute, then remove it with a stirring rod, rinse well, wring, and dry. Describe the results.

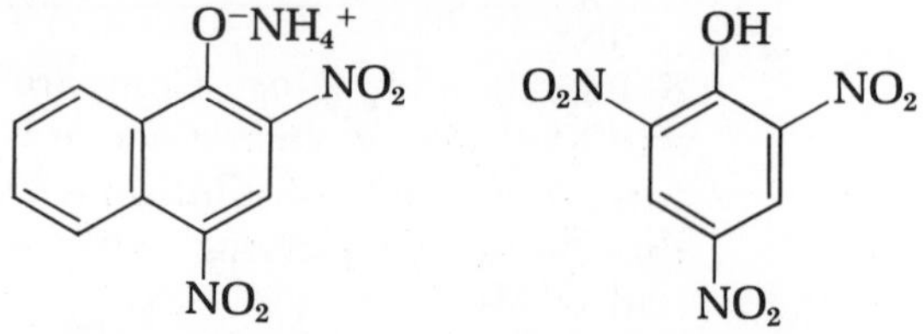

Martius Yellow   **Picric Acid**

Eosin

Dissolve 0.1 g of sodium eosin in 200 ml of water and dye a piece of test fabric by heating it with the solution for about 10 min.

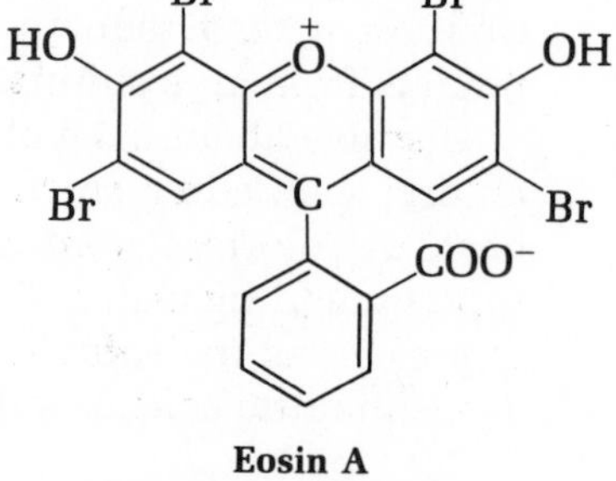

Eosin A
(λ_{max} 516,483 nm)

2. Substantive Dyes

Cotton and the rayons do not have the anionic and cationic carboxyl and amine groups of wool and silk and hence do not dye well with direct dyes, but they can be dyed with substances of rather high molecular weight showing colloidal properties; such dyes probably become fixed to the fiber by hydrogen bonding. Such a dye is Congo Red, a substantive dye.

Congo Red, a Benzidine Dye

Dissolve 0.1 g of Congo Red in 400 ml of water, add about 1 ml each of 10% solutions of sodium carbonate and sodium sulfate, heat to a temperature just below the boiling point, and introduce a piece of test fabric. At the end of 10 min remove the fabric and wash in warm water as long as the dye is removed. Place pieces of the dyed material in very dilute hydrochloric acid solution and observe the result. Rinse and wash the material with soap.

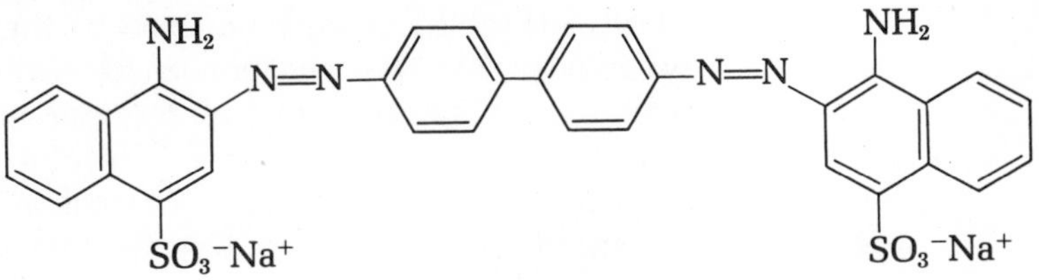

Congo Red
(λ_{max} 497 nm)

3. Mordant Dyes

One of the oldest known methods of producing wash-fast colors involves the use of metallic hydroxides, which form a link, or mordant (L. *mordere*, to bite), between the fabric and the dye. Other substances, such as tannic acid, also function as mordants. The color of the final product depends on both the dye used and the mordant. For instance, the dye Turkey Red is red with an aluminum mordant, violet with an iron mordant, and brownish-red with a chromium mordant. Important mordant dyes are those possessing a structure based on triphenylmethane, as exemplified by Crystal Violet and Malachite Green.

Synthesis of Crystal Violet, a Triphenylmethane Dye

Place 0.1 g of Michler's ketone, 5 drops of dimethylaniline, and 2 drops of phosphorus oxychloride in a test tube, and heat the tube in boiling water for 0.5 hr. Add 10 ml of water and stir. Add several drops of this solution to 20 ml of water and treat with a few drops of ammonium hydroxide solution. Let stand until the color has disappeared and then add dilute hydrochloric acid. Account for the color changes noted.

If the original solution is allowed to stand overnight, crystals of crystal violet should separate.

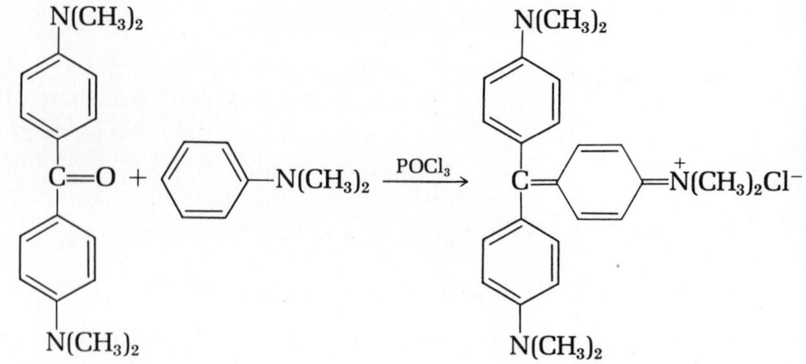

Michler's ketone   **Crystal Violet**
                                        (λ_{max} 591,540 nm)

Dyeing with a Triphenylmethane Dye and a Mordant

Mordant pieces of cotton cloth by allowing them to stand in a hot solution of 0.5 g of tannic acid in 500 ml of water for 5 min. The mordant must now be fixed to the cloth, otherwise it would wash out. For this purpose, transfer the cloth to a hot bath made from 0.2 g of tartar emetic (potassium antimonyl tartrate) in 200 ml of water. After 5 min, wring the cloth. A dye bath is prepared by dissolving 0.1 g of either Crystal Violet or Malachite Green in 200 ml of water (boiling). Dye the mordanted cloth in this bath for 5–10 min at a temperature just below the boiling point. Try further dyeings with the test fabric.

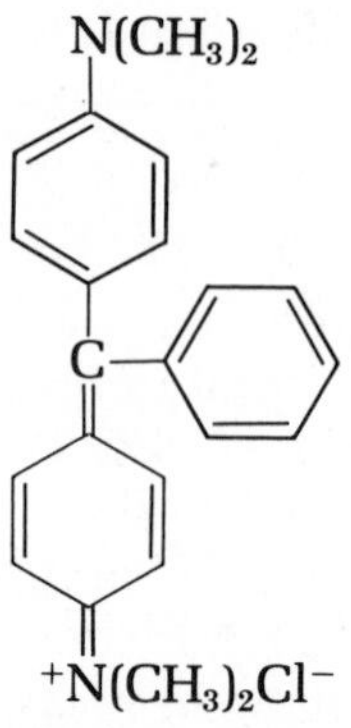

Malachite Green
(λ_{max} 617 nm)

Note: The stains on glass produced by triphenylmethane dyes can be removed with concd hydrochloric acid and washing with water, as HCl forms a di- or trihydrochloride more soluble in water than the original monosalt.

4. Developed Dyes

A superior method of applying azo dyes to cotton, patented in England in 1880, is that in which cotton is soaked in an alkaline solution of a phenol and then in an ice cold solution of a diazonium salt; the azo dye is developed directly on the fiber. The reverse process (ingrain dyeing) of impregnating cotton with an amine, which is then diazotized and developed by immersion in a solution of the phenol, was introduced in 1887. The first ingrain dye was Primuline Red, obtained by coupling the sulfur dye Primuline, after application to the cloth and diazotization, with 2-naphthol. Primuline (substantive to cotton) is a complex thiazole, prepared by heating p-toluidine with sulfur and then introducing a solubilizing sulfonic acid group.

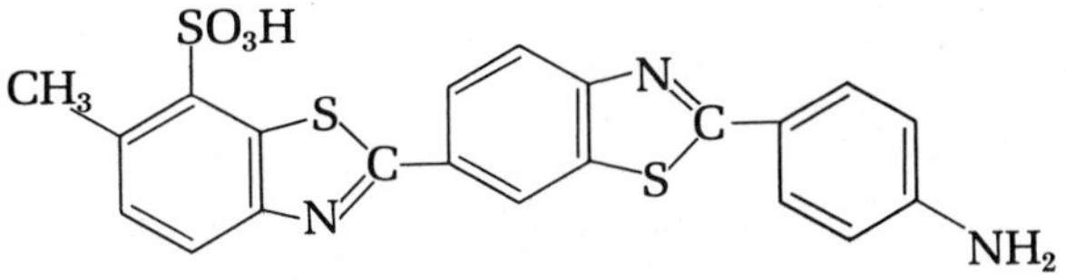

Primuline

Primuline Red

Dye three pieces of cotton cloth in a solution of 0.2 g of Primuline and 5 ml of sodium carbonate solution in 500 ml of water, at a temperature just below the boiling point for 15 min. Wash the cloth twice in about 500 ml of water. Prepare a diazotizing bath by dissolving 0.2 g of sodium nitrite in 500 ml of water containing a little ice and, just before using the bath, add 5 ml of concd hydrochloric acid. Allow the cloth dyed with Primuline to stay in this diazotizing bath for about 5 min. Now prepare three baths for the coupling reaction. Dissolve 0.1 g of 2-naphthol in 2 ml of 5% sodium hydroxide solution and dilute with 100 ml of water; prepare similar baths from phenol, resorcinol, Naphthol AS, or other phenolic substances.

Transfer the cloth from the diazotizing bath to a beaker containing about 500 ml of water and stir. Put one piece of cloth in each of the developing baths and allow them to stay for 5 min.

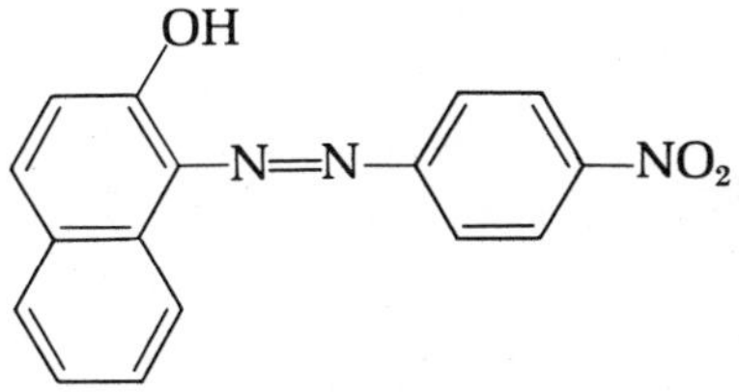

Para Red

Para Red, an Ingrain Color

Prepare a solution of p-nitrobenzenediazonium chloride as follows: dissolve 1.4 g of p-nitroaniline in a mixture of 30 ml of water and 6 ml of 10% hydrochloric acid by heating. Cool the solution in ice (the hydrochloride of the amine may crystallize), add all at once a solution of 0.7 g of sodium nitrite in about 5 ml of water and filter the solution by suction. The material to be dyed is first soaked in a solution prepared by suspending 0.5 g of 2-naphthol in 100 ml of water, stirring well and adding alkali, a drop at a time, until the naphthol all dissolves. This solution may also be painted onto the cloth. The cloth is then dried and dipped into the solution of the diazotized amine, after diluting the latter with about 300 ml of water.

Good results can be obtained by substituting Naphthol-AS for 2-naphthol; in this case, it is necessary to warm the Naphthol-AS with alkali and to break the lumps with a flattened stirring rod in order to bring the naphthol into solution.

5. Vat Dyes

Vat dyeing depends upon the fact that some dyes (e.g., indigo) can be reduced to a colorless, or leuco, derivative, which is soluble in dilute alkali. If fabric is immersed in this alkaline solution, the leuco compound is adsorbed by hydrogen bonding. On exposure to air the leuco compound is oxidized to the dye, which remains fixed to the cloth. Vat dyes are all quinonoid substances that are readily reduced to hydroquinonoid compounds reoxidizable by oxygen in the air.

The indigo so formed is very insoluble in all solvents. However, it is not covalently bound to the cotton, only adhering to the surface of the fiber. Hence, it is subject to removal by abrasion. This explains why the knees and other parts of blue jeans (dyed exclusively with indigo) subject to wear will gradually turn white.

Dyeing with a Vat Dye

Use 0.2 g of one of the following dyes: Indigo, Indanthrene Brilliant Violet, Indanthrene Yellow. If one of the dyes is available in

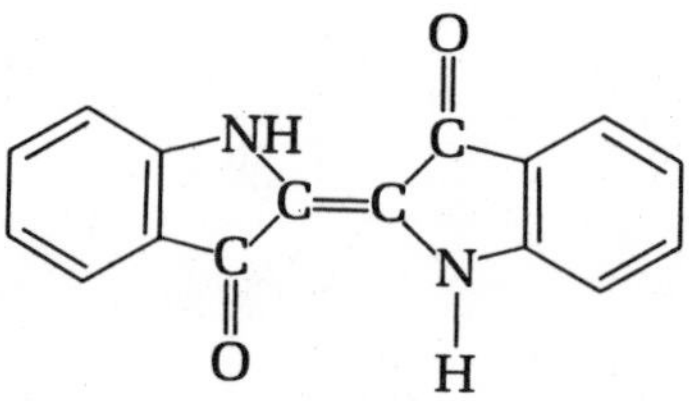

Indigo 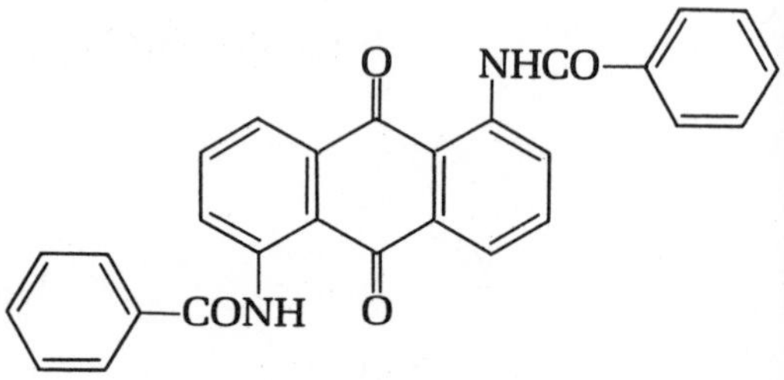   Indanthrene Yellow

the form of a paste use as much as will adhere to about 1 cm of the end of a stirring rod. Boil the dye with 100–200 ml of water, 5 ml of 10% sodium hydroxide solution, and about 1 g of sodium hydrosulfite until the dye is reduced. Introduce a piece of cloth and boil the solution gently for 10 min. Rinse the cloth well in water and allow it to dry.

6. Disperse Dyes

Fibers such as Dacron, acetate rayon, Nylon, and polypropylene are difficult to dye with conventional dyes because they contain so few polar groups. These fibers are dyed with substances which are insoluble in water but which at elevated temperatures (pressure vessels) are soluble in the fiber as true solutions. They are applied to the fiber in the form of a dispersion of finely divided dye, hence the name. The Cellitons are typical disperse dyes.

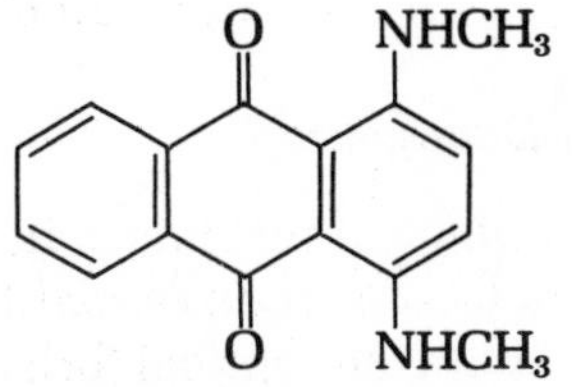

Celliton Fast Blue B 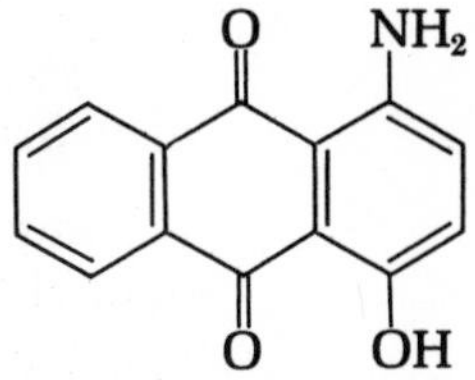   **Celliton Fast Pink B**

7. Fiber Reactive Dyes

Among the newest of the dyes are the fiber reactive compounds which form a covalent link to the hydroxyl groups of cellulose. The reaction involves an amazing and little understood nucleophilic displacement of a chloride ion from the triazine part of the molecule by the hydroxyl groups of cellulose yet the reaction occurs in aqueous solution.

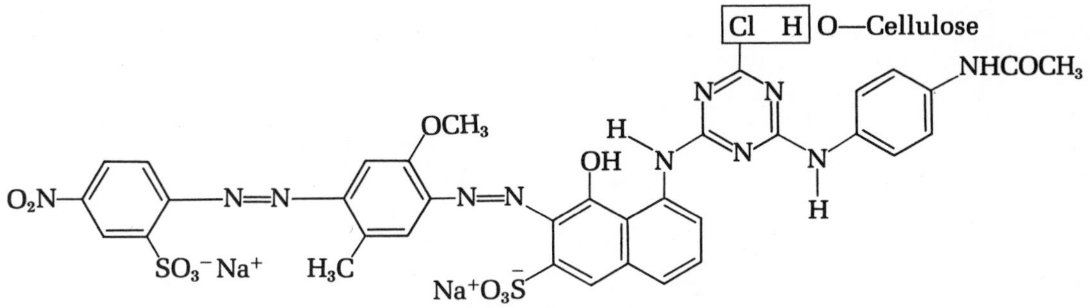

Chlorantin Light Blue 8G

8. Optical Brightners—Fluorescent White Dyes

Most modern detergents contain a blue-white fluorescent dye which is adsorbed on the cloth during the washing process. These dyes fluoresce, that is, absorb ultraviolet light and reemit light in the visible blue region of the spectrum. This blue color counteracts the pale yellow color of white goods, which develops because of a buildup of lipid soil. The modern day use of optical brighteners has replaced a past custom of using bluing (ferriferrocyanide).

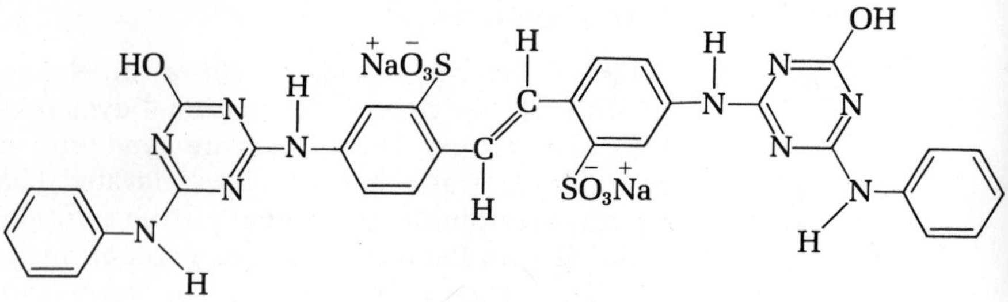

Blankophor B,
an optical brightener

Dyeing with Detergents

Immerse a piece of test fabric in a hot solution (0.5 g of detergent, 200 ml of water) of a commercial laundry detergent which you suspect may contain an optical brightener (*e.g., Tide* and *New Blue Cheer*) for 15 min. Rinse the fabric thoroughly, dry, and compare with an untreated fabric sample under an ultraviolet lamp.

54 Order of Reaction

KEYWORDS Tetraphenylcyclopentadienone
Maleic anhydride
Rate dependent on conc
  of dienophile?
CO given off?
Product structure?
Constant-temperature bath

Refluxing o-dichlorobenzene
Tetraethylene glycol, heat
  exchange fluid
Dewar condenser
Pyridinium bromide perbromide,
  aromatization

The second experiment of Chapter 51 involves a Diels-Alder reaction of tetraphenylcyclopentadienone with an acetylenic dienophile with loss of carbon monoxide and formation of an aromatic product. In the present experiment a pair of students is to study the reaction of tetraphenylcyclopentadienone with maleic anhydride at two concentrations of this dienophile and, from the results, to infer the nature of the adduct and to decide whether or not the reaction rate is dependent upon the concentration of the dienophile. Note that if the rate is indeed dependent upon concentration the time required for utilization of a given amount of colored dienone can be shortened by use of excess dienophile.

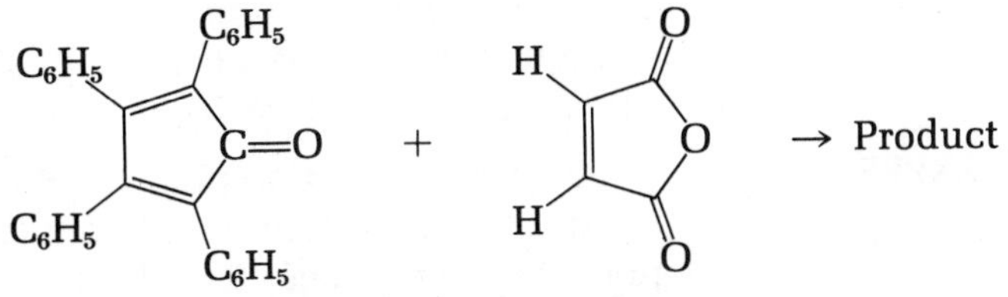

Tetraphenylcyclopentadienone
MW 384.45, mp 219°
(the diene)

Maleic anhydride
MW 98.06, mp 53°
(the dienophile)

For significant comparison of reaction times at two concentrations of one reactant, it is necessary that the two runs be made at the same temperature. This condition is not fulfilled by carrying out the reactions in a given refluxing solvent, for the *liquid* temperature will be higher in the more concentrated of the two solutions because of greater elevation of the boiling point. An elegant constant-temperature bath (Fig. 54.1) is constructed from a Kontes Dewar condenser (K457750) fitted with a 24/40 condenser and attached to a 125-ml round-bottomed flask containing 70 ml of o-dichlorobenzene (bp 179°).[1] The reaction vessel is a 20 × 150-mm test tube placed in a

[1] Ace Glass item.

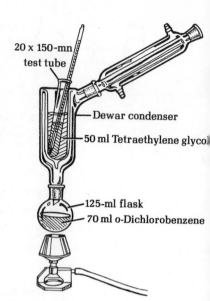

20 x 150-mn
test tube

— Dewar condenser

— 50 ml Tetraethylene glycol

— 125-ml flask
— 70 ml o-Dichlorobenzene

**Figure 54.1
Constant temperature
apparatus.**

*Constant-temperature
boiler*

heating bath containing 50 ml of the nonvolatile, high-boiling tetra-ethylene glycol as heat-exchange fluid.

You are to make runs at two concentrations of the dienophile that are sufficiently different to show whether or not the reaction rate is dependent upon the concentration of the dienophile. You are advised against use of only 1.0 equivalent of the dienophile, as the reaction will be inordinately slow.[2] For 1.0 g of tetraphenylcyclopentadienone and 5 ml of o-dichlorobenzene, it is suggested that in the first run you use 4 equivalents (1.02 g) of maleic anhydride. You can then decide on the amount to be used in the second run, for example, 3.0, 2.0, or 1.5 equivalents.

EXPERIMENT   Place 1.0 g of tetraphenylcyclopentadienone and 5 ml of o-dichlorobenzene in a 20 × 150-mm test tube, introduce a thermometer, note the time, place the tube in the heating bath of the o-dichlorobenzene boiler, and reflux vigorously until the purple solution acquires a constant temperature (which may be a few degrees below the true boiling point). The heatup period is about 18–20 min. A 1-g sample of maleic anhydride is then added and rinsed down the walls with a few drops of o-dichlorobenzene and the time is noted. Keep a time-temperature record and note particularly the time required to reach a stage which you think you can recognize and reproduce in a second run, for example, a pale yellow solution with reduction in the rate of bubbling. For isolation of the product, remove the test tube from the boiler, dilute the mixture with benzene, cool in ice, and use ligroin to wash and rinse out the tube.

[2] O. Grummit, *Org. Syn., Coll. Vol.* **3**, 807 (1945), mixed 0.094 mole of tetraphenylcyclopentadienone with 0.095 mole of maleic anhydride and 25 ml of bromobenzene (bp 155.5°) and refluxed the mixture for 3.5 hr; yield 88%.

QUESTIONS   1. *Is the reaction attended with elimination of carbon monoxide and what is the structure of the crystalline product? The reaction product is sparingly soluble in acetic acid, but a small sample (0.1 g) can be dissolved by refluxing it with this solvent. Addition to the hot solution of pyridinium hydrobromide perbromide (Chapter 14) will disclose unsaturation (and effect aromatization).*

2. *On the assumption that the reaction rate is dependent upon the concentration of the dienone, is the reaction:*

1st order, rate = k_1 [Dienone], or
2nd order, rate = k_2 [Dienone][Dienophile]?

Note: The heating chamber shown in Fig. 54.1 is large enough to hold five 13×100-mm test tubes, each of which can easily accommodate 5 ml of solution. Hence, as many as 5 runs on the scale suggested can be made at the same time.

55 Benzilic Acid Rearrangement

KEYWORDS

Base catalyzed rearrangement
Two step process or concerted mechanism?
Blue-violet pigment, structure?
Potassium benzilate
Cyclodehydration, mechanism
Carbonium ion intermediate

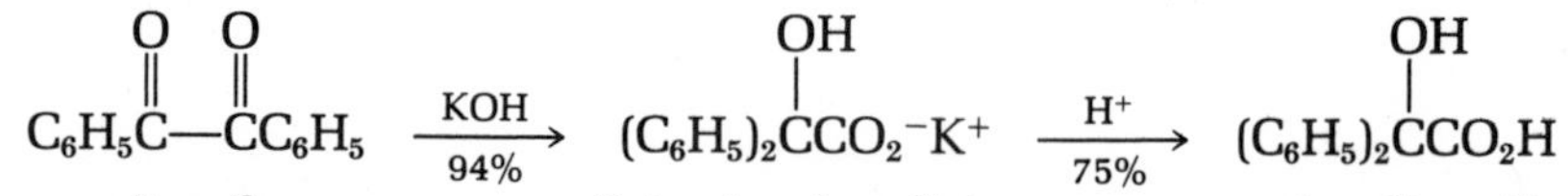

$$C_6H_5\overset{O}{\overset{\|}{C}}-\overset{O}{\overset{\|}{C}}C_6H_5 \xrightarrow[94\%]{KOH} (C_6H_5)_2\overset{OH}{\overset{|}{C}}CO_2{}^-K^+ \xrightarrow[75\%]{H^+} (C_6H_5)_2\overset{OH}{\overset{|}{C}}CO_2H$$

Benzil	**Potassium benzilate**	**Benzilic acid**
MW 210.22, mp 96°	MW 266.33	MW 228.24, mp 151° pK$_a$ 3.04

Potassium hydroxide effects rearrangement of benzil to give the potassium salt of benzilic acid, from which the free acid is obtained on acidification. In 1870 A. Jena[1] criticized formula **1** for benzilic acid proposed by Grimaux and formula **2** proposed by Kekulé. Jena then suggested formula **3** which indeed is closer to the formula now accepted.

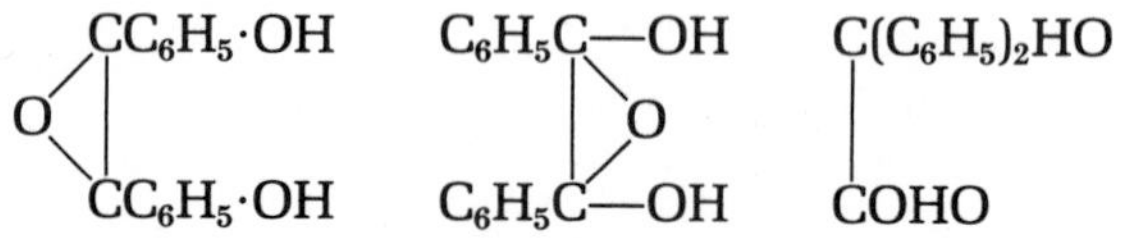

$$
\begin{matrix}
& CC_6H_5\cdot OH & & C_6H_5C-OH & & C(C_6H_5)_2HO \\
O & | & & | \quad | & & | \\
& CC_6H_5\cdot OH & & C_6H_5C-OH & & COHO \\
& \mathbf{1} & & \mathbf{2} & & \mathbf{3}
\end{matrix}
$$

In the first laboratory manual ever written for students,[2] Emil Fischer cited a preparative procedure for benzilic acid in which potassium hydroxide is treated with a little water, melted, and brought to 150°, and treated with benzil. Fischer stated that the benzilic acid prepared by this method contains benzoic acid, but that this contaminant can be removed by heating the product in a dish with water "bis der Geruch der Benzoësäure verschwunden ist" (but benzoic

[1] A. Jena, *Ann.*, **155**, 79 (1870).
[2] E. Fischer, "Anleitung zur Darstellung organischer Präparate," 8 Auflage, Braunschweig, 1908.

acid is odorless!). A better process, based upon the fact that benzilic acid is a stronger acid than benzoic acid (pK_a 4.17), is by fractional solution in sodium carbonate solution.

The present procedure calls for addition of a cold aqueous solution of potassium hydroxide to a hot solution of benzil in 95% ethanol, an operation which is attended by dramatic results: white needles are precipitated and the solution acquires an intense blue-violet color. On refluxing for a time the pigment gradually disappears and colorless needles of potassium benzilate separate in high yield. Scheuing[3] examined the crystalline product initially precipitated and found that it is a complex formed by addition of KOH to one of the carbonyl groups of benzil and that at 80° it rearranges to potassium benzilate:

$$C_6H_5\overset{\overset{O}{\|}}{C}-\overset{\overset{O}{\|}}{C}C_6H_5 \xrightarrow{KOH} C_6H_5\overset{\overset{O}{\|}}{C}-\underset{\underset{OH}{|}}{\overset{\overset{O^-K^+}{|}}{C}}C_6H_5 \rightarrow C_6H_5-\underset{\underset{C_6H_5}{|}}{\overset{\overset{OH}{|}}{C}}-\underset{\underset{O}{\|}}{\overset{\overset{O^-K^+}{|}}{C}}$$

<hr>
Intermediate adduct

Scheuing had expected to find evidence that the blue-violet pigment is an intermediate to potassium benzilate, but his observations led him to the view that such is not the case. As you carry out the experiment, see if you can form a judgment on this point. Also, as you view the beautiful blue-violet solution, you can reflect that the pigment has been known in solution for over a century but that the nature of the substance still awaits elucidation.

EXPERIMENT   1. Benzilic Acid

In a 25-ml Erlenmeyer dissolve 7.0 g of potassium hydroxide in 15 ml of water by heating on the hot plate and cool to room temperature. Place 7.0 g of benzil and 25 ml of methanol in a 100-ml round-bottomed flask and heat on the steam bath to dissolve the solid. Remove the flask from the heating bath, pour in the cold alkali solution, swirl, and observe the result. Then note the time and heat the mixture under reflux on the steam bath until the blue-violet color disappears. Cool in ice, collect the colorless needles of potassium benzilate, let dry, and take the weight. Dissolve the salt in the least amount of hot water, and acidify with hydrochloric acid (check the pH). Then add enough 95% ethanol to bring the free acid into solution at the boiling point and let the solution stand for crystallization.

2. Mechanism

In theory the rearrangement of benzil might proceed by either the

[3] G. Scheuing, *Ber.*, **56**, 252 (1923).

two-step process (a) or the concerted mechanism (b).

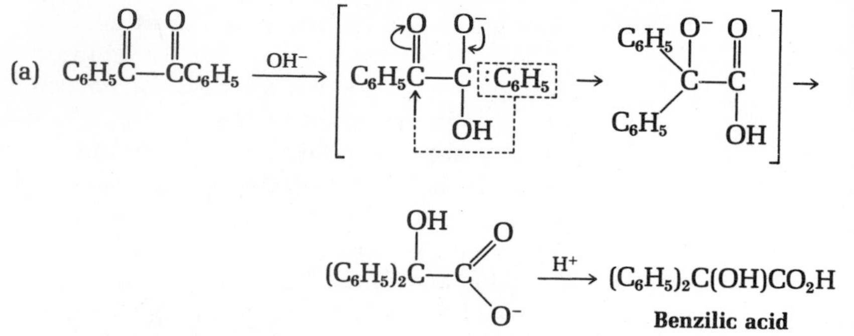

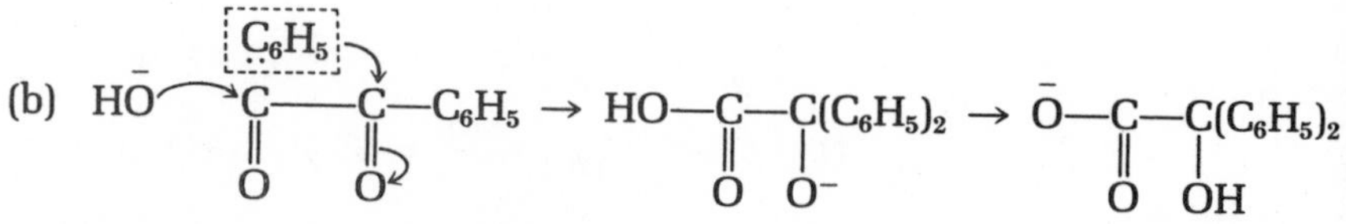

(a) $C_6H_5\overset{O}{\overset{\|}{C}}-\overset{O}{\overset{\|}{C}}C_6H_5 \xrightarrow{OH^-}$ $\left[C_6H_5\overset{O}{\overset{\|}{C}}-\underset{OH}{\overset{O^-}{\underset{|}{C}}}[C_6H_5] \rightarrow \underset{C_6H_5}{\overset{C_6H_5}{C}}\overset{O^-}{\underset{\|}{\overset{|}{C}}}-\overset{O}{\underset{OH}{\overset{\|}{C}}} \right] \rightarrow$

$(C_6H_5)_2\overset{OH}{\underset{|}{C}}-\overset{O}{\overset{\|}{C}}\underset{O^-}{\diagdown} \xrightarrow{H^+} (C_6H_5)_2C(OH)CO_2H$

Benzilic acid

(b) $\overset{-}{HO}\diagup\overset{\frown}{C}-\overset{[C_6H_5]\diagdown}{\underset{O^{\searrow}}{\overset{|}{C}}}-C_6H_5 \rightarrow HO-\overset{O}{\overset{\|}{C}}-\overset{O^-}{\underset{\|}{C}}(C_6H_5)_2 \rightarrow \overset{-}{O}-\overset{O}{\overset{\|}{C}}-\overset{OH}{\underset{\|}{C}}(C_6H_5)_2$

The intermediate mentioned previously and observed by you as a color, favors one of these two mechanisms. Can you devise still other experiments, both kinetic and nonkinetic, which could furnish further distinguishing evidence?

3. Preparation of 9-Fluorenecarboxylic acid

Catalysis by a Lewis acid

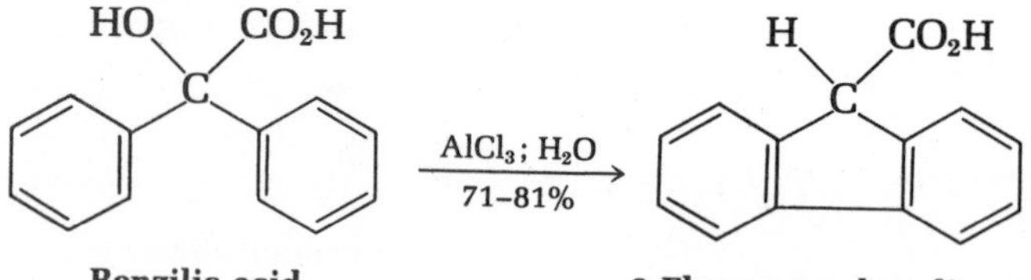

$\xrightarrow[71-81\%]{AlCl_3; H_2O}$

Benzilic acid
MW 228.24, mp 151°

9-Fluorenecarboxylic acid
MW 210.22, mp 223°

Formed from benzil by base-catalyzed rearrangement, benzilic acid when refluxed with aluminum chloride in benzene undergoes cyclodehydration with rearrangement to give 9-fluorenecarboxylic acid.[4] A standardized procedure[5] is available for carrying out the reaction, but this calls for use of a stirring motor and involves a tedious extraction with soda solution. It is suggested that instead of carrying out this standard procedure you work out a mechanism for the reaction. Note that rearrangements catalyzed by mineral acids or Lewis acids (for example, the Wagner-Meerwein rearrangement) usually proceed through a carbonium ion intermediate. Note also that triphenylcarbinol on reaction with hot phosphoric acid is converted into 9-phenylfluorene in 40–50% yield.[6]

[4] D. Vorländer and A. Pritszche, *Ber.*, **46**, 1793 (1913); R. T. Arnold, W. E. Parham and R. M. Dodson, *Am. Soc.*, **71**, 2439 (1949).
[5] H. J. Richter, *Org. Syn., Coll. Vol.* **4**, 482 (1963).
[6] A. Kliegl, *Ber.*, **38**, 284 (1905).

56 3,5-Dimethyl-pyrazole

KEYWORDS

Aromatic heterocycle synthesis
Pentanedione, hydrazine
Acidic and the basic properties of adduct
Dimedone
Picric acid salt 1 : 1 or 1 : 2?

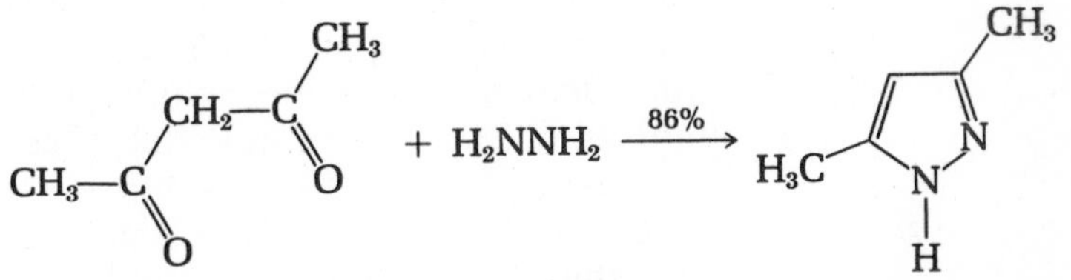

2,4-Pentanedione	Hydrazine	3,5-Dimethylpyrazole
MW 100.11,	MW 32.05,	MW 96.13,
bp 134°	64% in H₂O	mp 108°

The heterocycle 3,5-dimethylpyrazole is an interesting aromatic compound that can be prepared with finesse by the procedure described below. In case your results are satisfactory, you may care to compare the procedure used with that reported in *Organic Syntheses*[1] with respect to ease of experimentation and temperature control, reaction time, compensation for variation in the concentration of commercial hydrazine solution, ease of workup, yield and quality of product, and safety.

EXPERIMENTS

1. Preparation of 3,5-Dimethylpyrazole

High-yield condensation

Weigh 10.0 g of 2,4-pentanedione into a 125-ml Erlenmeyer, add 50 ml of distilled water, insert a thermometer, and make ready an ice-water bath. Measure 5.0 ml of a 64% solution of hydrazine in water (Eastman) in a 10-ml graduate and add the solution to the pentanedione, with a capillary dropping tube, in portions of about 0.5 ml. Keep the temperature at ca. 40° by cooling, as required. White plates of the product begin to separate after a few minutes.

A check to make sure that the hydrazine solution is of full strength

[1] H. Wiley and P. E. Hexner, *Org. Syn., Coll. Vol.*, **4**, 351 (1963).

321

can be made as follows: After the addition of the hydrazine is complete, adjust the temperature to 40°, add another 0.5-ml portion of the hydrazine solution and see if it causes a rise in temperature; if so, adjust to 40° and add another portion. When sure that the reaction is complete, cool well in ice, collect the product, and let dry in air.

Properties

Dimethylpyrazole has been chlorinated, nitrated, and sulfonated at the 4-position. The aromatic character of the compound, evident from these substitutions, provides the driving force for the formation of the pyrazole in high yield in an uncatalyzed condensation reaction. To see if dimethylpyrazole is acidic, place approximately equivalent small amounts in two small test tubes, add equivalent small amounts of water to each tube and a few drops of alkali to one tube. Shake, and see if the alkali promotes solution of the solid. Then test for basic character in the same way but with addition of a little hydrochloric acid to the other tube.

In case you find the substance basic, the next problem is to determine whether it combines with one or with two moles of an acid. Picric acid, MW 229.10, mp 122.5°, combines with the heterocycle to form a yellow salt (prisms, mp 165–167°). After running tests to find a suitable solvent or solvent pair, carry out an experiment which will determine whether the combining ratio of dimethylpyrazole to picric acid is 1:1 or 1:2.

2. Other Heterocycles from Hydrazine

Another example of the construction of a five-membered ring containing adjacent nitrogen atoms is the synthesis of indazole (**4**) by formylation of cyclohexanone, condensation with hydrazine to afford **3** and dehydrogenation.[2]

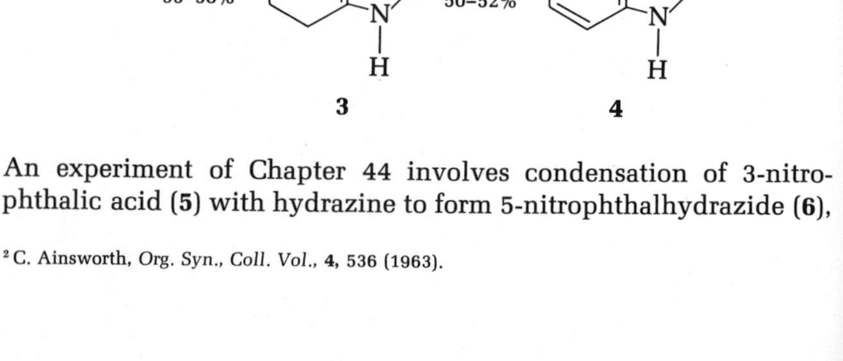

An experiment of Chapter 44 involves condensation of 3-nitrophthalic acid (**5**) with hydrazine to form 5-nitrophthalhydrazide (**6**),

---

[2] C. Ainsworth, *Org. Syn., Coll. Vol.*, **4**, 536 (1963).

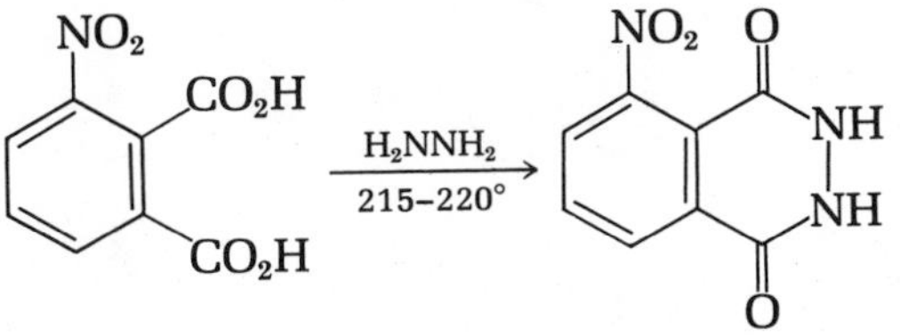

5   **6**

*Of interest for
further study*

which has a six-membered ring containing adjacent nitrogen atoms.
    Of particular interest is the observation of Stille and Ertz[3] that
condensation of hydrazine with 5,5-dimethyl-1,3-cyclohexanedione

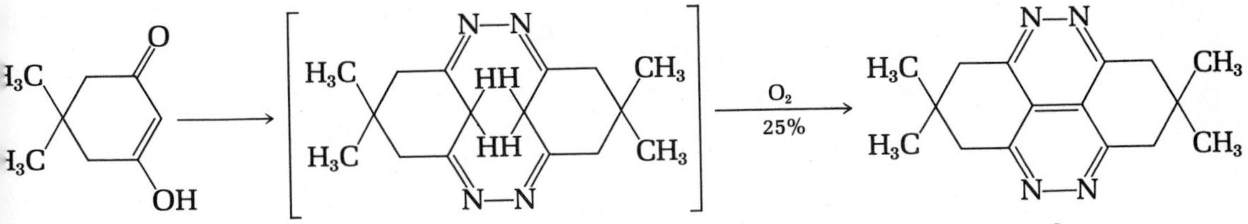

7   **8**   **9**

Mp 149°   Mp 277–278°, dec

(dimedone) as its enol **7** gives an initial product **8**, which apparently
is destabilized by repulsive interaction of internal hydrogens and
hence subject to air oxidation with aromatization to **9**, which can
be described as a derivative of 2,3,6,7-tetraazanaphthalene.[4] This
naphthalene analog, like naphthalene, combines with picric acid
in ethanol to give a 1 : 1 complex, mp 171–172°, dec. Inspection of
stereomodels of **8** and of **9** is interesting.

[3] J. K. Stille and R. Ertz, *Am. Soc.,* **86**, 661 (1964).
[4] In a procedure which is simplified, but which does not raise the yield above 25%, a mixture of
10 g of **1**, 30 ml of 95% hydrazine, and 50 ml of either ethylene glycol or N,N-dimethylformamide
is refluxed for 16 hr while passing in a stream of air.

57 Benzimidazole and a Problem

KEYWORDS   o-Phenylenediamine
Condensation with carboxylic acids
Practical grade
o-Phenylenediamine and crotonic acid, problem compound
Structure of product?
Stable to alkaline hydrolysis?

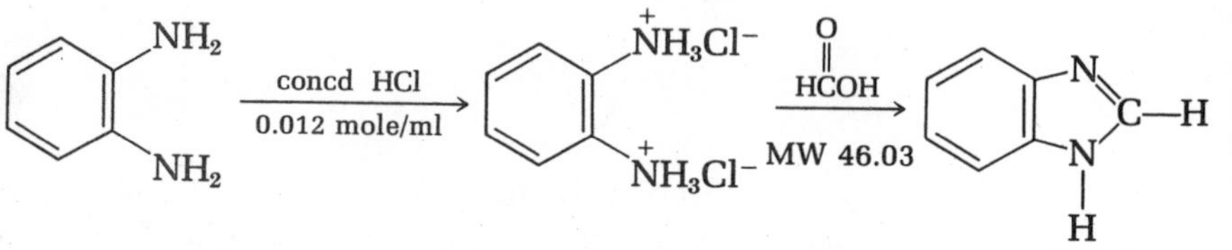

1
o-Phenylenediamine
MW 108.14,
mp 102°

2
MW 181.08,
mp about 260°, dec

3
Benzimidazole
MW 118.14,
mp 171–172°, pK$_a$ 5.48

Commercially available *o*-phenylenediamine is of practical grade, badly discolored, and contains considerable material insoluble in dilute mineral acid. A very small sample is conveniently purified by sublimation in an evacuated test tube. Commercial *o*-phenylenediamine that is to be converted into benzimidazole (**3**) by condensation with formic acid to effect elimination of two molecules of water can also be conveniently purified and used as the dihydrochloride (**2**).

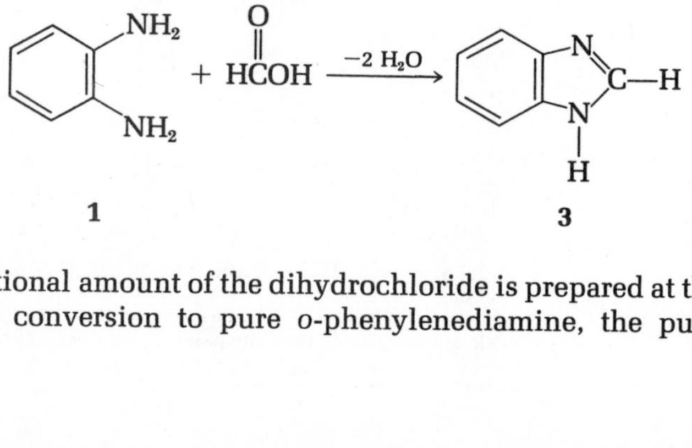

1

3

An additional amount of the dihydrochloride is prepared at the same time for conversion to pure *o*-phenylenediamine, the pure form being required for preparation of a problem compound in this chapter.

1. *o*-Phenylenediamine Dihydrochloride and Free Base

Heat a mixture of 10 g of practical grade *o*-phenylenediamine, 16 ml of concd hydrochloric acid, and 40 ml of water in a 125-ml Erlenmeyer to the boiling point. Note that considerable material remains undissolved. Suspend heating, add decolorizing charcoal for clarification, and filter by gravity into a 250-ml Erlenmeyer. The filtrate should be colorless, or nearly so. Add 47 ml of concd hydrochloric acid and swirl vigorously in an ice bath to promote full crystallization of the dihydrochloride as white or pinkish crystals. Collect, wash with a little acetone, and place on a tared filter paper on the drier. At constant weight, a typical yield is 12.4 g. Note that impurities in the starting material account for a considerable part of the difference between the theoretical yield (16.7 g) and the actual yield. Put aside half of the dihydrochloride (2) for experiment 3 and dissolve the remaining 6.2 g in 10 ml of distilled water by heating. Neutralize with ammonium hydroxide solution (29% NH_3), cool thoroughly to 0° to promote full crystallization of the colorless plates of product, and collect and dry the *o*-phenylenediamine; yield 2.95 g (82%).

Purification of a diamine through its dihydrochloride

2. Benzimidazole

Usual Reaction:

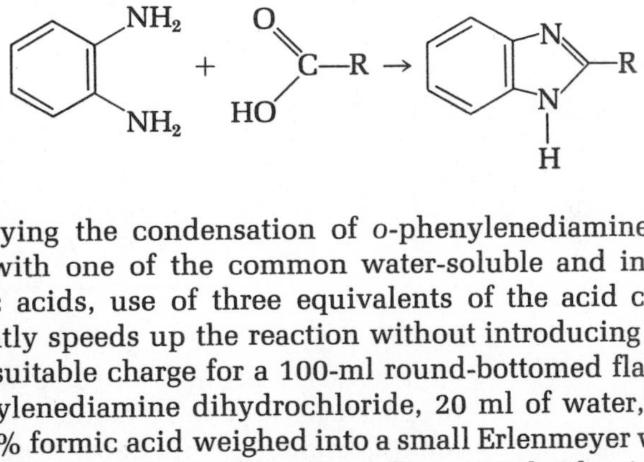

In studying the condensation of *o*-phenylenediamine dihydrochloride with one of the common water-soluble and inexpensive carboxylic acids, use of three equivalents of the acid component conveniently speeds up the reaction without introducing complications. A suitable charge for a 100-ml round-bottomed flask is 6.2 g of *o*-phenylenediamine dihydrochloride, 20 ml of water, and 4.8 g of 97–100% formic acid weighed into a small Erlenmeyer with use of a capillary dropping tube. Choose a reflux period either in the range 20–40 min, or 1–3 hr, or overnight, and so combine with other members of the class to acquire data for a time-yield curve. To run a reaction overnight without the hazard of a free flame, make ready a condenser with clamp and a hot plate at full heat, swirl the reaction flask over a free flame until the liquid begins to boil, note the time, attach the condenser, and rest the flask on the hot plate. Although there is little active boiling, because of the poor fit of the round flask to flat plate, if you suspend a thermometer on a nichrome wire with the bulb in the liquid you will note that it records a temperature a few degrees above 100°. (Why?) Another variation which might improve the yield in a given reaction period would be to replace some of the water—say half of it—by a high-boiling water-miscible solvent.

At the end of the reaction period chosen, cool the solution in an ice-water bath and add 10 ml of concd ammonia solution to neutralize the hydrogen chloride and some of the excess formic acid; this may cause some separation of a solid, probably a mixture of benzimidazole and its formate. With continued cooling, add more ammonia solution—about 1 ml at a time—until the mixture is basic to test paper. By heating over a free flame and addition of more water as required, bring all the white solid into solution, make sure the solution is basic, and let it stand; slow crystallization gives beautiful crystal sheaves.

Liberation of benzimidazole by addition of ammonia to the acidic reaction mixture shows that the heterocycle is basic. Tests will show that it is also acidic (pK$_a$ 5.48) and that it is stable to the hydrolytic action of alkalis, as well as that of acids.

In comparable small-scale reactions of o-phenylenediamine dihydrochloride with 3 equivalents of other acids, yields were as shown in Table 57.1.

Table 57.1   Reaction Time and Product of 9.05 g o-Phenylenediamine Dihydrochloride in 20 ml Water with 3 Equivalents of an Acid

Acid	Reflux time (hr)	Product (2-Alkylbenzimidazole)	Yield, %
Acetic	0.5	Methyl (mp 177°)	83
	2		90
	15		97
Propionic	2	Ethyl (mp 174.5°)	75
Isobutyric	0.5	Isopropyl (mp 232–233°)	17
	2		77.5
Phenylacetic	0.5	Benzyl (mp 187°)	37

Conversion of aliphatic acids into 2-alkylbenzimidazoles has been proposed as a general method for preparing solid derivatives for identification.[1] Comparison of the melting points of benzimidazole and naphthalene with their 2-substituted derivatives shows that there is little regularity here (Table 57.2).

Table 57.2   Melting Points of Derivatives

	2-H	2-CH$_2$	2-C$_2$H$_5$	2-(CH$_3$)$_2$CH	2-CH$_2$C$_6$H$_5$
Naphthalene	80°	24.4°	−19°	Liq. −21°	35.5°
Benzimidazole	171°	177°	174.5°	233°	187°

[1] R. Seka and R. H. Müller, *Monatsh.*, **57**, 97 (1931); W. O. Pool, H. J. Harwood, and A. W. Ralston, *Am. Soc.*, **59**, 178 (1937).

3. Problem Compound

The problem here is to condense o-phenylenediamine with an equivalent amount of the α,β-unsaturated acid crotonic acid, isolate a nicely crystalline product (mp 186°), submit it to such tests as you see fit, and deduce the structure.

What is the product?

$$\text{(o-phenylenediamine)} \quad + \quad CH_3CH{=}CHCO_2H \xrightarrow[-H_2O]{132°} \text{Problem Compound}$$

Crotonic acid
MW 86.09, mp 72°, pK_a 4.70   Mp 186°

The reaction is carried out conveniently by heating a mixture of the components in a 25×150-mm test tube mounted in the triethylene glycol-filled heating chamber of the constant temperature bath of Fig. 54.1, with the boiling flask charged with a xylene mixture, bp 138–140°. A mixture of 2.5 g of pure o-phenylenediamine and 2 g of crotonic acid is placed in the tube and this is connected, by means of a rubber stopper and glass and rubber tubing, to a suction pump turned on at full force. The tube is evacuated throughout the reaction both to promote elimination of the water formed and to inhibit air oxidation of the diamine or of the reaction product. After the tube has been evacuated, support it in the heating chamber with a clamp and note the time. Keep a record of time and temperature, and note the time at which droplets of water begin to condense on the upper walls of the test tube. Note that the bath temperature of about 138° is well above the melting point of both reactants but below the melting point of the organic product; also that water, the only other product, is largely eliminated as formed. There is thus a gradual transition from a fluid melt to a crystal magma and then a hard solid. A method of probing for the end point is to remove the tube, cool it in tap water, and look for signs of solidification; if there are no such signs, reheat and see if the appearance alters. A period of heating under vacuum for 5–6 hr is recommended.

The product is to be isolated by extraction with successive portions of boiling 95% ethanol. Remove the tube and cool slightly, insert a stirring rod with a flat end—useful for disintegrating a solid product, add about 10 ml of alcohol, and boil and stir to dissolve as much material as possible. Decant into a 125-ml Erlenmeyer and repeat the process of extraction until all the reaction mixture has been brought into a wine-red solution. Cool the hot solution in an ice bath to effect separation of prismatic needles of the product which, however, has a marked tendency to remain in supersaturated solution. Vigorous scratching of the flask walls with a stirring rod may be required to effect full crystallization. The product is obtained as slightly pinkish crystals, mp 184–185°. You may be able to secure a second crop by clarifying the mother liquor with decolorizing charcoal, removing most of the ethanol by distillation, and crystallizing the residue from methanol.

In characterizing the compound, determine if it is stable to the hydrolytic action of boiling alkali, as expected for 2-propenyl benzimidazole (1).

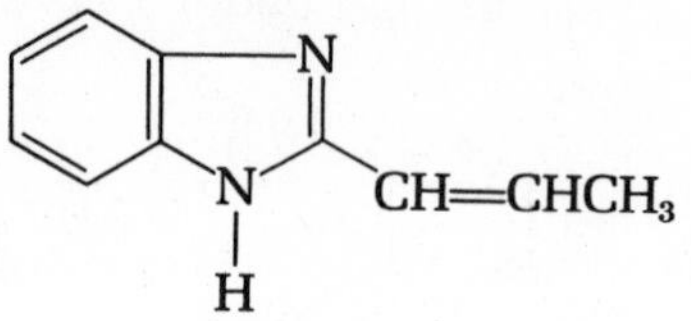

1

$C_{10}H_{10}N_2$

Place 0.5 g of the compound and 20 ml of 10% aqueous sodium hydroxide solution in a 100-ml boiling flask, reflux, and swirl to dislodge solid from the walls. After 20–25 min the solid is all dissolved. Cool the solution and acidify it with acetic acid. In case there is a precipitate, determine if it can be brought into solution by additon of dilute hydrochloric acid. What is the probable structure of the compound, mp 186°? What analytical evidence would serve to eliminate formula 1?

58 Lycopene and β-Carotene

KEYWORDS   Carotenoid, isoprene units
Chromophore, conjugated double bonds
Tomato, lycopene
Carrot, β-carotene
Photochemical air oxidation
Column chromatography

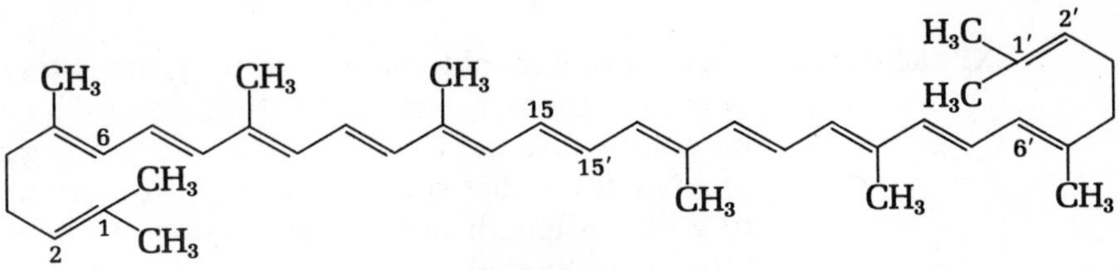

Lycopene ($C_{40}H_{56}$)
MW 536.85,
mp 173°, λ_{max}^{Hexane} 475 nm

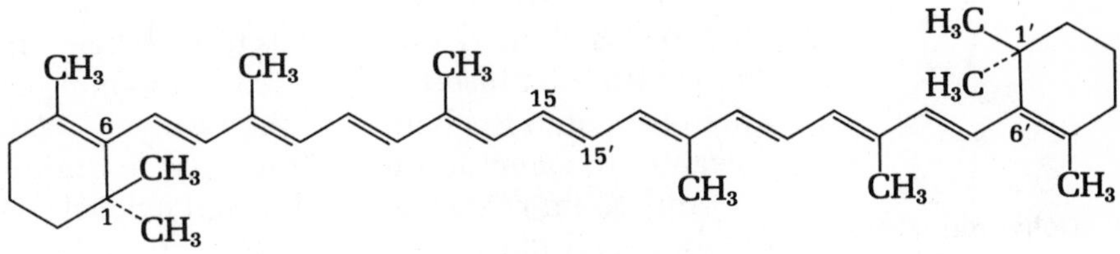

β-Carotene ($C_{40}H_{56}$)
Mp 183°, λ_{max}^{Hexane} 451 nm

Lycopene, the red pigment of the tomato, is a C_{40}-carotenoid made up of eight isoprene units. β-Carotene, the yellow pigment of the carrot, is an isomer of lycopene in which the double bonds at C_1—C_2 and C_1'—C_2' are replaced by bonds extending from C_1 to C_6 and from C_1' to C_6' to form rings. The chromophore in each case is a system of eleven all-*trans* conjugated double bonds; the closing of the two rings renders β-carotene less highly pigmented than lycopene.

The colored hydrocarbons are materials for an interesting experiment in thin layer chromatography (Chapter 23), but commercial samples are extremely expensive and subject to deterioration on storage. The isolation procedure in this chapter affords amounts of

329

pigments that are unweighable, except on a microbalance, but more than adequate for the thin layer experiment. It is suggested that half the students isolate lycopene and the other half β-carotene.

Lycopene and β-carotene from tomato paste and strained carrots

Fresh tomato fruit contains about 96% of water, and R. Willstätter and H. R. Escher isolated from this source 20 mg of lycopene per kg of fruit. They then found a more convenient source in commercial tomato paste, from which seeds and skin have been eliminated and the water content reduced by evaporation in vacuum to a content of 26% solids, and isolated 150 mg of lycopene per kg of paste. The expected yield in the present experiment is 0.75 mg.

As an interesting variation, try extraction of lycopene from commercial catsup

A jar of strained carrots sold as baby food serves as a convenient source of β-carotene. The German investigators isolated 1 g of β-carotene per kg of "dried" shredded carrots of unstated water content.

The following procedure calls for dehydration of tomato or carrot paste with ethanol and extraction with methylene chloride, an efficient solvent for lipids.

EXPERIMENT

A 5-g sample of tomato or carrot paste is transferred to the bottom of a 25 × 150-mm test tube with the aid of a paste dispenser,[1] made by connecting an 18-cm section of 11-mm glass tubing by means of a No. 00 rubber stopper and a short section of 5 mm glass tubing to a 50-cm length of $\frac{1}{8}$" rubber tubing. The open end of the glass tube is calibrated to deliver 5 g of paste and marked with a file scratch.

These carotenoids are very sensitive to photochemical air oxidation. Protect solutions and solid from undue exposure to light.

Thrust the dispenser into a can of fruit paste and "suck up" paste to the scratch mark on the glass tube. Then wipe the glass tube with a facial tissue, insert it into a 25 × 150-mm test tube, and blow out the paste onto the bottom of the receiver. Wash the dispenser tube clean with water and leave it upright to drain for the next user.

Dehydration of fruit paste

Add 10 ml of 95% ethanol, insert a cold finger condenser, heat to boiling, and reflux for 5 min. Then filter the hot mixture on a small Hirsch funnel. Scrape out the test tube with a spatula, let the tube drain thoroughly, and squeeze liquid out of the solid residue in the funnel with a spatula. Pour the yellow filtrate into a 125-ml Erlenmeyer flask. Then return the solid residue, with or without adhering filter paper, to the original test tube, add 10 ml of methylene chloride to effect an extraction, insert the condenser, and reflux the mixture for 3–4 min. Filter the yellow filtrate and add the filtrate to the storage flask. Repeat the extraction with two or three further 10-ml portions of methylene chloride, pour the combined extracts into a separatory funnel, add water and sodium chloride solution (to aid in layer separation), and shake. Run the colored lower layer through a cone of anhydrous sodium sulfate into a dry flask and evaporate to dryness (aspirator tube or rotary evaporator).

CH_2Cl_2
bp 41°
den 1.336

---

[1] To be supplied to each section of the class, along with a can opener.

The crude carotenoid is to be chromatographed on a 12-cm column of acid-washed alumina (Merck), prepared with petroleum ether (37–53°) as solvent. Run out excess solvent, or remove it from the top of the chromatography column with a suction tube, dissolve the crude carotenoid in a few milliliters of benzene, and then transfer the solution onto the chromatographic column with a capillary dropping tube. Elute the column with petroleum ether, discard the initial colorless eluate, and collect all yellow or orange eluates together in a 50-ml Erlenmeyer flask. Place a drop of solution on a microscope slide and evaporate the rest to dryness (rotary evaporator or aspirator tube). Examination of the material spotted on the slide may reveal crystallinity. Then put a drop of concd sulfuric acid

Color test

beside the spot and mix with a stirring rod. Compare the color of your test with that of a test on the other carotenoid.

Finally, dissolve the sample obtained by evaporating the petroleum ether in the least amount of methylene chloride, hold the Erlenmeyer in a slanting position for drainage into a small area, and transfer the solution with a capillary dropping tube to a 10×75-mm test tube or a 3-ml centrifuge tube. Add a boiling stone, hold the tube over a hot plate in a slanting position, and evaporate to dryness. Evacuate the tube at the suction pump, shake out the boiling stone, cork and label the test tube, and keep it in a dark place.

Look up the current prices of commercial lycopene[2] and β-carotene.[3] Note that β-carotene is in demand as a source of vitamin A and is manufactured by an efficient synthesis. No uses for lycopene have been found.

Notes for the instructor

[2] Pfaltz and Bauer, Flushing, N.Y., 11368.
[3] Aldrich Chemical Co., Milwaukee, Wis., 53233.

59 Quinones

KEYWORDS   Orange II, reduction
                     ortho-Quinone
                     Vitamin K_1, antihemorrhagic
                     Oxidation, chromic acid
                     1,4-Naphthoquinone, yellow

Reduction, hydroquinone
Phytol, allylic alcohol
Condensation
Phthiocol

In the first experiment of this chapter Orange II is reduced in aqueous solution with sodium hydrosulfite to water-soluble sodium sulfanilate and 1-amino-2-naphthol, which precipitates. This intermediate is purified as the hydrochloride and oxidized to 1,2-naphthoquinone.

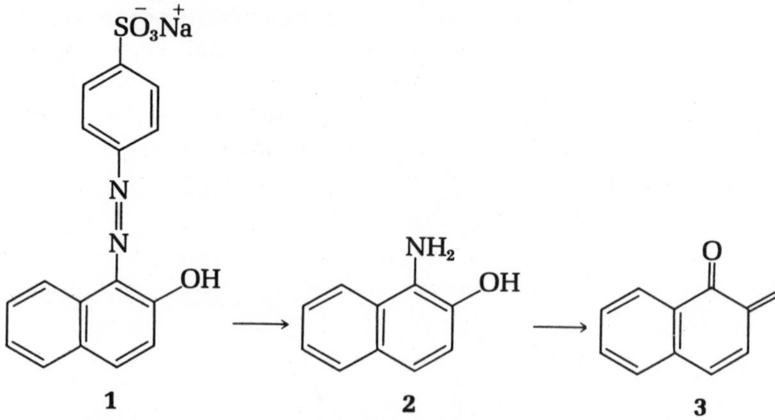

1
Orange II
MW 350.34

2
1-Amino-2-naphthol
MW 159.18

3
1,2-Naphthoquinone
MW 158.15, mp 145–147° dec

The next three experiments of this chapter are a sequence of steps for the synthesis of the antihemorrhagic Vitamin K_1 (or an analogue) starting with a coal-tar hydrocarbon.

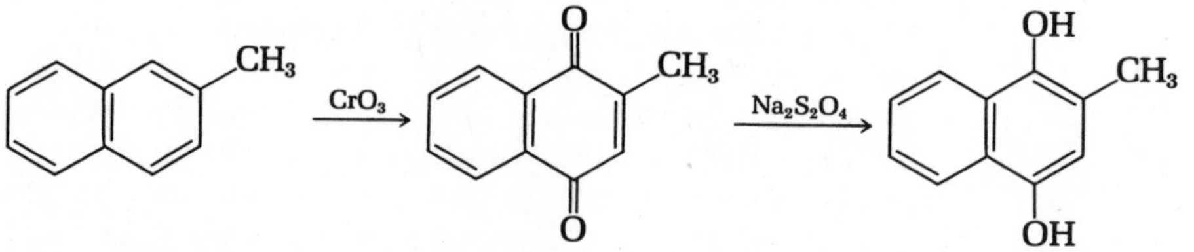

4
β-Methylnaphthalene
MW 142.19, mp 35°

5
2-Methyl-1,4-naphthoquinone
MW 172.17, mp 105–106°

6
2-Methyl-1,4-naphthohydroquinone
MW 174.19

2-Methylnaphthalene is oxidized with chromic acid to 2-methyl-1,4-naphthoquinone, the yellow quinone is purified and reduced to its hydroquinone by shaking an ethereal solution of the substance with aqueous hydrosulfite solution, the colorless hydroquinone is condensed with phytol, and the substituted hydroquinone oxidized to vitamin K_1.

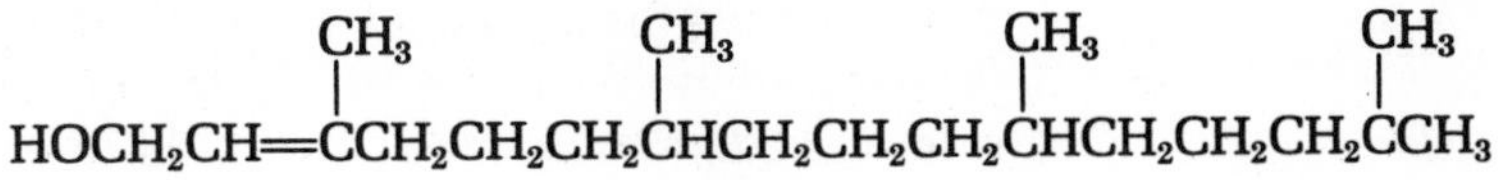

$$\underset{\text{CH}_3}{|}\quad \underset{\text{CH}_3}{|}\quad \underset{\text{CH}_3}{|}\quad \underset{\text{CH}_3}{|}$$

$$\text{HOCH}_2\text{CH}\!=\!\text{CCH}_2\text{CH}_2\text{CH}_2\text{CHCH}_2\text{CH}_2\text{CH}_2\text{CHCH}_2\text{CH}_2\text{CH}_2\text{CCH}_3$$

Phytol
MW 296.52, bp 145° at 0.03–0.04 torr

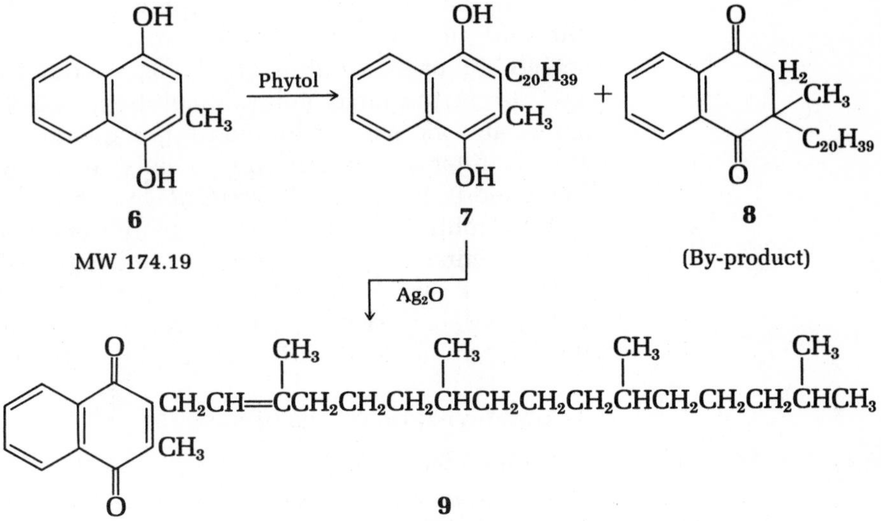

6
MW 174.19

7

8
(By-product)

Ag$_2$O

9
Vitamin K_1
MW 450.68

An additional or alternative experiment is conversion of 2-methyl-1,4-naphthoquinone through the oxide into the 3-hydroxy compound phthiocol, which has been isolated from human tubercle bacilli after saponification, probably as a product of cleavage of vitamin K_1.

EXPERIMENTS   1.   1,2-Naphthoquinone (3)

In a 125-ml Erlenmeyer flask dissolve 3.9 g of Orange II in 50 ml of water and warm the solution to 40–50°. Add 4.5 g of sodium hydrosulfite dihydrate and swirl until the red color is discharged and a cream-colored or pink precipitate of 1-amino-2-naphthol separates. To coagulate the product, heat the mixture nearly to boiling until it begins to froth, then cool in an ice bath, collect the product on a suction filter, and wash the residue with water. Prepare a solution of 1 ml of concd hydrochloric acid, 20 ml of water, and an estimated 50 mg of stannous chloride (antioxidant); transfer the precipitate of aminonaphthol to this solution and wash in material adhering to the funnel. Swirl, warm gently, and when all but a

Two short-time reactions

little fluffy material has dissolved, filter the solution by suction through a thin layer of decolorizing charcoal. Transfer the filtered solution to a clean flask, add 4 ml of concd hydrochloric acid, heat over a free flame until the precipitated aminonaphthol hydrochloride has been brought into solution, and then cool thoroughly in an ice bath. Collect the crystalline, colorless hydrochloride and wash it with a mixture of 1 ml of concd hydrochloric acid and 4 ml of water. Leave the air-sensitive crystalline product in the funnel while preparing a solution for its oxidation. Dissolve 5.5 g of ferric chloride crystals (FeCl$_3$ · 6H$_2$O) in 2 ml of concd hydrochloric acid and 10 ml of water by heating, cool to room temperature, and filter by suction.

Wash the crystalline aminonaphthol hydrochloride into a beaker, stir, add more water, and warm to about 35° until the salt is dissolved. Filter the solution quickly by suction from a trace of residue and stir in the ferric chloride solution. 1,2-Naphthoquinone separates at once as a voluminous precipitate and is collected on a suction filter and washed thoroughly to remove all traces of acid. The yield from pure, salt-free Orange II is 75%.

1,2-Naphthoquinone, highly sensitive and reactive, does not have a well-defined melting point but decomposes at about 145–147°. Suspend a sample in hot water and add concd hydrochloric acid. Dissolve a small sample in cold methanol and add a drop of aniline; the red product is 4-anilino-1,2-naphthoquinone.

2. 2-Methyl-1,4-naphthoquinone (5)

In the hood, clamp a separatory funnel in place to deliver into a 600-ml beaker which can be cooled in an ice bath when required. The oxidizing solution to be placed in the funnel is prepared by dissolving 50 g of chromic anhydride (CrO$_3$) in 35 ml of water and diluting the dark red solution with 35 ml of acetic acid.[1] In the beaker prepare a mixture of 14.2 g of 2-methylnaphthalene and 150 ml of acetic acid, and without cooling run in small portions of the oxidizing solution. Stir with a thermometer until the temperature rises to 60°. At this point ice cooling will be required to prevent a further rise in temperature. By alternate addition of reagent and cooling, the temperature is maintained close to 60° throughout the addition, which can be completed in about 10 min. When the temperature begins to drop spontaneously the solution is heated gently on the steam bath (85–90°) for one hour to complete the oxidation.

Dilute the dark green solution with water nearly to the top of the beaker, stir well for a few minutes to coagulate the yellow quinone, collect the product on a Büchner funnel, and wash it thoroughly with water to remove chromic acetate. The crude material can be

Reaction time: 1.25 hr

---

[1] The chromic anhydride is hygroscopic; weigh it quickly and do not leave the bottle unstoppered. The substance dissolves very slowly in acetic acid–water mixtures, and solutions are prepared by adding the acetic acid only after the substance has been completely dissolved in water.

crystallized from methanol (40 ml) while still moist (without filtering), and gives 6.5–7.3 g of satisfactory 2-methyl-1,4-naphthoquinone, mp 105–106°. The product is to be saved for the preparation of **6**, **9** and **11**. This quinone must be kept away from light, which converts it into a pale yellow, sparingly soluble polymer.

3. 2-Methyl-1,4-naphthohydroquinone (6)

Short-time reaction

In an Erlenmeyer flask dissolve 2 g of 2-methyl-1,4-naphthoquinone (**5**) in 35 ml of ether by warming on a steam bath, pour the solution into a separatory funnel, and shake with a fresh solution of 4 g of sodium hydrosulfite in 30 ml of water. After passing through a brown phase (quinhydrone) the solution should become colorless or pale yellow in a few minutes; if not, add more hydrosulfite solution. After removing the aqueous layer, shake the ethereal solution with 25 ml of saturated sodium chloride solution and 1–2 ml of saturated hydrosulfite solution to remove the bulk of the water.

Aspirator tube

Filter the ethereal layer by gravity through a filter paper moistened with ether and about one-third filled with sodium sulfate. Evaporate the filtrate on the steam bath until nearly all the solvent has been removed, cool, and add petroleum ether. The hydroquinone separating as a white or grayish powder is collected, washed with petroleum ether, and dried; yield 1.9 g (the substance has no sharp mp).

4. Vitamin K₁ (2-Methyl-3-phytol-1,4-naphthoquinone, 9)

Explanation

Phytol, being an allylic alcohol, is reactive enough to condense with 2-methyl-1,4-naphthohydroquinone (**6**) under mild conditions of acid catalysis as specified in the following procedure. Overheating must be avoided or the alcohol is dehydrated to phytadiene. The reaction mixture is diluted with water and extracted with ether and unchanged starting material removed by extraction with aqueous alkali containing hydrosulfite to keep the hydroquinones from being oxidized by air. The hydroquinone of Vitamin K₁ (**7**) has a methyl group adjacent to one hydroxyl group and the long phytol side chain adjacent to the other; it is a cryptophenol (hidden phenolic properties), insoluble in aqueous alkali. It is separated from the nonhydroxylic by-product **8** by crystallization from petroleum ether and oxidized to Vitamin K₁ (**9**).

Place 1.5 g of phytol[2] and 10 ml of dioxane in a 50-ml Erlenmeyer flask and warm to 50° on a hot plate. Prepare a solution of 1.5 g of 2-methyl-1,4-naphthohydroquinone (**6**) and 1.5 ml of boron trifluoride etherate in 10 ml of dioxane, and add this in portions with a capillary dropper in the course of 15 min with constant swirling and while maintaining a temperature of 50° (do not overheat).

Note for the instructor

[2] Suppliers: Matheson, Coleman, and Bell Co.; American Chlorophyll Division, Strong, Cobb and Co.; National Chlorophyll and Chemical Co. Geraniol can be used instead of phytol, reacting with 2-methyl-1,4-naphthohydroquinone to give a product similar in chemical and physical properties to the natural vitamin and with pronounced antihemorrhagic activity.

Continue in the same way for 20 min longer. Cool to 25°, wash the solution into a separatory funnel with 40 ml of ether, and wash the orange-colored ethereal solution with two 40-ml portions of water to remove boron trifluoride and dioxane. Extract the unchanged hydroquinone with a freshly prepared solution of 2 g of sodium hydrosulfite in 40 ml of 2% aqueous sodium hydroxide and 10 ml of a saturated sodium chloride solution (which helps break the resulting emulsion). Shake vigorously for a few minutes, during which time any red color should disappear and the alkaline layer should acquire a bright yellow vat color. After releasing the pressure through the stopcock, allow the layers to separate, keeping the funnel stoppered as a precaution against oxidation. Draw off the yellow liquor and repeat the extraction a second and a third time, or until the alkaline layer remains practically colorless. Separate the faintly colored ethereal solution, dry it over anhydrous sodium sulfate, filter into a tared flask, and evaporate the filtrate on the steam bath— eventually with evacuation at the aspirator. The total oil, which becomes waxy on cooling, amounts to 1.7–1.9 g.

Add 10 ml of petroleum ether (bp 20–40°) and boil and manipulate with a spatula until the brown mass has changed to a white paste. Wash the paste into small centrifuge tubes with 10–20 ml of fresh petroleum ether, make up the volume of paired tubes to the same point, cool well in ice, and centrifuge. Decant the brown supernatant liquor into the original tared flask, fill the tubes with fresh solvent, and stir the white sludge to an even suspension. Then cool, centrifuge, and decant as before. Evaporation of the liquor and washings gives 1.1–1.3 g of residual oil. Dissolve the portions of washed sludge of Vitamin K_1 hydroquinone in a total of 10–15 ml of absolute ether, and add a little decolorizing charcoal for clarification, if the solution is pink or dark. Add 1 g of silver oxide and 1 g of anhydrous sodium sulfate. Shake for 20 min, filter into a tared flask, and evaporate the clear yellow solution on the steam bath, removing traces of solvent at the water pump. Undue exposure to light should be avoided when the material is in the quinone form. The residue is a light yellow, rather mobile oil consisting of pure Vitamin K_1; yield 0.6–0.9 g. A sample for preservation is transferred with a capillary dropper to a small specimen vial wrapped in metal foil or black paper to exclude light.

To observe a characteristic color reaction, transfer a small bit of vitamin on the end of a stirring rod to a test tube, stir with 1 ml of alcohol, and add 1 ml of 10% alcoholic potassium hydroxide solution; the end pigment responsible for the red color is phthiocol.

5. Phthiocol (2-Methyl-3-hydroxy-1,4-naphthoquinone 11)

Dissolve 1 g of 2-methyl-1,4-naphthoquinone (5) in 10 ml of alcohol by heating, and let the solution stand while the second reagent is prepared by dissolving 0.2 g of anhydrous sodium carbonate in 5 ml of water and adding (cold) 1 ml of 30% hydrogen peroxide solution. Cool the quinone solution under the tap until

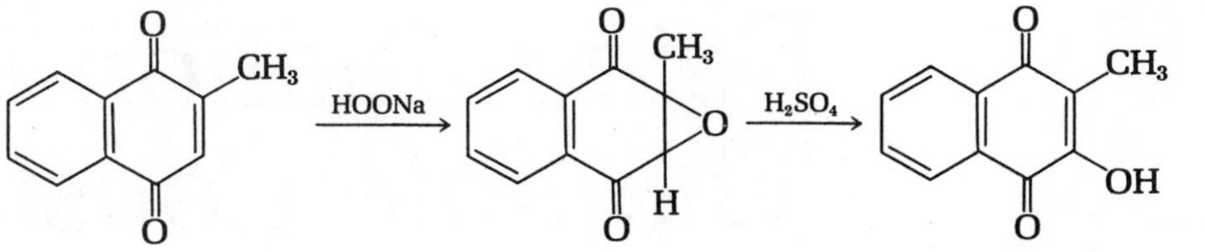

	5	**10**	**11**
		Oxide	**Phthiocol**
		MW 188.17, mp 93.5–94.5°	MW 188.17, mp 172–173°

crystallization begins, add the peroxide solution all at once, and cool the mixture. The yellow color of the quinone should be discharged immediately. Add about 100 ml of water, cool in ice, and collect the colorless, crystalline oxide, **10**; yield 0.97 g, mp 93.5–94.5° (pure: 95.5–96.5°).

To 1 g of **10** in a 25-ml Erlenmeyer flask add 5 ml of concd sulfuric acid; stir if necessary to produce a homogeneous deep red solution, and after 10 min cool this in ice and slowly add 20 ml of water. The precipitated phthiocol can be collected, washed, and crystallized by dissolving in methanol (25 ml), adding a few drops of hydrochloric acid to give a pure yellow color, treating with decolorizing charcoal, concentrating the filtered solution, and diluting to the saturation point. Alternatively, the yellow suspension is washed into a separatory funnel and the product extracted with a mixture of 25 ml each of benzene and ether. The organic layer is dried over anhydrous sodium sulfate and evaporated to a volume of about 10 ml for crystallization. The total yield of pure phthiocol (**11**) mp 172–173°, is 0.84–.88 g.

60 Reaction Kinetics: Determination of The Rate Constant and Order of a Reaction

---

KEYWORDS   Williamson ether synthesis   Potassium t-butoxide
                Rate constant, k BASIC computer program
                1st order reaction   Slope
                2nd order reaction   Time-zero

$$CH_3I + CH_3-\underset{\underset{CH_3}{|}}{\overset{\overset{CH_3}{|}}{C}}-O^-K^+ \xrightarrow{(CH_3)_3COH} CH_3-\underset{\underset{CH_3}{|}}{\overset{\overset{CH_3}{|}}{C}}-O-CH_3 + K^+I^-$$

The formation of methyl t-butyl ether from methyl iodide and potassium t-butoxide is an example of the Williamson ether synthesis. By analyzing this reaction as it proceeds you can determine the rate of the reaction as well as the order. In this way you also may be able to infer a mechanism for the reaction.

Known quantities of reactants are allowed to react for known lengths of time; from time to time, samples of the reaction mixture are analyzed quantitatively. You can follow either the disappearance of the reactants or the appearance of the products. Analysis of the reaction might include gas chromatography, infrared or nmr spectroscopy (methyl iodide decrease or ether increase), or reaction with aqueous silver nitrate to precipitate insoluble silver iodide. We shall use still another method,—titration of the unreacted base, potassium t-butoxide, with an acid. When using this method aliquots (known fractions) of the reaction mixture are removed and immediately diluted with ice water to stop the reaction. Unreacted t-butoxide ion will react with the water to form hydroxide ion and t-butanol; the base is titrated to a phenolphthalein end point with standard perchloric acid.

The rate of the reaction depends on the temperature. Ideally the reaction should be run in a constant temperature bath; otherwise wrap the reaction flask with a towel to shield it from drafts, keep it stoppered, and assure that all reactants and apparatus are at room temperature (which should be above 25.5°, the freezing point of t-butanol).

EXPERIMENT   *Procedure*

Pipet 200 ml of a 0.110 M solution of methyl iodide in t-butanol into a 300 ml Erlenmeyer flask, quickly followed by exactly 20.0 ml of ca. 0.5 M potassium t-butoxide in t-butanol (note the exact concentration of the potassium t-butoxide on the label). Swirl to insure thorough mixing, and place the flask in the constant temperature bath (note exact temperature). Immediately withdraw a 10-ml aliquot (1/22 of the reaction mixture) with a pipet and run it into 20 ml of an ice and water mixture in a 250-ml Erlenmeyer flask. Between aliquots rinse the pipet with water and a few drops of acetone and dry by sucking air through it with an aspirator. It is not necessary to carry out the titration immediately, so label the flask. During the first hour withdraw 10-ml aliquots at about 10-min intervals (note the exact time in seconds when the pipet is half empty), then at 20-min intervals during the second hour. By the end of the second hour well over half of the reactants will have reacted to give the product. Add a drop of phenolphthalein indicator to each quenched aliquot and titrate with standardized perchloric acid to a very faint pink color when seen against a white background. Note the exact concentration of the perchloric acid; it should be near 0.025 M.

Mathematics of Rate Studies

In studying a reaction you must first determine what the products are. It has previously been found that the reaction of methyl iodide and potassium t-butoxide gives only the ether, but the equation tells you nothing about how the rate of the reaction is affected by the relative concentrations of the reactants, nor does it tell you anything about the mechanism of the reaction.

(a) First Order Reaction   The rate of a first order reaction depends only upon the concentration of one reactant, A. For example, if

$$2A + 3B \rightarrow \text{Products}$$

and the reaction is first order in A, despite what the equation seems to say the rate equation is

$$\text{Rate} = \frac{-d[A]}{dt} = k[A]$$

where [A] is the concentration of A, $-d[A]/dt$ is the rate of disappearance of A with time, and k is a proportionality constant known as "the rate constant." Experimentally we determine

$$a = \text{concentration of A at time zero}$$
$$x = \text{concentration of A which has reacted by time } t$$
$$a - x = \text{concentration of A remaining at time } t$$

The rate expression is then

$$\frac{dx}{dt} = k(a - x)$$

which upon integration gives

$$-\ln(a - x) = kt - \ln(a)$$

Changing signs and changing from base e to base 10 logarithms give

$$\log(a - x) = \frac{-k}{2.303} t + \log(a)$$

This equation has the form of the equation for a straight line,

$$y = mx + b$$

If the reaction under study is first order, a straight line will be obtained when the concentration of A at various times is plotted against time. The slope of the line is $-k/2.303$, from which the value of the rate constant can be found. It will have units of (time)$^{-1}$.

(b) Second Order Reaction The rate of a second order reaction depends on the concentrations of two reactants. For example if

$$2A + 3B \rightarrow \text{Products}$$

the rate equation would be

$$\text{Rate} = -\frac{d[A]}{dt} = -\frac{d[B]}{dt} = k_2[A][B]$$

Experimentally we determine

$$a = \text{concentration of A at time zero}$$
$$b = \text{concentration of B at time zero}$$
$$x = \text{concentration of A (and concentration of B) which has reacted by time } t$$
$$a - x = \text{concentration of A remaining at time } t$$
$$b - x = \text{concentration of B remaining at time } t$$

The rate equation now becomes

$$\frac{dx}{dt} = k_2(a - x)(b - x)$$

which on integration gives

$$k_2t = \frac{1}{a-b} \ln \frac{b(a-x)}{a(b-x)}$$

which upon rearrangement and conversion to base ten logarithms gives

$$k_2t = \frac{2.303}{a-b} \log \frac{a-x}{b-x} + \frac{2.303}{a-b} \log \frac{b}{a}$$

The final term is a constant. Rearrangement into the form of an equation for a straight line gives

$$\log \frac{a-x}{b-x} = k_2 \left(\frac{a-b}{2.303}\right) t + \text{Constant}$$

If the reaction being studied is second order, a plot of $\log (a-x/b-x)$ against time will give a straight line with slope of $k_2(a-b/2.303)$. The value of the rate constant is given by

$$k_2 = \text{Slope} \left(\frac{2.303}{a-b}\right)$$

Treatment of Data

The fact that some reaction occurred before the first aliquot was withdrawn is immaterial. It is only necessary to know the concentrations of both reactants at the time of withdrawal of the first aliquot (which we shall call time-zero). The amount of base present at time-zero is determined directly by titration. The amount of methyl iodide present is equal to the concentration originally present after mixing minus that which reacted before the first aliquot was withdrawn. The amount which reacted before the first aliquot was withdrawn is equal to the amount of butoxide which reacted between the time of mixing and the time the first aliquot was withdrawn.

Make a rough plot of the concentration of methyl iodide and the concentration of potassium t-butoxide against time. If either gives a straight line relationship, the reaction is first order. If not, calculate the quantity $\log (a-x/b-x)$ for each aliquot and plot against time in seconds. If this plot gives a straight line (within your experimental error) for this reaction, you now have strong evidence that the reaction is second order, i.e., first order in each reactant. Obtain the rate constant, k, for this reaction at this temperature, from the slope of your plot. The units of k_2 are moles sec^{-1} l^{-1}.

Computer Analysis of Data

The repeated calculation of $\log (a-x/b-x)$ for each aliquot from titration data is, to say the least, tedious. A computer program can do this simple job. After a little experience you will be able to write a program that would take as input the concentrations of all reactants,

ml of acid, the time, and then print out k. There are standard com
puter programs ("library programs") available which will accep
your various values of ln $(a - x/b - x)$ and the corresponding time
calculate the mathematically "best" straight line through the poin
(the "least squares" line) and print out the slope and intercept of th
line.

A program written in the BASIC computer language to calculat
ln $(a - x/b - x)$ for various values of x looks like this:

    10   READ A, B

    20   DATA 0.098, 0.479

    30   READ X

    40   DATA 0.007, 0.019, 0.025, 0.037, 0.049, 0.058, 0.062

    50   LET Y = LOG((A − X)/(B − X))

    60   PRINT A,B,X,Y

    70   GO TO 10

    80   END

In this example, a = 0.098, b = 0.479 and x = 0.007, 0.019, etc. Th
computer will READ for "a" the value 0.098, because that is the firs
number it finds in the data printed in line 20. It reads 0.479 for "b
because that is the second number in line 20. The next comman
(line 30) says READ X and the computer reads the number 0.00
(line 40). We are trying to calculate ln $(a - x/b - x)$ so we set thi
equal to Y. The LET statement tells the computer to carry out th
calculation substituting 0.098 for A, 0.479 for B, and 0.007 for X
LOG means natural logarithms to the computer so it carries out th
computation and prints out the values of A, B, and X, which w
supplied, and Y, which it calculates. The next command (line 70
says go back to line 10 and start all over. This the computer does, bu
this time it automatically substitutes 0.019 in the equation (A an
B remaining the same), ignoring 0.007 which it has already used
Again it carries out the computation, prints out A, B, X, and Y, an
keeps going through the cycle until it runs out of data in line 40.

61 *Synthesis of Carpanone*

KEYWORDS

Lignan
Stereospecific synthesis
Five asymmetric carbons
Biosynthetic-like synthesis

Williamson ether synthesis
Claisen rearrangement
Oxidative dimerization
Diels-Alder cyclization

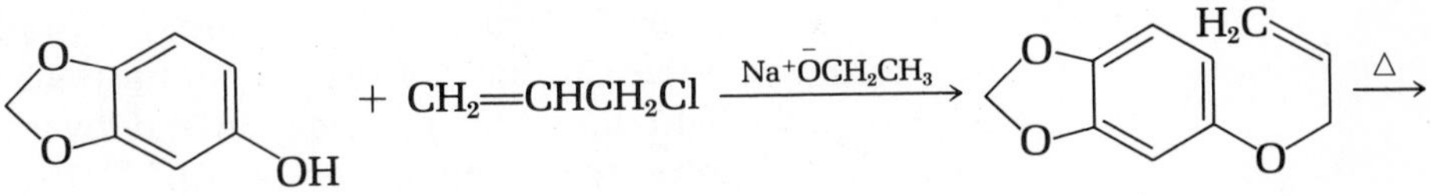

1
MW 138.12, mp 63–64°

2
MW 178.19

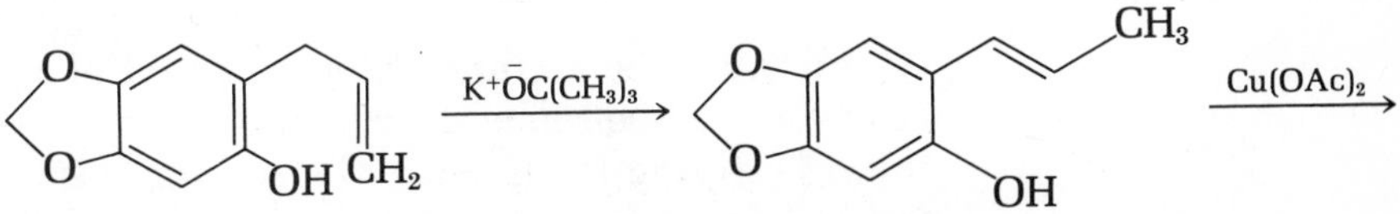

3
MW 178.19

4
MW 178.19

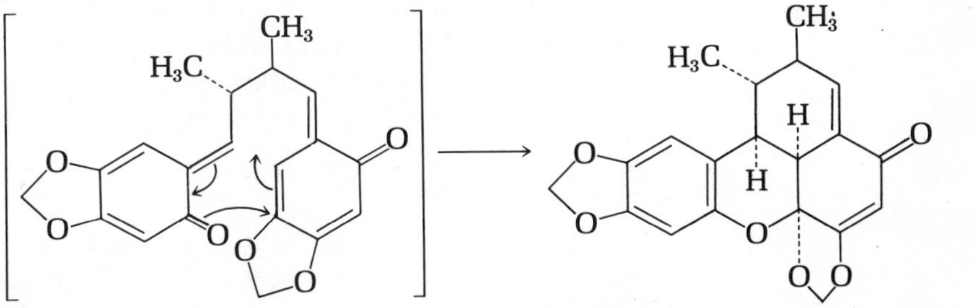

Carpanone, MW 354.36, mp 210–212°
Carpanone · CCl₄, MW 508.20, mp 185°

Carpanone is a lignan from a tree, native to the South Pacific island of Bougainville. Lignans, which can be extracted from wood with organic solvents, are related to the lignins, the insoluble, non-carbohydrate polymers which bind cellulose fibers together in wood.

The structure of carpanone was determined primarily by an analysis of its nmr spectrum.[1] The synthesis[2] might, at first glance, seem a formidable task. Carpanone has no element of symmetry and five contiguous asymmetric carbon atoms. A rational total synthesis of a compound of this type is the ultimate test of its structure determination. In the course of such a synthesis, new organic reactions might be invented to circumvent particular problems, and the synthesis might provide clues to biosynthetic pathways. In this case, the ease with which the five asymmetric centers are formed in the last reaction, all with the correct stereochemistry, may mirror the way this compound is formed in nature.

The first reaction, addition of allyl chloride to sesamol (1), is an example of the Williamson ether synthesis. The phenyl allyl ether, (2), undergoes the Claisen rearrangement on thermolysis to 2-allylsesamol (3), which, in turn, is isomerized by strong base to the conjugated olefin, 2-propenylsesamol (4). Under catalysis by cupric ion 2-propenylsesamol undergoes oxidative dimerization and cyclization, via a Diels-Alder type reaction, to carpanone, with the formation of five asymmetric centers. A study of models will clarify this reaction, as well as the stereochemistry of the product.

EXPERIMENTS

1. Sesamol allyl ether (2)

Add 0.84 g of sodium pellets to 25 ml of anhydrous ethanol in a 125-ml round-bottomed flask equipped with condenser and calcium chloride drying tube. After the sodium has dissolved, add 5.0 g of sesamol in 6 ml of anhydrous ethanol and then, with swirling, 3.4 g of allyl chloride. Reflux the solution 2 hr, cool to room temperature, pour into 200 ml of water, and extract three times with 35-ml portions of diethyl ether. Wash the combined ether extracts with 100-ml portions of 5% aqueous sodium hydroxide, 5% aqueous sulfuric acid, water, and saturated sodium chloride solutions. Dry the ether solution of the product over 5 g of anhydrous magnesium sulfate, remove the drying agent by filtration or careful decantation, and evaporate the diethyl ether on the steam bath (boiling chip, aspirator tube), or on a rotary evaporator. The product is a brown liquid.

2. 2-Allylsesamol (3)

Place the crude sesamol allyl ether (2) from the previous experiment in a 50-ml round-bottomed flask equipped with a condenser. Heat the flask in an oil bath at 170° for 2 hr. This is most easily done

[1] G. C. Brophy, J. Mohandas, M. Slaytor, S. Sternhall, T. R. Watson and L. A. Wilson, *Tetrahedron Lett.* 5159 (1969).
[2] O. L. Chapman, M. R. Engel, J. P. Springer, and J. C. Clardy, *J. Amer. Chem. Soc.*, **93**, 6696 (1971).

in a large Petri dish containing cottonseed oil and a thermometer placed atop an electric hot plate. The oil bath is large enough for several reactions to be carried out at the same time. The product will solidify to a dark-brown mass, on cooling to room temperature. Remove a small sample of the product and sublime it. This is done by heating a few mg of the compound in an inclined evacuated test tube. Crystals of the product should appear on the cool upper portions of the tube. The crude product is, however, pure enough to carry on to the next step.

3. 2-Propenylsesamol (4)

Equip a 150-ml three-necked flask with a gas inlet tube, a thermometer, a Teflon-covered magnetic stirring bar, and a condenser leading to a bubbler (Fig. 61.1). Add to the flask 50 ml of dimethyl

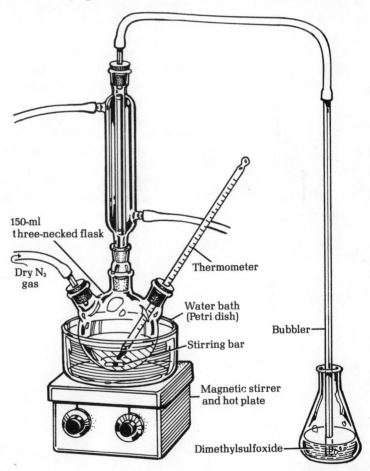

Figure 61.1
Inert atmosphere
reaction flask.

sulfoxide, 5 g of potassium t-butoxide[3] and the 2-allyl sesamol (3) prepared in the previous experiment. Pass a slow stream of nitro-

[3] Potassium t-butoxide is very sensitive to moisture. Weigh rapidly and keep container tightly closed. Alternatively, the laboratory instructor may prepare a stock solution containing 5 g of t-butoxide per 50 ml of dimethyl sulfoxide. Use 50 ml of this solution.

gen through the flask for 5 min and then reduce the flow to just a few bubbles per minute. Start the stirrer and heat the reaction mixture in a water bath at 80° for 30 min. Cool the mixture to room temperature and then pour into a 250-ml beaker filled with ice. Stir the slurry and make the solution acidic to Congo Red paper with concd hydrochloric acid (about 10 ml). Extract the suspension with three 35-ml portions of diethyl ether, wash the combined ether extracts three times with 100-ml portions of water and once with 100 ml of saturated aqueous sodium chloride solution. Dry the ether extracts over anhydrous magnesium sulfate, remove the drying agent, and evaporate to give 2-propenylsesamol, which will solidify on standing.

4. Carpanone (5)

Dissolve 1.0 g of the 2-propenylsesamol (4) prepared in the previous step in a solution of 5 ml of methanol and 10 ml of water. Add 3 g of sodium acetate and then a solution of 1 g of cupric acetate in 5 ml of water dropwise with stirring. Stir for an additional 5 min and filter the resulting suspension. Dilute the filtrate with twice its volume of water and extract with three 35-ml portions of ether. Wash the combined ether extracts with two 50-ml portions of saturated sodium chloride solution and dry the extracts over anhydrous magnesium sulfate. Remove the drying agent by filtration and evaporate the ether. Take up the residue in 2 ml of warm carbon tetrachloride, stopper the flask, and set aside to crystallize. If crystals have not formed overnight, add a few seed crystals and let the solution stand for several hours. Decant the solvent from the crystals with a micropipette and allow them to dry. Since each molecule of carpanone crystallizes with a molecule of carbon tetrachloride the theoretical yield is 1.4 g, mp 185°.

Caution. Carbon tetrachloride vapor should not be inhaled

62 Vacuum Distillation

KEYWORDS Claisen distilling head
Pot
Bumping, hair-fine capillary
Changing fractions
Mercury manometer
Vigreaux distilling column, flooding
Hold up
Fraction collector, cow
Vacuum distillation head
Water aspirator vacuum, 15 torr
Rotary oil pump, 0.1 torr
Mercury diffusion pump, 10^{-8} torr
McLeod gauge
Dry ice trap
Distilling solids
Sublimation

Many substances cannot be distilled satisfactorily in the ordinary way, either because they boil at such high temperatures that decomposition occurs or because they are sensitive to oxidation. In such cases purification can be accomplished by distillation at diminished pressure. A few of the many pieces of apparatus for vacuum distillation are described on the following pages. Round-bottomed pyrex ware and thick-walled suction flasks are not liable to collapse, but even so safety glasses should be worn at all times when carrying out this type of distillation.

2. Vacuum Distillation Assemblies

A typical vacuum distillation apparatus is illustrated in Fig. 62.1. It is constructed of a round-bottomed flask (often called the "pot") containing the material to be distilled, a Claisen distilling head fitted with a hair-fine capillary mounted through a rubber tubing sleeve, and a thermometer with the bulb extending below the side arm opening. The condenser fits into a vacuum adapter which is connected to the receiver and, via heavy-walled rubber tubing, to a mercury manometer and thence to the trap and water aspirator.

Prevention of bumping

Liquids usually bump vigorously when boiled at reduced pressure and most boiling stones lose their activity in an evacuated system; it is therefore essential to make special provision for controlling the bumping. This is done by allowing a fine stream of air bubbles to be drawn into the boiling liquid through a glass tube drawn to a hair-fine capillary. The capillary should be so fine that even under vacuum only a few bubbles of air are drawn in each

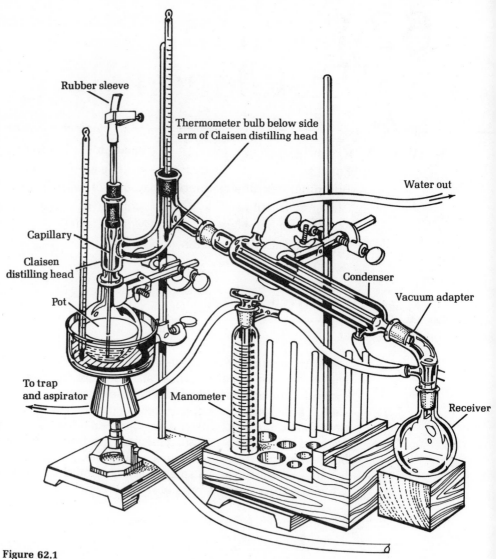

**Figure 62.1
Vacuum distillation
apparatus.**

Labels on figure:
Rubber sleeve
Thermometer bulb below side arm of Claisen distilling head
Water out
Capillary
Claisen distilling head
Pot
Condenser
Vacuum adapter
To trap and aspirator
Manometer
Receiver

second; smooth boiling will be promoted and the pressure will remain low. The capillary should extend to the very bottom of the flask and it should be slender and flexible so that it will whip back and forth in the boiling liquid. Another method used to prevent bumping, when small quantities of material are being distilled, is to introduce sufficient glass wool into the flask to fill a part of the space above the liquid.

The capillary is made in three operations. First, a six-inch length of 6-mm glass tubing is rotated and heated in a small area over a *very hot* flame to collapse the glass and thicken the side walls as seen in Fig. 62.2a. The tube is removed from the flame, allowed to cool slightly and then drawn into a thick-walled, coarse diameter capillary (Fig. 62.2b). This coarse capillary is heated at point X over the

*Making a hair-fine
capillary*

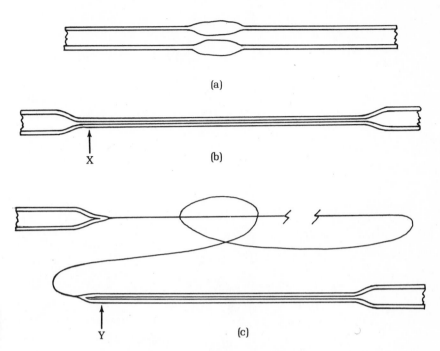

(a)

(b)

X

(c)

Y

**Figure 62.2
Capillary for vacuum
distillation.**

wing top of a Bunsen burner turned 90°. When the glass is very soft, but not so soft as to collapse the tube entirely, the tubing is lifted from the flame and without hesitation drawn smoothly and rapidly into a hair-fine capillary by stretching the hands as far as they will reach (5 or 6 ft)(Fig. 62.2c). The two capillaries so produced can be snapped off to the desired length. To ascertain that there is indeed a hole in the capillary, place the end beneath a low viscosity liquid such as acetone or ether and blow in the large end. A stream of very small bubbles should be seen. Should the right-hand capillary of Fig. 62.2c break when in use, it can be fused to a scrap of glass (for use as a handle) and heated again at point Y (Fig. 62.2c). In this way the capillary can be redrawn many times.

The pot is heated with a heating bath rather than a free flame to promote even boiling and make possible accurate determination of the boiling point. The bath is filled with a suitable heat transfer liquid (water, cottonseed oil, silicone oil, or molten metal) and heated to a temperature about 20° higher than that at which the substance in the flask distils. The bath temperature is kept constant throughout the distillation. The surface of the liquid in the flask should be below that of the heating medium, for this condition lessens the tendency to bump. Heating of the flask is begun only after the system has been evacuated to the desired pressure; otherwise the liquid might boil too suddenly on reduction of the pressure.

Heating baths

To change fractions the following must be done in sequence: Remove the source of heat, release the vacuum, change the receiver, restore the vacuum to the former pressure, resume heating.

Changing fractions

The pressure of the system is measured with a closed-end mercury manometer. The manometer (Fig. 62.3) is connected to the system

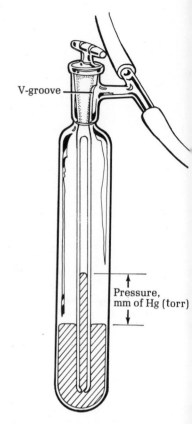

V-groove

Pressure,
mm of Hg (torr)

**Figure 62.3
Closed end mercury
manometer.**

Mercury manometer

by turning the stopcock until the V-groove in the stopcock is aligned with the side arm. To avoid contamination of the manometer it should be connected to the system only when a reading is being made. [The pressure in the system can be monitored continuously by observing the rubber-tubing pressure gauge on the trap (Fig. 2.11)]. The pressure, in torr, is given by the height, in mm, of the central mercury column above the reservoir of mercury and represents the difference in pressure between the nearly perfect vacuum in the center tube (closed at the top, open at the bottom) and the large volume of the manometer, which is at the pressure of the system.

A better vacuum distillation apparatus is shown in Fig. 62.4. The distillation neck of the Claisen adapter is longer than other adapters and has a series of indentures made from four directions, so that the points nearly meet in the center. These indentations increase the surface area over which rising vapor can come to equilibrium with descending liquid and it then serves as a fractionating column (a Vigreaux column). A column packed with a metal sponge has a great tendency to become filled with liquid (flood) at reduced pressure. The apparatus illustrated in Fig. 62.4 also has a fraction collector, which allows the removal of a fraction without disturbing the vacuum in the system. While the receiver is being changed the distillate collects in the small reservoir A. The clean receiver is

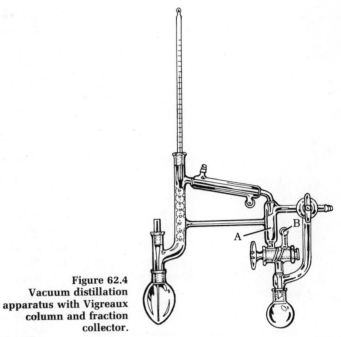

**Figure 62.4
Vacuum distillation
apparatus with Vigreaux
column and fraction
collector.**

evacuated by another aspirator at tube B before being connected
again to the system.

If only a few milliliters of a liquid are to be distilled, the apparatus
shown in Fig. 62.5 has the advantage of low hold-up, that is, not

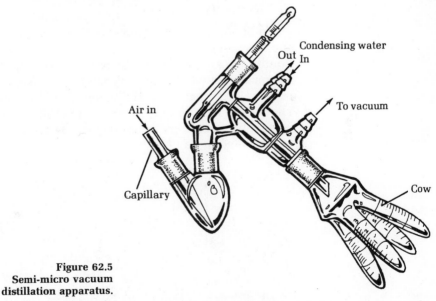

**Figure 62.5
Semi-micro vacuum
distillation apparatus.**

much liquid is lost wetting the surface area of the apparatus. The
fraction collector illustrated is known as a "cow." Rotation of the
cow about the standard taper joint will allow four fractions to be
collected without interrupting the vacuum.

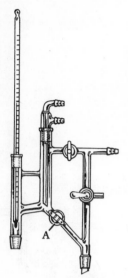

A distillation head of the type shown in Fig. 62.6 allows fractions to be removed without disturbing the vacuum, and it also allows control of the reflux ratio (Chapter 5) by manipulation of the condenser and stopcock A. These can be adjusted to remove all material which condenses or only a small fraction, with the bulk of the liquid being returned to the distilling column to establish equilibrium between descending liquid and ascending vapor. In this way liquids with small boiling-point differences can be separated.

2. The Water Aspirator in Vacuum Distillation

A water aspirator in good order gives a vacuum nearly corresponding to the vapor pressure of the water flowing through it. Polypropylene aspirators give good service and are not subject to corrosion as are the brass ones. If a manometer is not available, and the assembly is free of leaks and the trap and lines clean and dry, an approximate estimate of the pressure can be made by measuring the water temperature and reading the pressure from Table 62.1.

Figure 62.6
Vacuum distillation head.

Table 62.1   Vapor Pressure of Water at Different Temperatures

t	p	t	p	t	p	t	p
0°	4.58	20°	17.41	24°	22.18	28°	28.10
5°	6.53	21°	18.50	25°	23.54	29°	29.78
10°	9.18	22°	19.66	26°	24.99	30°	31.55
15°	12.73	23°	20.88	27°	26.50	35°	41.85

3. The Rotary Oil Pump

To obtain pressures below 10 torr a mechanical vacuum pump of the type illustrated in Fig. 62.7a is used. A pump of this type in good condition can give pressures as low as 0.1 torr. These low pressures are measured with a tilting type McLeod gauge (Fig. 62.7b). When a reading is being made the gauge is tilted to the vertical position shown and the pressure is read as the difference between the heights of the two columns of mercury. Between readings the gauge is rotated clockwise 90°.

Figure 62.7
(a) Rotary oil pump,
(b) Tilting McLeod gauge.

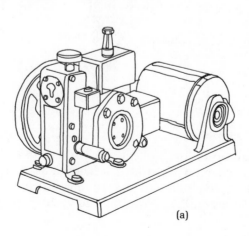

(a)

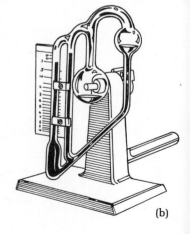

(b)

Never use a mechanical vacuum pump before placing a mixture of dry ice and acetone in a Dewar flask (Fig. 62.7c) around the trap and

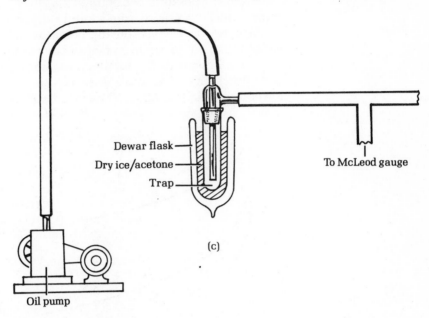

Dewar flask

Dry ice/acetone

Trap

To McLeod gauge

(c)

Figure 62.7
(c) Vacuum system.   Oil pump

never pump corrosive vapors (e.g., HCl gas) into the pump. With care, it will give many years of good service. The dry ice trap condenses organic vapors and water vapor, both of which would otherwise contaminate the vacuum pump oil and exert enough vapor pressure to destroy a good vacuum.

For an exceedingly high vacuum (5×10^{-8} torr) a high-speed three-stage mercury diffusion pump is used (Fig. 62.8).

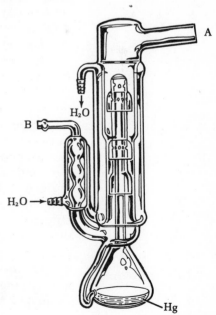

Figure 62.8
High speed three-stage
mercury diffusion pump
capable of producing a
vacuum of 5×10^{-8}
torr. Mercury is boiled
in the flask, the vapor
rises in the center tube,
is deflected downward
in the inverted cups and
entrains gas molecules
which diffuse in from
the space to be evac-
uated, A. The mercury
condenses to a liquid
and is returned to the
flask, the gas molecules
are removed at B by
an ordinary rotary
vacuum pump.

A

H_2O

B

$H_2O \rightarrow$

Hg

4. Relationship between Boiling Point and Pressure

It is not possible to calculate the boiling point of a substance at some reduced pressure from a knowledge of the boiling temperature at 760 torr, for the relationship between boiling point and pressure varies from compound to compound and is somewhat unpredictable. It is true, however, that boiling point curves for organic substances have much the same general disposition, as illustrated by the two lower curves in Fig. 62.9. These are similar and do not differ greatly

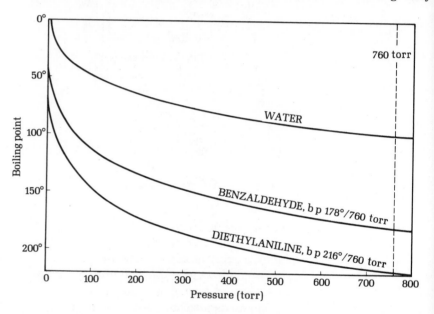

Figure 62.9
Boiling point curves.

from the curve for water. For substances boiling in the region 150–250° at 760 torr, the boiling point at 20 torr is 100–120° lower than at 760 torr. Benzaldehyde, which is very sensitive to air oxidation at the normal boiling point of 178°, distils at 76° at 20 torr, and the concentration of oxygen in the rarefied atmosphere is just $^{20}/_{760}$, or 3%, of that in an ordinary distillation.

The curves all show a sharp upward inclination in the region of very low pressure. The lowering of the boiling point attending a reduction in pressure is much more pronounced at low than at high pressures. A drop in the atmospheric pressure of 10 torr lowers the normal boiling point of an ordinary liquid by less than a degree, but a reduction of pressure from 20 to 10 torr causes a drop of about 15° in the boiling point. The effect at pressures below 1 mm is still more striking, and with development of practical forms of the highly efficient mercury vapor pump, distillation at a pressure of a few thousandths or ten thousandths of a millimeter has become a standard operation in many research laboratories. High vacuum distillation, that is at a pressure below 1 torr, affords a useful means of purifying extremely sensitive or very slightly volatile substances. Table 62.2 indicates the order of magnitude of the reduction in boil-

Table 62.2   Distillation of a (Hypothetical) Substance at Various Pressures

Method		Pressure (torr)	Bp
Ordinary distillation		760	250°
Water pump {	summer	25	144°
	winter	15	134°
Oil pump {	poor condition	10	124°
	good condition	3	99°
	excellent condition	1	89°
Mercury vapor pump		0.01	30°

ing point attainable by operating in different ways and illustrates the importance of keeping vacuum pumps in good repair.

The boiling point of a substance at various pressures can be estimated from a pressure-temperature nomograph such as the one shown in Fig. 62.10. If the boiling point of a substance at 760 torr

Figure 62.10
Pressure-temperature
nomograph.

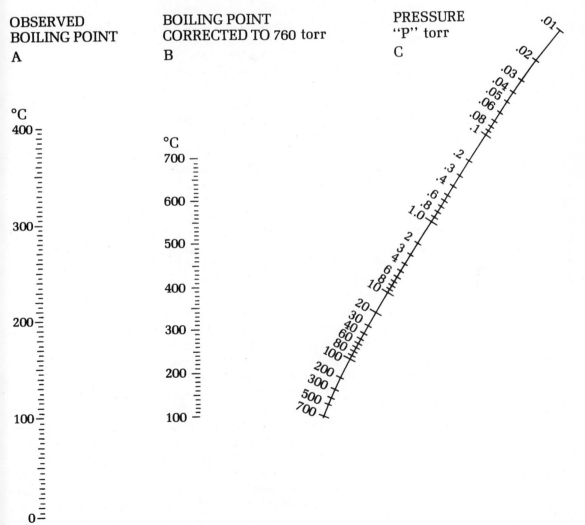

is known, e.g., 300° (Column B), and the new pressure measured, e.g., 10 torr (Column C), then a straight line connecting these values on Columns B and C when extended intersects Column A to give the observed bp of 160°. Conversely, a substance observed to boil at 50° (Column A) at 1.0 torr (Column C) will boil at approximately 212° at atmospheric pressure (Column B).

Vacuum distillation is not confined to the purification of substances liquid at ordinary temperatures but often can be used to advantage for solid substances. The operation is conducted for a different purpose and by a different technique. A solid is seldom distilled to cause a separation of constituents of different degrees of volatility but rather to purify the solid. It is often possible in one vacuum distillation to remove foreign coloring matter and tar without appreciable loss of product, whereas several wasteful crystallizations might be required to attain the same purity. It is often good practice to distil a crude product and then to crystallize it. Time is saved in the latter operation because the hot solution usually requires neither filtration nor clarification. The solid must be dry and a test should be made to determine if it will distil without decomposition at the pressure of the pump available. That a compound lacks the required stability at high temperatures is sometimes indicated by the structure, but a high melting point should not be taken as an indication that distillation will fail. Substances melting as high as 300° have been distilled with success at the pressure of an ordinary rotary vacuum pump.

It is not necessary to observe the boiling point in distillations of this kind because the purity and identity of the distillate can be checked by melting point determinations. The omission of the customary thermometer simplifies the technique. A simple and useful assembly is shown in Fig. 62.11. A rather stout capillary tube carrying an adjustable vent at the top is fitted into the neck of the two-bulb flask by means of a rubber stopper and the suction pump is connected through a trap at the other bulb. It is not necessary to insert in the mouth a rubber stopper of just the right size; a somewhat larger stopper may be put on backwards as shown, as it will be held in place by atmospheric pressure. The same scheme can be used for the other stopper. Water cooling is unnecessary. If some cooling of the receiving bulb is required it is best to use an air blast. Since the connection between the distilling and the receiving flask is of glass, any material that solidifies and tends to plug the side arm can be melted with a free flame. A heating bath should not be used; it is best to heat the flask with a rather large flame. Hold the burner in the hand and play the flame in a rotary motion around the side walls of the flask. This allows less bumping than when the flask is heated from the bottom. If there is much frothing at the start of the heating, direct the flame upon the upper walls and the neck of the flask. If the liquid froths over into the receiving bulb, the flask is tilted to such a position that this bulb can drain through the con-

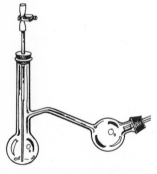

Figure 62.11
Two-bulb flask for the distillation of solids.

necting tube back into the distillation bulb when suitably warmed.

At the end of the distillation the vacuum is broken by carefully opening the pinchcock, and the contents of the receiving bulb melted and poured out. This method of emptying the bulb is sometimes inadvisable because the hot, molten material may be susceptible to air oxidation. In such a case, the material is allowed to solidify and cool completely before the vacuum is broken. The solid is then chipped out with a clean knife or with a strong nickel spatula and the last traces recovered with the solvent to be used in the crystallization. The tar usually remaining in the distillation bulb is best removed by adding small quantities of concd nitric and sulfuric acids, mixing the layers well, and heating the mixture while the flask is supported by a clamp under the hood. After cooling and pouring the acid mixture down the drain, loose char is removed with water and a brush and the process repeated.

6. Sublimation

When the quantity of solid is small or the thermal stability is such as to preclude distillation, sublimation is used to purify the solid. In an apparatus such as that illustrated in Fig. 62.12a, the solid to be sublimed is placed in the lower flask and connected via a lubricant free rubber "O" ring to the condenser, which in turn is connected to a vacuum pump. The lower flask is immersed in an oil bath at the appropriate temperature and the product sublimed and condensed on the cool walls of the condenser. The parts of the apparatus are gently separated and the condenser inverted; the vacuum connection serves as a convenient funnel for product removal. For large scale work the sublimator of Fig. 62.12b is used. The inner

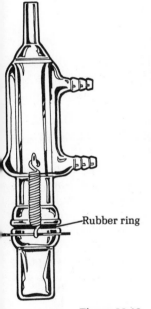

—Rubber ring

Figure 62.12
(a) Mallory sublimator.

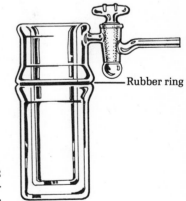

—Rubber ring

Figure 62.12
(b) Large vacuum sub-
limator.

well is filled with a coolant (ice or dry ice). The sublimate clings to this cool surface, from which it can be removed by scraping and dissolving in an appropriate solvent.

63 *Qualitative Organic Analysis*

Identification and characterization of the structures of unknown substances are an important part of organic chemistry. It is sometimes possible to establish the structure of a compound on the basis of spectra alone (IR, UV and nmr), but these spectra must usually be supplemented with other information about the unknown: physical state, elementary analysis, solubility, and confirmatory tests for functional groups. Conversion of the unknown to a solid derivative of known melting point will often provide final confirmation of structure.

PROCEDURES

1. Physical State

Check for sample purity. Distill or recrystallize as necessary. Constant bp and sharp mp are indicators, but beware of azeotropes and eutectics. Check homogeneity by TLC, gas or paper chromatography.

Note the color. Common colored compounds include nitro and nitroso compounds (yellow), α-diketones (yellow), quinones (yellow to red), azo compounds (yellow to red), and polyconjugated olefins and ketones (yellow to red). Phenols and amines are often brown to dark-purple because of traces of air oxidation products.

Note the odor. Some liquid and solid amines are recognizable by their fishy odors; esters are often pleasantly fragrant. Alcohols, ketones, aromatic hydrocarbons, and aliphatic olefins have characteristic odors. On the unpleasant side are thiols, isonitriles, and low MW carboxylic acids.

Make an ignition test. Heat a small sample on a spatula; first hold the sample near the side of a microburner to see if it melts normally and then burns. Heat it in the flame. If a large ashy residue is left after ignition, the unknown is probably a metal salt. Aromatic compounds often burn with a smoky flame.

2. Spectra

Obtain nuclear magnetic resonance and infrared spectra following the procedures of Chapters 10 and 11.[1] If these spectra indicate the

---

[1] Kits of unknowns with IR and nmr spectra may be purchased from the Aldrich Chemical Co. The properties of all unknowns in kits A, B and C are included in the tables at the end of this chapter.

presence of conjugated double bonds, aromatic rings, or conjugated carbonyl compounds obtain the ultraviolet spectrum following the procedures of Chapter 15. Interpret the spectra as fully as possible by reference to the sources cited at the end of the various spectroscopy chapters.

3. Elementary Analysis, Sodium Fusion[2]

Explanation

This method for detection of nitrogen, sulfur, and halogen in organic compounds depends upon the fact that fusion of substances containing these elements with sodium[3] yields NaCN, Na_2S, and NaX (X = Cl, Br, I). These products can, in turn, be readily identified. The method has the advantage that the most usual elements other than C, H, and O present in organic compounds can all be detected following a single fusion, although the presence of sulfur sometimes interferes with the test for nitrogen. Unfortunately, even in the absence of sulfur the test for nitrogen is sometimes unsatisfactory (nitro compounds in particular).

Caution! Manipulate sodium with a knife and forceps; never touch it with the fingers. Wipe it free of kerosene with a dry towel or filter paper; return scraps to the bottle or destroy scraps with methyl or ethyl alcohol, never with water. Safety glasses!

Place a 3-mm cube of sodium[4] (30 mg, no more)[5] in a 10×75-mm pyrex test tube and support the tube in a vertical position (Fig. 63.1).

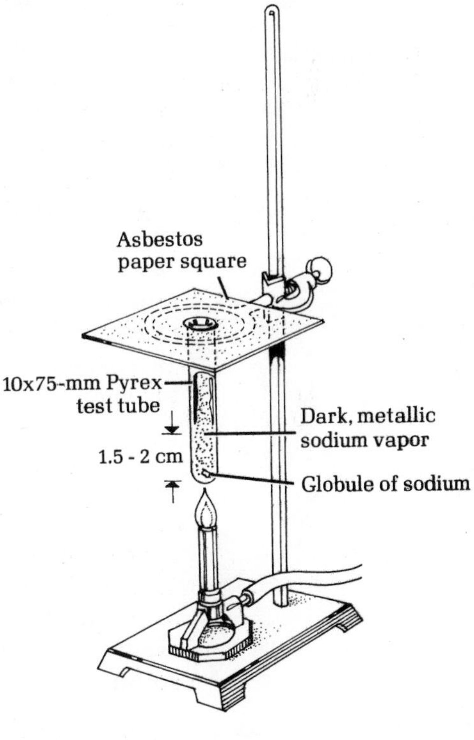

Asbestos paper square

10x75-mm Pyrex test tube

1.5 - 2 cm

Dark, metallic sodium vapor

Globule of sodium

Figure 63.1
Sodium fusion, just prior to addition of sample.

[2] S. H. Tucker, *J. Chem. Ed.*, **22**, 212 (1945).
[3] Metallic potassium, which is more reactive, is also used; F. Feigl. *Spot Tests in Organic Analysis*, 7th ed., Elsevier, 1966. A method of fusion with potassium carbonate and magnesium is described by R. H. Baker and C. Barkenbus, *Ind. Eng. Chem., Anal. Ed.*, **9**, 135 (1937).
[4] Sodium spheres 1/16″ to 1/4″ (CB1035) from Matheson, Coleman, and Bell are convenient.
[5] A dummy 3 mm cube of rubber can be attached to the sodium bottle to indicate the correct amount.

Have a microburner with small flame ready to move under the tube,
place an estimated 20 mg of solid on a spatula or knife blade, put
the burner in place, and heat until the sodium first melts and then
vapor rises 1.5–2.0 cm in the tube. Remove the burner and at once
drop the sample onto the hot sodium. If the substance is a liquid add
2 micro drops of it. If there is a flash or small explosion the fusion
is complete; if not, heat briefly to produce a flash or a charring.
Then let the tube cool to room temperature, be sure it is cold, add a
micro drop of methanol, and let it react (heat effect). Repeat until
10 micro drops have been added. With a stirring rod break up the
char to uncover sodium. When you are sure that all sodium has
reacted, empty the tube into a 13 × 100-mm test tube, hold the small
tube pointing away from you or a neighbor, and pipette into it 1 ml
of water. Boil and stir the mixture and pour the water into the larger
tube; repeat with 1 ml more water. Then transfer the solution with a
capillary dropper to a 2.5-cm funnel (fitted with a fluted filter paper)
resting in a second 13 × 100-mm test tube. Portions of the alkaline
filtrate are used for the tests that follow:

Do not use CHCl₃ or CCl₄ as samples in sodium fusion

(a) Nitrogen.   The test is done by boiling the alkaline solution with
ferrous sulfate and then acidifying. Sodium cyanide reacts with
ferrous sulfate to produce ferrocyanide, which combines with ferric
salts, inevitably formed by air oxidation in the alkaline solution,
to give Prussian Blue, $NaFe^{+3}Fe^{+2}(CN)_6$. Ferrous and ferric hydroxides precipitate along with the blue pigment but dissolve on acidification.

Run each test on a known and an unknown

Place 50 mg of powdered ferrous sulfate (this is a large excess)
in a 13 × 100-mm test tube, add 0.5 ml of the alkaline solution from
the fusion, heat the mixture gently with shaking to the boiling point,
and then—without cooling—acidify with dilute sulfuric acid
(hydrochloric acid is unsatisfactory). A deep blue precipitate indicates the presence of nitrogen. If the coloration is dubious, filter
through a 2.5-cm funnel and see if the paper shows blue pigment.

(b) Sulfur.   (1) Dilute one drop of the alkaline solution with 1 ml
of water and add a drop of sodium nitroprusside; a purple coloration
indicates the presence of sulfur. (2) Prepare a fresh solution of
sodium plumbite by adding 10% sodium hydroxide solution to
0.2 ml of 0.1 M lead acetate solution until the precipitate just dissolves, and add 0.5 ml of the alkaline test solution. A black precipitate or a colloidal brown suspension indicates the presence of
sulfur.

Differentiation of the halogens

(c) Halogen.   Acidify 0.5 ml of the alkaline solution from the fusion
with dilute nitric acid (indicator paper) and, if nitrogen or sulfur
has been found present, boil the solution (hood) to expel HCN or
H_2S. On addition of a few drops of silver nitrate solution, halide
ion is precipitated as silver halide. Filter with minimum exposure
to light on a 2.5-cm funnel, wash with water, and then with 1 ml of

concd ammonia solution. If the precipitate is white and readily soluble in ammonium hydroxide solution it is AgCl; if it is pale yellow and difficultly soluble it is AgBr; if yellow and insoluble it is AgI. Fluorine is not detected in this test since silver fluoride is soluble in water.

4. Beilstein Test

Heat the tip of a copper wire in a burner flame until no further coloration of the flame is noticed. Allow the wire to cool slightly then dip it into the unknown (solid or liquid) and again heat it in the flame. A green flash is indicative of chlorine, bromine, and iodine; fluorine is not detected since copper fluoride is not volatile. The Beilstein test is very sensitive; halogen-containing impurities may give misleading results. Run the test on a compound known to contain halogen for comparison with your unknown.

It is considered good practice to run tests on knowns in parallel with unknowns for all qualitative organic reactions. In this way, interpretations of positive reactions are clarified and defective test reagents can be identified and replaced.

5. Solubility Tests

Like dissolves like; a substance is most soluble in that solvent to which it is most closely related in structure. This statement serves as a useful classification scheme for all organic molecules. The solubility measurements are done at room temperature with 1 micro drop of a liquid, or 5 mg of a solid (finely crushed), and 0.2 ml of solvent. The mixture should be rubbed with a rounded stirring rod and shaken vigorously. Lower members of a homologous series are easily classified, higher members become more like the hydrocarbons from which they are derived.

If a *very* small amount of the sample fails to dissolve when added to some of the solvent, it can be considered insoluble, and, conversely, if several portions dissolve readily in a small amount of the solvent, the substance is obviously soluble.

If an unknown seems to be more soluble in dilute acid or base than in water, the observation can be confirmed by neutralization of the solution; the original material will precipitate if it is less soluble in a neutral medium.

If both acidic and basic groups are present, the substance may be amphoteric and therefore soluble in both acid and base. Aromatic aminocarboxylic acids are amphoteric, like aliphatic ones, but they do not exist as inner salts. They are soluble in both dilute hydrochloric acid and sodium hydroxide, but not in bicarbonate solution. Aminosulfonic acids exist as inner salts; they are soluble in alkali but not in acid.

The solubility tests are not infallible and many borderline cases are known.

Carry out the tests according to the scheme of Fig. 63.2 and tentatively assign the unknown to one of the groups, I–X.

6. Classification Tests

After the unknown is assigned to one of the solubility groups (Fig. 63.2) on the basis of solubility tests, the possible type should be further narrowed by application of classification tests, e.g., for alcohols, or methyl ketones, or esters.

7. Complete Identification—Preparation of Derivatives

Once the unknown has been classified by functional group, the physical properties should be compared with those of representative members of the group (see Tables at the end of this chapter). Usually, several possibilities present themselves, and the choice can be narrowed by preparation of derivatives. Select derivatives that distinguish most clearly between the possibilities.

Notes to Solubility Tests

1. Groups I, II, III (soluble in water). Test the solution with pH paper. If the compound is not easily soluble in cold water, treat it as water-insoluble but test with indicator paper.

2. If the substance is insoluble in water but dissolves partially in 5% sodium hydroxide, add more water; the sodium salts of some phenols are less soluble in alkali than in water. If the unknown is colored, be careful to distinguish between the *dissolving* and the *reacting* of the sample. Some quinones (colored) *react* with alkali and give highly colored solutions. Some phenols (colorless) *dissolve and then* become oxidized to give colored solutions. Some compounds (e.g., benzamide) are hydrolyzed with such ease that careful observation is required to distinguish them from acidic substances.

3. Nitrophenols (yellow), aldehydophenols, and polyhalophenols are sufficiently strongly acidic to react with sodium bicarbonate.

4. Oxygen- and nitrogen-containing compounds form oxonium and ammonium ions in concd sulfuric acid and dissolve. The test distinguishes phenyl ethers from hydrocarbons.

5. On reduction in the presence of hydrochloric acid these compounds form water-soluble amine hydrochlorides. Dissolve 0.5 g of stannous chloride in 1 ml of concd hydrochloric acid, add 0.1 g of the unknown, and warm. The material should dissolve with the disappearance of the color and give a clear solution when diluted with water.

6. Most amides can be hydrolyzed by short boiling with 10% sodium hydroxide solution; the acid dissolves with evolution of ammonia. Reflux 0.2 g of the sample and 10% sodium hydroxide solution for 15–20 min under a cold finger condenser. Test for the evolution of

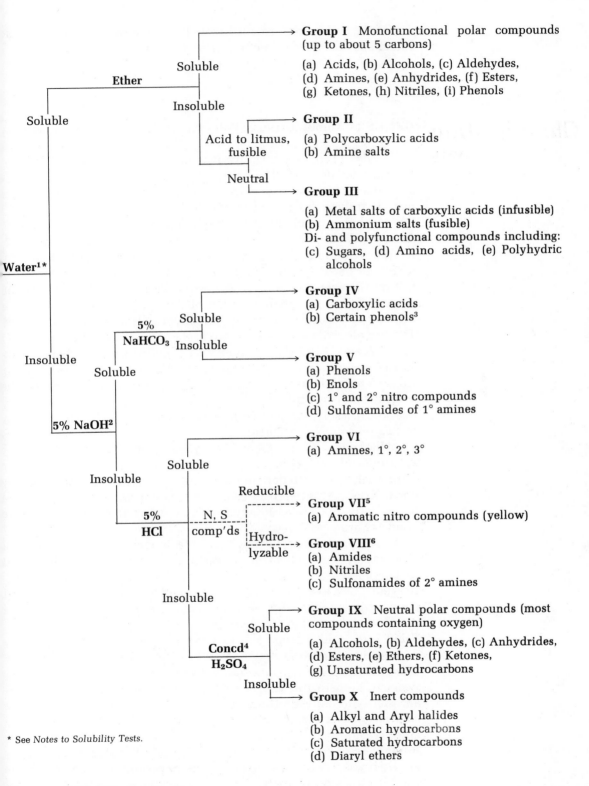

Figure 63.2   Solubility Classification.

The content of the classification chart:

Water[1]*

Soluble → **Ether**
- Soluble → **Group I** Monofunctional polar compounds (up to about 5 carbons)
  - (a) Acids, (b) Alcohols, (c) Aldehydes, (d) Amines, (e) Anhydrides, (f) Esters, (g) Ketones, (h) Nitriles, (i) Phenols
- Insoluble
  - Acid to litmus, fusible → **Group II**
    - (a) Polycarboxylic acids
    - (b) Amine salts
  - Neutral → **Group III**
    - (a) Metal salts of carboxylic acids (infusible)
    - (b) Ammonium salts (fusible)
    - Di- and polyfunctional compounds including:
    - (c) Sugars, (d) Amino acids, (e) Polyhydric alcohols

Insoluble → **5% NaOH[2]**
- Soluble → **5% NaHCO₃**
  - Soluble → **Group IV**
    - (a) Carboxylic acids
    - (b) Certain phenols[3]
  - Insoluble → **Group V**
    - (a) Phenols
    - (b) Enols
    - (c) 1° and 2° nitro compounds
    - (d) Sulfonamides of 1° amines
- Insoluble → **5% HCl**
  - Soluble → **Group VI**
    - (a) Amines, 1°, 2°, 3°
  - Insoluble
    - N, S comp'ds
      - Reducible → **Group VII[5]**
        - (a) Aromatic nitro compounds (yellow)
      - Hydrolyzable → **Group VIII[6]**
        - (a) Amides
        - (b) Nitriles
        - (c) Sulfonamides of 2° amines
    - **Concd[4] H₂SO₄**
      - Soluble → **Group IX** Neutral polar compounds (most compounds containing oxygen)
        - (a) Alcohols, (b) Aldehydes, (c) Anhydrides, (d) Esters, (e) Ethers, (f) Ketones, (g) Unsaturated hydrocarbons
      - Insoluble → **Group X** Inert compounds
        - (a) Alkyl and Aryl halides
        - (b) Aromatic hydrocarbons
        - (c) Saturated hydrocarbons
        - (d) Diaryl ethers

* See *Notes to Solubility Tests.*

ammonia, which confirms the elementary analysis for nitrogen and establishes the presence of a nitrile or amide.

Classification Tests

GROUP I Monofunctional Polar Compounds (up to ca. 5 carbons)

(a) Acids (Table 63.1; Derivatives, page 374)

No classification test is necessary. Carboxylic and sulfonic acids are detected by testing aqueous solutions with litmus. Acid halides may hydrolyze during the solubility test.

(b) Alcohols (Table 63.2; Derivatives, page 375)

Jones Oxidation. Follow the procedure of Section 2, Chapter 16 for the oxidation of cholesterol with chromic acid but using one tenth the listed quantities of reagents and solvent. Run a control test on the acetone solvent. A positive test is formation of a green color within 5 sec upon addition of the orange-yellow reagent to a primary or secondary alcohol. Aldehydes also give a positive test. Tertiary alcohols are readily dehydrated to olefins with acid.

Ceric Nitrate Test. (For alcohols with ten or fewer carbons.) Dissolve 4 or 5 drops (or 25–30 mg) of the unknown in water or dioxane. Add 0.5 ml of the reagent and shake the mixture. Alcohols cause the reagent to change from yellow to red.

Reagent: Dissolve 90 g of ceric ammonium nitrate, $Ce(NH_4)_2(NO_3)_6$, in 225 ml of 2 N nitric acid.

(c) Aldehydes (Table 63.3; Derivatives, page 376)

2,4-Dinitrophenylhydrazones. All aldehydes and ketones readily form bright yellow to dark red 2,4-dinitrophenylhydrazones. Yellow derivatives are formed from isolated carbonyl groups and orange-red to red derivatives from aldehydes or ketones conjugated with double bonds or aromatic rings.

Add 1 or 2 drops (20 mg) of the unknown to 2 ml of ethanol followed by 3 ml of 2,4-dinitrophenylhydrazine reagent. Shake and let sit for a few minutes. A yellow to red precipitate is a positive test.

Reagent: Dissolve 3 g of 2,4-dinitrophenylhydrazine in 15 ml of concd sulfuric acid. Add this solution, with stirring, to a mixture of 20 ml of water and 70 ml of ethanol.

Schiff Test. Add 3 drops of the unknown to 2 ml of Schiff's reagent. A magenta color will appear within ten minutes with aldehydes. Compare the color of your unknown with that of a known aldehyde.

Reagent: Prepare 100 ml of a 0.1 percent aqueous solution of p-rosaniline hydrochloride (fuchsia). Add 4 ml of a saturated aqueous solution of sodium bisulfite. After 1 hr add 2 ml of concd hydrochloric acid.

Bisulfite Test. Follow the procedure on page 134, Chapter 22. Nearly all aldehydes and most methyl ketones form solid, water soluble bisulfite addition products.

Tollens Test. Follow the procedure on page 180, Chapter 31. A positive test, deposition of a silver mirror, is given by most aldehydes, but not by ketones.

(d) Amines   (Tables 63.5 and 63.6; Derivatives, pages 377, 379)

Hinsberg Test. Follow the procedure on page 173, Chapter 30, using benzenesulfonyl chloride to distinguish between primary, secondary, and tertiary amines.

(e) Anhydrides and Acid Halides   (Table 63.7; Derivatives, page 379)

Anhydrides and acid halides will react with water to give acidic solutions, detectable with litmus paper.

Acidic Ferric Hydroxamate Test.[6] With ferric chloride alone a number of substances give a color which can interfere with this test. Dissolve 2 or 3 drops (or about 30 mg) of the unknown in 1 ml of ethanol and add 1 ml of 1 N hydrochloric acid followed by 1 drop of 10% aqueous ferric chloride solution. If any color except yellow appears you will find it difficult to interpret the results from the following test.

Add 2 or 3 drops (or about 30 mg) of the unknown to 0.5 ml of a 1 N solution of hydroxylamine hydrochloride in alcohol. Add 2 drops of 6 M hydrochloric acid to the mixture, warm it slightly for 2 min, and boil it for a few seconds. Cool the solution and add 1 drop of 10 percent ferric chloride solution. A red-blue color is a positive test.

(f) Esters   (Table 63.8, page 380. Derivatives prepared from component acid and alcohol obtained on hydrolysis)

Esters, unlike anhydrides and acid halides, do not react with water to give acidic solutions and do not react with acidic hydroxylamine hydrochloride. They do, however, react with alkaline hydroxylamine.

Alkaline Ferric Hydroxamate Test. First test the unknown with ferric chloride alone. (See under Group I(e), Acidic Ferric Hydroxamate Test.)

To a solution of two drops (or about 30 mg) of the unknown in 1 ml of 0.5 N hydroxylamine hydrochloride in ethanol add four drops of 20% sodium hydroxide solution. Heat the solution to boiling, cool slightly, and add 2 ml of 1 N hydrochloric acid. If cloudiness develops add up to 2 ml of ethanol. Add 10% ferric chloride solution dropwise with swirling. A red-blue color is a positive test. Compare your unknown with a known ester.

(g) Ketones   (Table 63.14; Derivatives, page 383)

2,4-Dinitrophenylhydrazone. See under Group I(c), Aldehydes. All ketones react with 2,4-dinitrophenylhydrazine reagent.

Idoform Test for Methyl Ketones. Follow the procedure on page 135, Chapter 22. A positive iodoform test is given by substances contain-

$$\overset{O}{\overset{\|}{}}$$

ing the CH_3C- group or by compounds easily oxidized to this group, e.g., CH_3COR, CH_3CHOHR, CH_3CH_2OH, CH_3CHO, $RCOCH_2COR$.

[6] R. E. Buckles and C. J. Thelen, *Anal. Chem.,* **22,** 676 (1950).

The test is negative for compounds of the structure CH_3COOR CH_3CONHR, and other compounds of similar structure which give acetic acid on hydrolysis. It is also negative for $CH_3COCH_2CO_2R$ CH_3COCH_2CN, $CH_3COCH_2NO_2$.

Bisulfite Test. Follow the procedure on page 134, Chapter 22. Aliphatic methyl ketones and unhindered cyclic ketones form bisulfite addition products. Methyl aryl ketones, such as acetophenone $C_6H_5COCH_3$, fail to react.

(h) Nitriles (Table 63.15, page 384. Derivatives prepared from the carboxylic acid obtained by hydrolysis)

At high temperature nitriles (and amides) are converted to hydroxamic acids by hydroxylamine: $RCN + 2\ H_2NOH \rightarrow RCONHOH + NH_3$. The hydroxamic acid forms a red-blue complex with ferric ion. The unknown must first give a negative test with hydroxylamine at lower temperature (Group I(f), Alkaline Hydroxamic Acid Test) before trying this test.

Hydroxamic Acid Test for Nitriles (and Amides). To 2 ml of a 1 M hydroxylamine hydrochloride solution in propylene glycol add one drop (30 mg) of the unknown dissolved in the minimum amount of propylene glycol. Then add one ml of 1 N potassium hydroxide in propylene glycol and boil the mixture for 2 min. Cool the mixture and add 0.25 to 0.5 ml of 10% ferric chloride solution. A red-blue color is a positive test for almost all nitrile and amide groups, although benzanilide fails to give a positive test.

(i) Phenols (Table 63.17, page 385)

Ferric Chloride Test. Dissolve one drop (30 mg) of the unknown compound in a ml of water or water-alcohol mixture and add up to three drops of 1% ferric chloride solution. A red, blue, green or purple color is a positive test.

A more sensitive test for phenols[7] consists of dissolving or suspending 30 mg of the unknown in 1 ml of chloroform and adding 2 drops of a solution made by dissolving 1 g of ferric chloride in 100 ml of chloroform. Addition of a drop of pyridine, with stirring, will produce a color if phenols or enols are present.

GROUP II   Water-soluble Acidic Salts, Insoluble in Ether

(a) Amine salts [Table 63.5 (1° and 2° amines); Table 63.6 (3° amines)] The free amine can be liberated by addition of base and extraction into ether. Following evaporation of the ether the Hinsberg test, Group I(d), can be applied to determine if the compound is a primary, secondary, or tertiary amine.

The acid ferric hydroxamate test, Group I(d), can be applied directly to the amine salt.

---

[7] S. Soloway and S. H. Wilen, *Anal. Chem.*, **24**, 979 (1952).

GROUP III   Water-soluble Neutral Compounds, Insoluble in Ether

(a) Metal Salts of Carboxylic Acids   (Table 63.1, carboxylic acids; Derivatives, page 374)

The free acid can be liberated by addition of acid and extraction into an appropriate solvent, after which the carboxylic acid can be characterized by mp or bp before proceeding to prepare a derivative.

(b) Ammonium Salts   (Table 63.1, Carboxylic acids; Derivatives, page 374)

Ammonium salts on treatment with alkali liberate ammonia, which can be detected by its odor and the fact it will turn damp red litmus, blue. A more sensitive test utilizes the cupric ion, which is blue in the presence of ammonia [see Group VIII a(i)]. Ammonium salts will not give a positive hydroxamic acid test (Ih) as given by amides.

(c) Sugars. See page 180, Chapter 31, for Tollens test and phenyl-osazone formation.

(d) Amino Acids. See Chapter 34, paper chromatography of amino acids and color reaction with ninhydrin.

(e) Polyhydric Alcohols   (Table 63.2; Derivatives, page 375)

Periodic Acid Test for vic-Glycols.[8]   Vicinal glycols (hydroxyl groups on adjacent carbon atoms) can be detected by reaction with periodic acid.

To 2 ml of periodic acid reagent add one drop (no more) of concentrated nitric acid and shake. Then add one drop or a small crystal of the unknown. Shake for 15 sec, and add 1 or 2 drops of 5% aqueous silver nitrate solution. Instantaneous formation of a white precipitate is a positive test.

Reagent: Dissolve 0.5 g of paraperiodic acid (H_5IO_6) in 100 ml of water.

In addition to 1,2-glycols a positive test is given by α-hydroxy aldehydes, α-hydroxy ketones, α-hydroxy acids, and α-amino alcohols, as well as 1,2-diketones.

GROUP IV   Certain Carboxylic Acids, Certain Phenols, and Sulfonamides of 1° Amines

(a) Carboxylic Acids. Solubility in both 5% sodium hydroxide and sodium bicarbonate is usually sufficient to characterize this class of compounds. Addition of mineral acid should regenerate the carboxylic acid. The neutralization equivalent can be obtained by titrating a known quantity of the acid (*ca.* 0.15 g) dissolved in water-ethanol with 0.1 N sodium hydroxide to a phenolphthalein end point.

(b) Phenols. Negatively substituted phenols such as nitrophenols,

[8] R. L. Shriner, R. C. Fuson, and D. Y. Curtin, *The Systematic Identification of Organic Compounds,* 5th ed., John Wiley & Sons, Inc., New York, 1964, p. 145.

aldehydophenols, and polyhalophenols are sufficiently acidic to dissolve in 5% sodium bicarbonate. See Group I(i), page 366, for the ferric chloride test for phenols; however this test is not completely reliable for these acidic phenols.

GROUP V   Acidic Compounds, Insoluble in Bicarbonate

(a) Phenols. See Group I(i).
(b) Enols. See Group I(i).
(c) 1° and 2° Nitro Compounds (Table 63.16; Derivatives, page 384).
Ferrous Hydroxide Test. In a small test tube add 1 drop (or 20 mg) of the unknown to 1.5 ml of a freshly prepared 5% aqueous solution of ferrous ammonium sulfate. Add one drop of 3 N sulfuric acid and 1 ml of 2 N methanolic potassium hydroxide solution. Quickly stopper the tube and shake it. The appearance of a red-brown precipitate of ferric hydroxide within 1 min is a positive test. Almost all nitro compounds give a positive test within 30 sec.
(d) Sulfonamides of 1° amines. An extremely sensitive test for sulfonamides (Feigl, *Spot Tests in Organic Analysis*) consists of placing a drop of a suspension or solution of the unknown on sulfonamide test paper followed by a drop of 0.5% hydrochloric acid. A red color is a positive test for sulfonamides.

The test paper is prepared by dipping filter paper into a mixture of equal volumes of a 1% aqueous solution of sodium nitrite and a 1% methanolic solution of N,N-dimethyl-1-naphthylamine. Allow the filter paper to dry in the dark. Handle the naphthylamine with great care. It is carcinogenic.

GROUP VI   Basic Compounds, Insoluble in Water, Soluble in Acid

(a) Amines. See Group I(d).

GROUP VII   Reducible, Neutral N- and S-Containing Compounds

(a) Aromatic Nitro Compounds. See Group V(c).

GROUP VIII   Hydrolyzable, Neutral N- and S-Containing Compounds (Identified through the acid and amine obtained on hydrolysis)

(a) Amides. Unsubstituted amides are detected by the hydroxamic acid test, Group I(h).
(1) Unsubstituted Amides. Upon hydrolysis, unsubstituted amides liberate ammonia, which can be detected by reaction with cupric ion (Group III(b)).

To 2 ml of 20% sodium hydroxide solution add 50 mg of the unknown. Cover the mouth of the test tube with a piece of filter paper moistened with a few drops of 10% copper sulfate solution. Boil for 1 min. A blue color on the filter paper is a positive test for ammonia.

(2) Substituted Amides. The identification of substituted amides is not easy. There are no completely general tests for the substituted amide groups and hydrolysis is often difficult.

Hydrolyze the amide by refluxing 0.5 g with 5 ml of 20% sodium hydroxide for 20 min. Isolate the primary or secondary amine produced, by extraction into ether, and identify as described under Group I(d). Liberate the acid by acidification of the residue and isolate by filtration or extraction and characterize by bp or mp and the mp of an appropriate derivative.

(3) Anilides. Add 100 mg of the unknown to 3 ml of concd sulfuric acid. Carefully stopper the test tube with a rubber stopper and shake vigorously. **Caution.** Add 50 mg of finely powdered potassium dichromate. A blue-pink color is a positive test for an anilide which does not have substituents on the ring (e.g., acetanilide).

(b) Nitriles. See Group I(h).

(c) Sulfonamides. See Group V(d), page 368.

GROUP IX   Neutral Polar Compounds, Insoluble in Dilute Hydrochloric Acid, Soluble in Concentrated Sulfuric Acid. (Most compounds containing oxygen.)

(a) Alcohols. See Group I(b).

(b) Aldehydes. See Group I(c).

(c) Anhydrides. See Group I(e).

(d) Esters. See Group I(f).

(e) Ethers (Table 63.9, page 381)

Ethers are very unreactive. Care must be used to distinguish ethers from those hydrocarbons which are soluble in concd sulfuric acid. *Ferrox Test.*[9] In a dry test tube grind together, with a stirring rod, a crystal of ferric ammonium sulfate (or ferric chloride) and a crystal of potassium thiocyanate. Ferric hexathiocyanatoferriate will adhere to the stirring rod. In a clean tube place 3 or 4 drops of a liquid unknown or a saturated benzene solution of a solid unknown and stir with the rod. The salt will dissolve if the unknown contains oxygen to give a red to red-purple color, but it will not dissolve in hydrocarbons or halocarbons. Diphenyl ether does not give a positive test.

Alkyl ethers are generally soluble in concd hydrochloric acid; alkyl aryl and diaryl ethers are not soluble.

(f) Ketones. See Group I(g).

(g) Unsaturated Hydrocarbons (Table 63.12, page 382)

Bromine in Carbon Tetrachloride. Dissolve 3–4 drops (40–60 mg) of the unknown in 2 ml of carbon tetrachloride. Add a 2% solution of bromine in carbon tetrachloride dropwise with shaking. If more than 2 drops of bromine solution are required to give a permanent red color, unsaturation is indicated.

[9] D. Davidson, *Ind. Eng. Chem., Anal. Ed.,* **12,** 40 (1940).

Potassium Permanganate Solution. Dissolve 2 or 3 drops (30 mg) of the unknown in reagent grade acetone and add a 1% aqueous solution of potassium permanganate dropwise with shaking. If more than one drop of reagent is required to give a purple color to the solution, unsaturation or an easily oxidized functional group is present. Run parallel tests on pure acetone and, as usual, a compound known to be an alkene.

GROUP X   Inert Compounds. Insoluble in Concentrated Sulfuric Acid

(a) Alkyl and Aryl Halides (Table 63.10, page 381)
Alcoholic Silver Nitrate. Add one drop of the unknown (or a saturated solution of 20 mg of unknown in ethanol) to 1 ml of a saturated solution of silver nitrate. A precipitate which forms within 2 min is a positive test for an alkyl bromide, or iodide, or a tertiary alkyl chloride, as well as allyl halides.

If no precipitate forms within 2 min, heat the solution to boiling. A precipitate of silver chloride will form from primary and secondary alkyl chlorides. Aryl halides and vinyl halides will not react.

(b) Aromatic Hydrocarbons (Table 63.13; Derivatives, page 382)
Aromatic hydrocarbons are best identified and characterized by ultraviolet and nmr spectroscopy, but the Friedel-Crafts reaction produces a characteristic color with certain aromatic hydrocarbons.
Friedel-Crafts Test. Heat a test tube containing about 0.1 g of anhydrous aluminum chloride in a hot flame to sublime the salt up onto the sides of the tube. Add a solution of about 20 mg of the unknown dissolved in 0.1 ml of chloroform to the cool tube in such a way that it comes into contact with the sublimed aluminum chloride. Note the color that appears.

Nonaromatic compounds fail to give a color with aluminum chloride, benzene and its derivatives give orange or red colors, naphthalenes a blue or purple color, biphenyls a purple color, phenanthrene a purple color, and anthracene a green color.

(c) Saturated Hydrocarbons
Saturated hydrocarbons are best characterized by nmr and IR spectroscopy, but they can be distinguished from aromatic hydrocarbons by the Friedel-Crafts test (Group X(b)).

(d) Diaryl Ethers
Because they are so inert, diaryl ethers are difficult to detect and may be mistaken for aromatic hydrocarbons. They do not give a positive Ferrox Test, Group IX(e), for ethers, and do not dissolve in concd sulfuric acid. Their infrared spectra, however, are characterized by an intense C—O single-bond, stretching vibration in the region 1270–1230 cm^{-1}.

Derivatives

1. Acids (Table 63.1)
p-Toluidides and Anilides. Reflux a mixture of the acid (0.5 g) and thionyl chloride (2 ml) under a cold finger condenser for 0.5 hr.

Cool the reaction mixture and add 1 g of aniline or *p*-toluidine in 30 ml of benzene. Warm the mixture on the steam bath for 2 min, transfer the benzene solution to a separatory funnel, and wash with 5-ml portions of water, 5% hydrochloric acid, 5% sodium hydroxide and water. The benzene is filtered through a cone of anhydrous sodium sulfate and evaporated; the derivative is recrystallized from water or ethanol-water.

Amides. Reflux a mixture of the acid (0.5 g) and thionyl chloride (2 ml) under a cold finger condenser for 0.5 hr. Pour the cool reaction mixture into 7 ml of ice-cold concd ammonia. Stir until reaction is complete, collect the product by filtration, and recrystallize it from water or water-ethanol.

2. Alcohols   (Table 63.2)

3,5-Dinitrobenzoates. Gently boil 0.5 g of 3,5-dinitrobenzoyl chloride and 0.5 ml of the alcohol for 5 min. Cool the mixture, pulverize any solid that forms, and add 5 ml of 2% sodium carbonate solution. Continue to grind and stir the solid with the sodium carbonate solution (to remove 3,5-dinitrobenzoic acid) for about a minute, filter, and wash the crystals with water. Dissolve the product in about 10 ml of hot ethanol, add water to the cloud point, and allow crystallization to proceed. Wash the 3,5-dinitrobenzoate with water-alcohol and dry.

Phenylurethans. Mix 0.5 ml of anhydrous alcohol (or phenol) and 0.5 ml of phenyl isocyanate (or α-naphthyl isocyanate) and heat on the steam bath for 5 min. (If the unknown is a phenol add 2 or 3 drops of pyridine to the reaction mixture.) Cool, add about 6 ml of ligroin or carbon tetrachloride, heat to dissolve the product, filter hot to remove a small amount of diphenylurea which usually forms, and cool the filtrate in ice, with scratching, to induce crystallization.

3. Aldehydes   (Table 63.3)

Semicarbazones. See page 134, Chapter 22. Use 0.5 ml of the stock solution and an estimated 1 millimole of the unknown aldehyde (or ketone).

3,4-Dinitrophenylhydrazones. See page 133, Chapter 22. Use 10 ml of the stock solution of 0.1 M 2,4-dinitrophenylhydrazine and an estimated 1 millimole of the unknown aldehyde (or ketone).

4. Primary and Secondary Amines   (Table 63.5)

Benzamides. Add about 0.5 g of benzoyl chloride in small portions with vigorous shaking and cooling to a suspension of 1 millimole of the unknown amine in 1 ml of 10% aqueous sodium hydroxide solution. After about 10 min of shaking the mixture is made pH 8 (pH paper) with dilute hydrochloric acid. The lumpy product is removed by filtration, washed thoroughly with water, and recrystallized from ethanol-water.

Picrates. Add a solution of 0.15 g of the unknown in 5 ml of ethanol (or 5 ml of a saturated solution of the unknown) to 5 ml of a saturated solution of picric acid (2,4,6-trinitrophenol, a strong acid) in ethanol and heat the solution to boiling. Cool slowly, remove the picrate by filtration, and wash with a small amount of ethanol. Recrystallization is not usually necessary; in the case of hydrocarbon picrates the product is often too unstable to be recrystallized.

Acetamides. Reflux about 1 millimole of the unknown with 0.4 ml of acetic anhydride for 5 min, cool, and dilute the reaction mixture with 5 ml of water. Initiate crystallization by scratching, if necessary. Remove the crystals by filtration and wash thoroughly with dilute hydrochloric acid to remove unreacted amine. Recrystallize the derivative from alcohol-water. Amines of low basicity, e.g., p-nitroaniline, should be refluxed for 30 to 60 min with 2 ml of pyridine as a solvent. The pyridine is removed by shaking the reaction mixture with 10 ml of 2% sulfuric acid solution; the product is isolated by filtration and recrystallized.

5. Tertiary Amines   (Table 63.6)

Picrates. See under Primary and Secondary Amines.

Methiodides. Reflux 0.3 ml of the amine and 0.3 ml of methyl iodide under a cold finger condenser for 5 min on the steam bath. Cool, scratch to induce crystallization, and recrystallize the product from ethyl alcohol or ethyl acetate.

6. Anhydrides and Acid Chlorides   (Table 63.7)

Acids. Reflux 0.2 g of the acid chloride or anhydride with 5 ml of 5% sodium carbonate solution for 20 min or less. Extract unreacted starting material with 5 ml of ether, if necessary, and acidify the reaction mixture with dil sulfuric acid to liberate the carboxylic acid.

Amides. Since the acid chloride (or anhydride) is already present, simply mix the unknown (0.5 g) and 7 ml of ice-cold concentrated ammonia until reaction is complete, collect the product by filtration, and recrystallize it from water or ethanol-water.

Anilides. Reflux 0.2 g of the acid halide or anhydride with 0.5 g of aniline in 10 ml of benzene for 10 min. Transfer the benzene solution to a separatory funnel and wash with 5 ml portions each of water, 5% hydrochloric acid, 5% sodium hydroxide and water. The benzene solution is filtered through a cone of anhydrous sodium sulfate and evaporated; the anilide is recrystallized from water or ethanol-water.

7. Aryl Halides   (Table 63.11)

Nitration. Add 2 ml of concd sulfuric acid to 0.5 g of the aryl halide (or aromatic compound) and stir. Add 2 ml of concd nitric acid dropwise with stirring and shaking while cooling the reaction mixture in water. Then heat and shake the reaction mixture in a water bath at about 50° for 15 min, pour into 10 ml of cold water, and collect the product by filtration. Recrystallize from methanol to constant mp.

To nitrate unreactive compounds use fuming nitric acid in place of concd nitric acid.

Sidechain Oxidation Products. Dissolve 1 g of sodium dichromate in 3 ml of water and add 2 ml of concd sulfuric acid. Add 0.25 g of the unknown and boil for 30 min. Cool, add 2 or 3 ml of water and then remove the carboxylic acid by filtration. Wash the crystals with water and recrystallize from methanol-water.

8. Hydrocarbons: Aromatic   (Table 63.13)

Nitration. See preceding, under Aryl Halides.

Picrates. See preceding, under Primary and Secondary Amines.

9. Ketones   (Table 63.14)

Semicarbazones and 2,4-dinitrophenylhydrazones. See preceding directions under Aldehydes.

10. Nitro Compounds   (Table 63.16)

Reduction to Amines. Place 0.5 g of the unknown in a 25-ml round-bottomed flask, add 1 g of tin, and then—in portions—10 ml of 10% hydrochloric acid. Reflux for 30 min, add 5 ml of water, then add slowly, with good cooling, sufficient 40% sodium hydroxide solution to dissolve the tin hydroxide. Extract the reaction mixture with three 5-ml portions of ether, dry the ether extract over anhydrous sodium sulfate, and evaporate the ether to leave the amine. Determine the bp or mp of the amine and then convert it into a benzamide or acetamide as described under Primary and Secondary Amines.

11. Phenols   (Table 63.17)

α-Naphthylurethan. Follow the procedure for preparation of a phenylurethan under Alcohols, page 371.

Bromo Derivative. In a 10-ml Erlenmeyer flask dissolve 0.8 g of potassium bromide in 5 ml of water. *Carefully* add 0.5 g of bromine. In a separate flask dissolve 0.1 g of the phenol in 1 ml of methanol and add 1 ml of water. Add about 1.5 ml of the bromine solution with swirling (hood); continue the addition of bromine until the yellow color of unreacted bromine persists. Add 3 to 4 ml of water to the reaction mixture and shake vigorously. Remove the product by filtration and wash well with water. Recrystallize from methanol-water.

Table 63.1   Acids

Bp	Mp	Name of Compound	Derivatives p-Toluidide[a] Mp	Anilide[b] Mp	Amide[c] Mp
101		Formic acid	53	47	43
118		Acetic acid	126	106	
139		Acrylic acid	141	104	85
141		Propionic acid	124	103	81
162		n-Butyric acid	72	95	115
163		Methacrylic acid		87	102
165		Pyruvic acid	109	104	124
185		Valeric acid	70	63	
186		2-Methylvaleric acid	80	95	79
194		Dichloroacetic acid	153	118	98
202–203		Hexanoic acid	75	95	101
237		Octanoic acid	70	57	107
254		Nonanoic acid	84	57	99
	31–32	Decanoic acid	78	70	108
	43–45	Lauric acid	87	78	100
	47–49	Bromoacetic acid		131	91
	47–49	Hydrocinnamic acid	135	92	105
	54–55	Myristic acid	93	84	103
	54–58	Trichloroacetic acid	113	97	141
	61–62	Chloroacetic acid	162	137	121
	61–62.5	Palmitic acid	98	90	106
	67–69	Stearic acid	102	95	109
	68–69	3,3-Dimethylacrylic acid		126	107
	71–73	Crotonic acid		118	158
	77–78.5	Phenylacetic acid	136	118	156
	98–102	Azelaic acid (nonanedioic)	164 (di)	107 (mono) 186 (di)	
	103–105	o-Toluic acid	144	125	142
	108–110	m-Toluic acid	118	126	94
	119–121	dl-Mandelic acid	172	151	133
	122–123	Benzoic acid	158	163	130
	127–128	2-Benzoylbenzoic acid		195	165
	129–130	2-Furoic acid	107	123	143
	131–134	Sebacic acid	201	122 (mono) 200 (di)	170 (mono) 210 (di)
	134–135	trans-Cinnamic acid	168	153	147
	134–136	Maleic acid	142 (di)	198 (mono) 187 (di)	260 (di)
	135–137	Malonic acid	86 (mono) 253 (di)	132 (mono) 230 (di)	
	138–140	2-Chlorobenzoic acid	131	118	
	140–142	3-Nitrobenzoic acid	162	155	143
	144–148	Anthranilic acid	151	131	109
	147–149	Diphenylacetic acid	172	180	167
	152–153	Adipic acid	239	151 (mono) 241 (di)	125 (mono) 220 (di)
	153–154	Citric acid	189 (tri)	199 (tri)	210 (tri)
	157–159	4-Chlorophenoxyacetic acid		125	133
	158–160	Salicylic acid	156	136	142
	163–164	Trimethylacetic acid		127	178
	164–166	5-Bromosalicylic acid		222	232
	166–167	Itaconic acid		190	191 (di)

Table 63.1 Acids (cont.)

Bp	Mp	Name of Compound	Derivatives		
			p-Toluidide[a]	Anilide[b]	Amide[c]
			Mp	Mp	Mp
	171–174	d-Tartaric acid		180 (mono)	171 (mono)
				264 (di)	196 (di)
	179–182	3,4-Dimethoxybenzoic acid		154	164
	180–182	4-Toluic acid	160	145	160
	182–185	4-Anisic acid	186	169	167
	187–190	Succinic acid	180 (mono)	143 (mono)	157 (mono)
			255 (di)	230 (di)	260 (di)
	201–203	3-Hydroxybenzoic acid	163	157	170
	203–206	3,5-Dinitrobenzoic acid		234	183
	210–211	Phthalic acid	150 (mono)	169 (mono)	149 (mono)
			201 (di)	253 (di)	220 (di)
	214–215	4-Hydroxybenzoic acid	204	197	162
	225–227	2,4-Dihydroxybenzoic acid		126	228
	236–239	Nicotinic acid	150	132	128
	239–241	4-Nitrobenzoic acid	204	211	201
	299–300	Fumaric acid		233 (mono)	270 (mono)
				314 (di)	266 (di)
	>300	Terephthalic acid		334	

[a] For preparation, see page 370.
[b] For preparation, see page 370.
[c] For preparation, see page 371.

Table 63.2 Alcohols

Bp	Mp	Name of Compound	Derivatives	
			3,5-Dinitro-benzoate[a]	Phenyl-urethan[b]
			Mp	Mp
65		Methanol	108	47
78		Ethanol	93	52
82		2-Propanol	123	88
83		t-Butanol	142	136
96–98		Allyl alcohol	49	70
97		1-Propanol	74	57
98		2-Butanol	76	65
102		2-Methyl-2-butanol	116	42
104		2-Methyl-3-butyn-2-ol	112	
108		2-Methyl-1-propanol	87	86
114–115		Propargyl alcohol		
114–115		3-Pentanol	101	48
116–119		1-Butanol	64	61
118–119		2-Pentanol	62	
123		3-Methyl-3-pentanol	96(62)	43
129		2-Chloroethanol	95	51
130		2-Methyl-1-butanol	70	31
132		4-Methyl-2-pentanol	65	143
136–138		1-Pentanol	46	46
139–140		Cyclopentanol	115	132
140		2,4-Dimethyl-3-pentanol		95
146		2-Ethyl-1-butanol	51	

Table 63.2   Alcohols (cont.)

Bp	Mp	Name of Compound	3,5-Dinitro-benzoate[a] Mp	Phenyl urethar[Mp
151		2,2,2-Trichloroethanol	142	87
157		1-Hexanol	58	42
160–161		Cyclohexanol	113	82
170		Furfuryl alcohol	80	45
174–181		2-Octanol	32	oil
176		1-Heptanol	47	60(68
178		Tetrahydrofurfuryl alcohol	83	61
183–184		2,3-Butanediol		201 (di
183–186		2-Ethyl-1-hexanol		34
187		1,2-Propanediol		153 (di
194–197		Linaloöl		66
195		1-Octanol	61	74
196–198		Ethylene glycol	169	157 (di
200–215		1,3-Butanediol		122
203–205		Benzyl alcohol	113	77
204		1-Phenyl-1-ethanol	95	92
219–221		2-Phenethyl alcohol	108	78
230		1,4-Butanediol		183 (di
231		1-Decanol	57	59
259		4-Methoxybenzyl alcohol		92
	33–35	Cinnamyl alcohol	121	90
	38–40	1-Tetradecanol	67	74
	48–50	1-Hexadecanol	66	73
	58–60	1-Octadecanol	77 (66)	79
	66–67	Benzhydrol	141	139
	147	Cholesterol		168

[a] For preparation, see page 371.
[b] For preparation, see page 371.

Table 63.3   Aldehydes

Bp	Mp	Name of Compound	Semicarbazone[a] Mp	2,4-Dinitrophenyl hydrazone[b] Mp
21		Acetaldehyde	162	168
46–50		Propionaldehyde	89(154)	148
63		i-Butyraldehyde	125(119)	187(183)
75		Butyraldehyde	95(106)	123
90–92		2-Methylbutyraldehyde	103	120
98		Chloral	90	131
104		Crotonaldehyde	199	190
117		2-Ethylbutyraldehyde	99	95(130)
153		Heptaldehyde	109	108
162		2-Furaldehyde	202	212(230)
163		2-Ethylhexanal	254	114(120)
195		Phenylacetaldehyde	153	121(110)
197		Salicylaldehyde	231	248
204–205		4-Tolualdehyde	234(215)	232

Table 63.3   Aldehydes (cont.)

Bp	Mp	Name of Compound	Derivatives	
			Semicarbazone[a]	2,4-Dinitrophenyl-hydrazone[b]
			Mp	Mp
209–215		2-Chlorobenzaldehyde	146(229)	213
247		2-Ethoxybenzaldehyde	219	
248		4-Anisaldehyde	210	253
250–252		trans-Cinnamaldehyde	215	255
	33–34	1-Naphthaldehyde	221	
	37–39	2-Anisaldehyde	215	254
	42–45	3,4-Dimethoxybenzaldehyde	177	261
	44–47	4-Chlorobenzaldehyde	230	254
	57–59	3-Nitrobenzaldehyde	246	293
	81–83	Vanillin	230	271

[a] For preparation see page 371.
[b] For preparation see page 371.

Table 63.4   Amides

Bp	Mp	Name of Compound
153		N,N-Dimethylformamide
164–166		N,N-Dimethylacetamide
210		Formamide
243–244		N-Methylformanilide
	26–28	N-Methylacetamide
	79–81	Acetamide
	109–111	Methacrylamide
	113–115	Acetanilide
	116–118	2-Chloroacetamide
	127–129	Isobutramide
	128–129	Benzamide
	130–133	Nicotinamide
	177–179	4-Chloroacetanilide

Table 63.5   Primary and Secondary Amines

Bp	Mp	Name of Compound	Derivatives		
			Benzamide[a]	Picrate[b]	Acetamide[c]
			Mp	Mp	Mp
33–34		i-Propylamine	71	165	
46		t-Butylamine	134	198	
48		n-Propylamine	84	135	
53		Allyl amine		140	
55		Diethylamine	42	155	
63		s-Butylamine	76	139	
64–71		i-Butylamine	57	150	

Table 63.5   Primary and Secondary Amines (cont.)

Bp	Mp	Name of Compound	Derivatives Benzamide[a] Mp	Picrate[b] Mp	Acetamide Mp
78		n-Butylamine	42		
84		Di-i-propylamine		140	
87–88		Pyrrolidine	oil	112	
105–110		Dipropylamine		75	
106		Piperidine	48	152	
118		Ethylenediamine	244(di)	233	172(di)
129		Morpholine	75	146	
137–139		Di-i-butylamine		121	86
145–146		Furfurylamine		150	
149		N-Methylcyclohexylamine	85	170	
159		Di-n-butylamine		59	
182–185		Benzylamine	105	199	60
184		Aniline	163	198	114
196		N-Methylaniline	63	145	102
199–200		2-Toluidine	144	213	110
203–204		3-Toluidine	125	200	65
205		N-Ethylaniline	60	138(132)	54
208–210		2-Chloroaniline	99	134	87
210		2-Ethylaniline	147	194	111
216		2,6-Dimethylaniline	168	180	177
218		2,4-Dimethylaniline	192	209	133
218		2,5-Dimethylaniline	140	171	139
221		N-Ethyl-m-toluidine	72		
225		2-Anisidine	60(84)	200	85
230		3-Chloroaniline	120	177	72(78)
231–233		2-Phenetidine	104		79
241		4-Chloro-2-methylaniline	142		140
242		3-Chloro-4-methylaniline	122		105
250		4-Phenetidine	173	69	137
256		Dicyclohexylamine	153(57)	173	103
	35–38	N-Phenylbenzylamine	107	48	58
	41–44	4-Toluidine	158	182	147
	49–51	2,5-Dichloroaniline	120		132
	52–54	Diphenylamine	180	182	101
	57–60	4-Anisidine	154	170	130
	57–60	2-Aminopyridine	165(di)	216(223)	
	60–62	N-Phenyl-1-naphthylamine	152		115
	62–64	4-Bromoaniline	204	180	168
	62–65	2,4,5-Trimethylaniline	167		162
	64–66	3-Phenylenediamine	125(mono) 240(di)	184	87(mono) 191(di)
	68–71	4-Chloroaniline	192	178	179(172)
	71–73	2-Nitroaniline	110(98)	73	92
	97–99	2,4-Diaminotoluene	224(di)		224(di)
	100–102	2-Phenylenediamine	301	208	185
	104–107	2-Methyl-5-nitroaniline	186		151
	107–109	2-Chloro-4-nitroaniline	161		139
	112–114	3-Nitroaniline	157(150)	143	155(76)
	115–116	4-Methyl-2-nitroaniline	148		99
	117–119	4-Chloro-2-nitroaniline			104
	120–122	2,4,6-Tribromoaniline	198(204)		232

Table 63.5 Primary and Secondary Amines (cont.)

Bp	Mp	Name of Compound	Derivatives		
			Benzamide[a]	Picrate[b]	Acetamide[c]
			Mp	Mp	Mp
	131–133	2-Methyl-4-nitroaniline			202
	138–140	2-Methoxy-4-nitroaniline	149		
	138–142	4-Phenylenediamine	128(mono)		162(mono)
			300(di)		304(di)
	148–149	4-Nitroaniline	199	100	215
	162–164	4-Aminoacetanilide			304
	176–178	2,4-Dinitroaniline	202(220)		120

[a] For preparation see page 371.
[b] For preparation see page 371.
[c] For preparation see page 372.

Table 63.6 Tertiary Amines

Bp	Name of Compound	Derivatives	
		Picrate[a]	Methiodide[b]
		Mp	Mp
85–91	Triethylamine	173	280
115	Pyridine	167	117
128–129	2-Picoline	169	230
143–145	2,6-Lutidine	168(161)	233
144	3-Picoline	150	92(36)
145	4-Picoline	167	149
155–158	Tri-n-propylamine	116	207
159	2,4-Lutidine	180	113
183–184	N,N-Dimethylbenzylamine	93	179
216	Tri-n-butylamine	105	186
217	N,N-Diethylaniline	142	102
237	Quinoline	203	133(72)

[a] For preparation see page 372.
[b] For preparation see page 372.

Table 63.7 Anhydrides and Acid Chlorides

Bp	Mp	Name of Compound	Derivatives			
			Acid[a]		Amide[b]	Anilide[c]
			Bp	Mp	Mp	Mp
52		Acetyl chloride	118		82	114
77–79		Propionyl chloride	141		81	106
102		Butyryl chloride	162		115	96
138–140		Acetic anhydride	118		82	114
167		Propionic anhydride	141		81	106
198–199		Butyric anhydride	162		115	96
198		Benzoyl chloride		122	130	163
225		3-Chlorobenzoyl chloride		158	134	122
238		2-Chlorobenzoyl chloride		142	142	118
	32–34	cis-1,2-Cyclohexanedicarboxylic anhydride		192		
	35–37	Cinnamoyl chloride		133	147	151

Table 63.7   Anhydrides and Acid Chlorides (cont.)

Bp	Mp	Name of Compound	Derivatives		
			Acid[a]	Amide[b]	Anilide[c]
			Bp / Mp	Mp	Mp
	39–40	Benzoic anhydride	122	130	163
	54–56	Maleic anhydride	130	181(mono)	173(mono)
				266(di)	187(di)
	72–74	4-Nitrobenzoyl chloride	241	201	211
	119–120	Succinic anhydride	186	157(mono)	148(mono)
				260(di)	230(di)
	131–133	Phthalic anhydride	206	149(mono)	170(mono)
				220(di)	253(di)
	254–258	Tetrachlorophthalic anhydride	250		
	267–269	1,8-Naphthalic anhydride	274		250–282(di

[a] For preparation see page 372.
[b] For preparation see page 372.
[c] For preparation see page 372.

Table 63.8   Esters

Bp	Mp	Name of Compound	Bp	Mp	Name of Compound
34		Methyl formate	169–170		Methyl acetoacetate
52–54		Ethyl formate	180–181		Dimethyl malonate
72–73		Vinyl acetate	181		Ethyl acetoacetate
77		Ethyl acetate	185		Diethyl oxalate
79		Methyl propionate	198–199		Methyl benzoate
80		Methyl acrylate	206–208		Ethyl caprylate
85		i-Propyl acetate	208–210		Ethyl cyanoacetate
93		Ethyl chloroformate	212		Ethyl benzoate
94		Isopropenyl acetate	217		Diethyl succinate
98		Isobutyl formate	218		Methyl phenylacetate
98		t-Butyl acetate	218–219		Diethyl fumarate
99		Ethyl propionate	222		Methyl salicylate
99		Ethyl acrylate	225		Diethyl maleate
100		Methyl methacrylate	229		Ethyl phenylacetate
101		Methyl trimethylacetate	234		Ethyl salicylate
102		n-Propyl acetate	268		Dimethyl suberate
106–113		s-Butyl acetate	271		Ethyl cinnamate
120		Ethyl butyrate	282		Dimethyl phthalate
127		n-Butyl acetate	298–299		Diethyl phthalate
128		Methyl valerate	298–299		Phenyl benzoate
130		Methyl chloracetate	340		Dibutyl phthalate
131–133		Ethyl isovalerate		56–58	Ethyl p-nitrobenzoate
142		n-Amyl acetate		88–90	Ethyl p-aminobenzoate
142		Isoamyl acetate		94–96	Methyl p-nitrobenzoate
143		Ethyl chloroacetate		95–98	n-Propyl p-hydroxybenzoate
154		Ethyl lactate		116–118	Ethyl p-hydroxybenzoate
168		Ethyl caproate (ethyl hexanoate)		126–128	Methyl p-hydroxybenzoate

Table 63.9 Ethers

Bp	Mp	Name of Compound	Bp	Mp	Name of Compound
32		Furan	215		4-Bromoanisole
33		Ethyl vinyl ether	234–237		Anethole
65–67		Tetrahydrofuran	259		Phenylether
94		n-Butyl vinyl ether	273		2-Nitroanisole
54		Anisole	298		Benzyl ether
74		4-Methylanisole		50–52	4-Nitroanisole
75–176		3-Methylanisole		56–60	1,4-Dimethoxybenzene
98–203		4-Chloroanisole		73–75	2-Methoxynaphthalene
06–207		1,2-Dimethoxybenzene			

Table 63.10 Halides

Bp	Name of Compound	Bp	Name of Compound
34–36	2-Chloropropane	100–105	1-Bromobutane
40–41	Dichloromethane	105	Bromotrichloromethane
44–46	Allyl chloride	110–115	1,1,2-Trichloroethane
57	1,1-Dichloroethane	120–121	1-Bromo-3-methylbutane
59	2-Bromopropane	121	Tetrachloroethylene
68	Bromochloromethane	123	3,4-Dichloro-1-butene
68–70	2-Chlorobutane	125	1,3-Dichloro-2-butene
69–73	Iodoethane	131–132	1,2-Dibromoethane
70–71	Allyl bromide	140–142	1,2-Dibromopropane
71	1-Bromopropane	142–145	1-Bromo-3-chloropropane
72–74	2-Bromo-2-methylpropane	146–150	Bromoform
74–76	1,1,1-Trichloroethane	147	1,1,2,2-Tetrachloroethane
81–85	1,2-Dichloroethane	156	1,2,3-Trichloropropane
87	Trichloroethylene	161–163	1,4-Dichlorobutane
88–90	2-Iodopropane	167	1,3-Dibromopropane
90–92	1-Bromo-2-methylpropane	177–181	Benzyl chloride
91	2-Bromobutane	197	(2-Chloroethyl) benzene
94	2,3-Dichloro-1-propene	219–223	Benzotrichloride
95–96	1,2-Dichloropropane	238	1-Bromodecane
96–98	Dibromomethane		

Table 63.11 Aryl Halides

Bp	Mp	Name of Compound	Nitration Product[a]		Oxidation Product[b]	
			Position	Mp	Name	Mp
132		Chlorobenzene	2,4	52		
156		Bromobenzene	2,4	70		
157–159		2-Chlorotoluene	3,5	63	2-Chlorobenzoic acid	141
162		4-Chlorotoluene	2	38	4-Chlorobenzoic acid	240
172–173		1,3-Dichlorobenzene	4,6	103		
178		1,2-Dichlorobenzene	4,5	110		
196–203		2,4-Dichlorotoluene	3,5	104	2,4-Dichlorobenzoic acid	164
201		3,4-Dichlorotoluene	6	63	3,4-Dichlorobenzoic acid	206
214		1,2,4-Trichlorobenzene	5	56		
279–281		1-Bromonaphthalene	4	85		
	51–53	1,2,3-Trichlorobenzene	4	56		
	54–56	1,4-Dichlorobenzene	2	54		

Table 63.11 Aryl Halides (cont.)

Bp	Mp	Name of Compound	Derivatives			
			Nitration Product[a]		Oxidation Product[b]	
			Position	Mp	Name	Mp
66–68		1,4-Bromochlorobenzene	2	72		
87–89		1,4-Dibromobenzene	2,5	84		
138–140		1,2,4,5-Tetrachlorobenzene	3	99		
			3,6	227		

[a] For preparation, see page 372.
[b] For preparation, see page 373.

Table 63.12 Hydrocarbons: Alkenes

Bp	Name of Compound	Bp	Name of Compound
34	Isoprene	149–150	1,5-Cyclooctadiene
83	Cyclohexene	152	dl-α-Pinene
116	5-Methyl-2-norbornene	160	Bicyclo[4.3.0]nona-3,7-diene
122–123	1-Octene	165–167	(−)-β-Pinene
126–127	4-Vinyl-1-cyclohexene	165–169	α-Methylstyrene
132–134	2,5-Dimethyl-2,4-hexadiene	181	1-Decene
141	5-Vinyl-2-norbornene	181	Indene
143	1,3-Cyclooctadiene	251	1-Tetradecene
145	4-Butylstyrene	274	1-Hexadecene
145–146	Cycloctene	349	1-Octadecene
145–146	Styrene		

Table 63.13 Hydrocarbons: Aromatic

Bp	Mp	Name of Compound	Melting Point of Derivatives		
			Nitro[a]		Picrate[b]
			Position	Mp	Mp
80		Benzene	1,3	89	84
111		Toluene	2,4	70	88
136		Ethylbenzene	2,4,6	37	96
138		p-Xylene	2,3,5	139	90
138–139		m-Xylene	2,4,6	183	91
143–145		o-Xylene	4,5	118	88
145		4-t-Butylstyrene	2,4	62	
145–146		Styrene			
152–154		Cumene	2,4,6	109	
163–166		Mesitylene	2,4	86	97
			2,4,6	235	
165–169		α-Methylstyrene			
168		1,2,4-Trimethylbenzene	3,5,6	185	97
176–178		p-Cymene	2,6	54	
189–192		4-t-Butyltoluene			
197–199		1,2,3,5-Tetramethylbenzene	4,6	181(157)	
203		p-Di-i-propylbenzene			
204–205		1,2,3,4-Tetramethylbenzene	5,6	176	92
207		1,2,3,4-Tetrahydronaphthalene	5,7	95	
240–243		1-Methylnaphthalene	4	71	142
	34–36	2-Methylnaphthalene	1	81	116
	50–51	Pentamethylbenzene	6	154	131

Table 63.13   Hydrocarbons: Aromatic (cont.)

Bp	Mp	Name of Compound	Melting Point of Derivatives		
			Nitro[a]		Picrate[b]
			Position	Mp	Mp
	69–72	Biphenyl	4,4′	237(229)	
	80–82	1,2,4,5-Tetramethylbenzene	3,6	205	
	80–82	Naphthalene	1	61(57)	149
	90–95	Acenaphthene	5	101	161
	99–101	Phenanthrene			144(133)
	112–115	Fluorene	2	156	87(77)
			2,7	199	
	214–217	Anthracene			138

[a] For preparation see page 373.
[b] For preparation see page 373.

Table 63.14   Ketones

Bp	Mp	Name of Compound	Derivatives	
			Semi-carbazone[a]	2,4-Dinitro-phenylhydrazone[b]
			Mp	Mp
56		Acetone	187	126
80		2-Butanone	136,186	117
88		2,3-Butanedione	278	315
100–101		2-Pentanone	112	143
102		3-Pentanone	138	156
106		Pinacolone	157	125
114–116		4-Methyl-2-pentanone	132	95
124		2,4-Dimethyl-3-pentanone	160	88,95
128–129		5-Hexen-2-one	102	108
129		4-Methyl-3-penten-2-one	164	205
130–131		Cyclopentanone	210	146
133–135		2,3-Pentanedione	122(mono) 209(di)	209
145		4-Heptanone	132	75
145		5-Methyl-2-hexanone	147	95
145–147		2-Heptanone	123	89
146–149		3-Heptanone	101	
156		Cyclohexanone	166	162
162–163		2-Methylcyclohexanone	195	137
169		2,6-Dimethyl-4-heptanone	122	66,92
169–170		3-Methylcyclohexanone	180	155
173		2-Octanone	122	58
191		Acetonylacetone	185(mono) 224(di)	257(di)
202		Acetophenone	198	238
216		Phenylacetone	198	156
217		Isobutyrophenone	181	163
218		Propiophenone	182	191
220–222		n-Butyrophenone	188	
226		4-Methylacetophenone	205	258
231–232		2-Undecanone	122	63
232		4-Chloroacetophenone	204	236
235		Benzylacetone	142	127

Table 63.14   Ketones (cont.)

Bp	Mp	Name of Compound	Semi-carbazone[a] Mp	2,4-Dinitro-phenylhydrazone[b] Mp
	35–37	4-Chloropropiophenone	176	223
	35–39	4-Phenyl-3-buten-2-one	187	227
	36–38	4-Methoxyacetophenone	198	228
	48–49	Benzophenone	167	238
	53–55	2-Acetonaphthone	235	262
	60	Desoxybenzoin	148	204
	76–78	3-Nitroacetophenone	257	228
	78–80	4-Nitroacetophenone		257
	82–85	9-Fluorenone		283
	134–136	Benzoin	206	245
	147–148	4-Hydroxypropiophenone		240

[a] For preparation, see page 373.
[b] For preparation, see page 373.

Table 63.15   Nitriles

Bp	Mp	Name of Compound	Bp	Mp	Name of Compound
77		Acrylonitrile	212		3-Tolunitrile
83–84		Trichloroacetonitrile	217		4-Tolunitrile
97		Propionitrile	233–234		Benzyl cyanide
107–108		Isobutyronitrile	295		Adiponitrile
115–117		n-Butyronitrile		30.5	4-Chlorobenzyl cyanide
174–176		3-Chloropropionitrile		32–34	Malononitrile
191		Benzonitrile		38–40	Stearonitrile
205		2-Tolunitrile		46–48	Succinonitrile
				71–73	Diphenylacetonitrile

Table 63.16   Nitro Compounds

Bp	Mp	Name of Compound	Amine Obtained by Reduction of Nitro Groups Bp	Mp	Acetamide[a] Mp	Benzamide[b] Mp
210–211		Nitrobenzene	184		114	160
225		2-Nitrotoluene	200		110	146
225		2-Nitro-m-xylene	215		177	168
230–231		3-Nitrotoluene	203		65	125
245		3-Nitro-o-xylene	221		135	189
245–246		4-Ethylnitrobenzene	216		94	151
	34–36	2-Chloro-6-nitrotoluene	245		157(136)	173
	36–38	4-Chloro-2-nitrotoluene		21	139(131)	
	40–42	3,4-Dichloronitrobenzene		72	121	
	43–50	1-Chloro-2,4-dinitrobenzene		91	242(di)	178(di)
	52–54	4-Nitrotoluene		45	147	158
	55–56	1-Nitronaphthalene		50	159	160
	83–84	1-Chloro-4-nitrobenzene		72	179	192
	88–90	m-Dinitrobenzene		63	87(mono) 191(di)	125(mono) 240(di)

[a] For preparation, see page 373.
[b] For preparation, see page 373.

Table 63.17 Phenols

Bp	Mp	Name of Compound	Derivatives: α-Naphthylurethan[a] Mp	Derivatives: Bromo[b] Mp
175–176		2-Chlorophenol	120	48(mono) 76(di)
181	42	Phenol	133	95(tri)
202	32–34	*p*-Cresol	146	49(di) 108(tetra)
203		*m*-Cresol	128	84(tri)
228–229		3,4-Dimethylphenol	141	171(tri)
	32–33	*o*-Cresol	142	56(di)
	42–43	2,4-Dichlorophenol		68
	42–45	4-Ethylphenol	128	
	43–45	4-Chlorophenol	166	
	44–46	2,6-Dimethylphenol	176	79
	44–46	2-Nitrophenol	113	117(di)
	49–51	Thymol	160	55
	62–64	3,5-Dimethylphenol		166(tri)
	64–68	4-Bromophenol	169	95(tri)
	68–71	2,5-Dimethylphenol	173	178(tri)
	92–95	2,3,5-Trimethylphenol	174	
	95–96	1-Naphthol	152	105(di)
	98–101	4-*t*-Butylphenol	110	50(mono) 67(di)
	104–105	Catechol	175	192(tetra)
	109–110	Resorcinol	275	112(tri)
	112–114	4-Nitrophenol	150	142(di)
	121–124	2-Naphthol	157	84
	133–134	Pyrogallol	173	158(di)

[a] For preparation, see page 373.
[b] For preparation, see page 373.

64 Crown Ethers— 18-Crown-6

KEYWORDS   Macrocyclic polyether, crown ether
Host-guest chemistry, molecular complex
Naked anions, powerful nucleophiles
Triethylene glycol

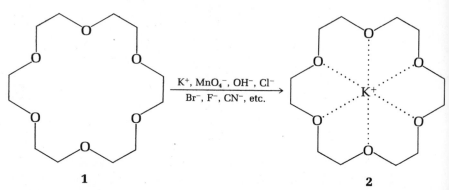

1

18-Crown-6
(1,4,7,10,13,16-Hexaoxacyclooctadecane)
MW 264.32, mp 37–38°

2

Potassium ion complex
(soluble in organic solvents)

Host-guest chemistry

Macrocyclic polyethers of the type exemplified by 18-crown-6 have the extraordinary ability to form stable molecular complexes with cations, owing to the efficient coordination of the cation by the ether oxygens.[1] The better the cation (the guest) fits into the cavity of the crown ether (the host), the more stable the complex. The particular crown ether to be synthesized in this experiment, 18-crown-6 (18-membered ring, 6 oxygen atoms), has a cavity diameter estimated to be between 2.6A and 3.2A. It will, therefore, complex most strongly with Rb^+, NH_4^+, and K^+, which have ionic diameters of 2.94A, 2.86A, and 2.66A, respectively. It will not, however, complex with Na^+ (diameter of Na^+, 2.2A). When a complex is formed, the cation will be in the hole and the anion will be associated with the cation; the complex is soluble in organic solvents.

---

[1] For a discussion of the complexes see C. J. Pederson and H. K. Frensdorff, *Angew. Chem., Internat. Edit.*, **11**, 16 (1972).

The anion is not highly solvated in organic solvents, as it would be in water, and consequently this "naked" anion is highly reactive. For instance, the 18-crown-6 complex with potassium fluoride is a highly reactive nucleophile in benzene solutions. Primary alkyl fluorides can be prepared from primary alkyl bromides in >90% yield employing this potassium fluoride complex.[2] Under the usual conditions (e.g., KF in diethylene glycol) the yields are about 40%. The crown ether complex of potassium hydroxide will displace the chloride ion from aryl chlorides to form phenols in organic solvents. In many of these reactions the crown ether can be used in catalytic amounts to effect solubility of the solid inorganic compound in an aprotic solvent like benzene. As such, it functions as a phase transfer catalyst between the solid and liquid phases [compare the phase transfer catalyst used to prepare dichlorocarbene (Chapter 42)]. 18-Crown-6 is the simplest member of this interesting new class of compounds, the crown ethers. Other crown ethers form complexes with ions of different sizes; asymmetric crown ethers will selectively complex one enantiomer in a racemic mixture, facilitating simple resolutions of, for example, amino acids.[3] A study of such host-guest chemistry may shed some light on the complexation-decomplexation reactions between enzymes and their substrates. The crown ethers bear a resemblance to the macrocyclic antibiotics in their ability to complex metal ions, and to effect ion transport and are, therefore, of interest to biochemists.

The procedure to be followed in this experiment involves initial conversion of the inexpensive solvent triethylene glycol to the corresponding 1,8-dichloro compound, **4**, followed by condensation of this halide with an equimolar amount of triethylene glycol in the presence of potassium hydroxide (an example of the Williamson ether synthesis) to give the product, 18-crown-6 (**1**).

$$HOCH_2CH_2OCH_2CH_2OCH_2CH_2OH \quad + \quad 2\ SOCl_2 \quad + \quad 2\ \text{(pyridine)}$$

3

Triethylene glycol
MW 150.17, bp 285°

Thionyl chloride
MW 118.97, bp 79°
den 1.631

Pyridine
MW 79.1, bp 115°

$$2\ SO_2 \quad + \quad 2\ \text{(pyridinium)}\ Cl^- \quad + \quad ClCH_2CH_2OCH_2CH_2OCH_2CH_2Cl$$

4

1,8-Dichloro-3,6-dioxaoctane
MW 187.08, bp 108–110° at 8 torr

[2] C. L. Liotta and H. P. Harris, *J. Amer. Chem. Soc.*, **96**, 2250 (1974).
[3] R. C. Helgeson, K. Koga, J. M. Timko, and D. J. Cram, *J. Amer. Chem. Soc.*, **95**, 3023 (1973).

EXPERIMENTS

1. 1,8-Dichloro-3,6-dioxaoctane (4)

To a mixture of 10g of triethylene glycol, 70 ml of benzene, and 11.9 g of pyridine in a 250-ml, round-bottomed flask add 10.8 ml of thionyl chloride dropwise with swirling.[4] Reflux the mixture in the hood (SO_2 evolution) for 30 min, then cool the reaction mixture, pour it into a separatory funnel, and shake it with 25 ml of 10% hydrochloric acid. Separate the layers, filter the organic layer through a cone of anhydrous sodium sulfate, and evaporate the benzene on a rotary evaporator (any residual water will evaporate as an azeotrope with the benzene). The residual light-yellow 1,8-dichloro-3,6-dioxaoctane (4) [also known as 1,2-bis(2-chloroethoxy)ethane] should weigh 12.9 g (100% yield).

2. 18-Crown-6 (1)

$$ClCH_2CH_2OCH_2CH_2OCH_2CH_2Cl \ + \ HOCH_2CH_2OCH_2CH_2OCH_2CH_2OH$$

4   **3**

MW 187.08   MW 150.17

$\downarrow$ KOH

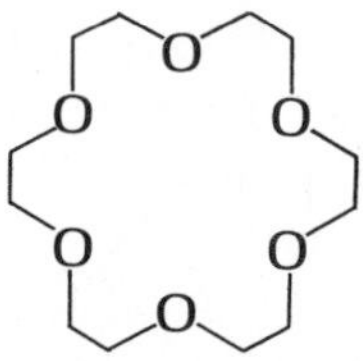

1

In a 25-ml Erlenmeyer flask held with a clamp add 2.2 g of potassium hydroxide pellets to 2.26 g of triethylene glycol. Heat the mixture to boiling over a microburner to dissolve all of the potassium hydroxide. Boil for one minute after the potassium hydroxide dissolves, then allow the mixture to cool slightly while rotating the flask to coat the wall and bottom with a uniformly thick layer of glycol-KOH paste. While the flask is still quite warm add 2.8 g of the dichloroether, **4**, prepared in the previous experiment. With a stainless steel spatula scrape and stir the mixture in the flask. After approximately one minute a vigorous exothermic reaction will begin. Be prepared to lower the flask into a pan of water to control the reaction. Continue to scrape and stir the paste until it begins to cool, then warm the mixture to boiling over a microburner for 3 min to complete the reaction. Cool the mixture under the tap, add 20 ml of methylene chloride, and remove the potassium chloride by filtration on a small Büchner funnel, using 5 ml more of methylene chloride to complete the transfer. Dry the methylene chloride solution over anhydrous sodium sulfate, remove the drying agent

Do not scale up this reaction. It is vigorously exothermic.

[4] Conveniently and safely dispensed from a burette in the hood.

Caution! The toxicological effects of the macrocyclic polyethers are uncertain. Do not inhale the vapors of 18-Crown-6 ether or allow it to contact the skin.[7]

by filtration, and evaporate the methylene chloride to leave crude 18-crown-6 (**1**) as a light brown oil. This oil can be used directly for the test that follows or several preparations can be pooled and distilled (bp 130–133° at 0.3 torr) to give a semisolid oil.

The semisolid oil can also be purified as its acetonitrile complex.[5] To do this, dissolve each gram of crude crown ether in 2.5 ml of hot acetonitrile. Cool the solution to room temperature and then in an ice-salt mixture before collecting the complex on a Büchner or Hirsch funnel. The complex melts at 63.5–65.5° and can be converted to the pure crown ether by pumping off the acetonitrile at 0.5 to 1.0 torr, while heating the complex at 40° in a small round-bottomed flask. The resulting 18-crown-6 has mp 36.5–38°.

3. Potassium permanganate complex of 18-crown-6[6]

In each of four 13 × 100-mm test tubes place small quantities (about 10 mg) of finely powdered (mortar and pestle) potassium permanganate. Add 3 ml of benzene to each tube and then, to the first, add two drops of your crude 18-crown-6 prepared in Section 2; to the second, add two drops of the dichloro compound (**4**) prepared in Section 1; and to the third, add two drops of triethylene glycol. To the fourth add nothing. Stopper the tubes, shake them vigorously for three min, and allow the permanganate to settle for one min before recording your observations. Add further small portions to each tube and repeat. Finally, add a drop of cyclohexene to the first tube and note any apparent reaction.

[5] C. L. Liotta, H. P. Harris, F. L. Cook, G. W. Gokel, and D. J. Cram, *J. Org. Chem.*, **39**, 2445 (1974).
[6] D. J. Sam and H. E. Simons, *J. Amer. Chem. Soc.*, **94**, 4024 (1972).
[7] See also *C&EN*, Jan. 27, 1975, p. 5.

Index

4 5 6 7 8 9 10

MOLECULAR WEIGHT CHANGE ON
CONVERTING A FUNCTIONAL GROUP TO A DERIVATIVE

Name of Derivative	Change in Converting Functional Group	Molecular Wt. Change	
		Formula	Mol. Wt.
Acetyl	—H → —COCH$_3$	C$_2$H$_2$O	42.038
Allophanate	—H → —CONHCONH$_2$	C$_2$H$_2$O$_2$N$_2$	86.050
Benzal	=H$_2$ → =CHC$_6$H$_5$	C$_7$H$_4$	88.110
Benzenesulfonyl	—H → —SO$_2$C$_6$H$_5$	C$_6$H$_4$O$_2$S	140.162
Benzhydryl	—CO$_2$H → —CO$_2$CH(C$_6$H$_5$)$_2$	C$_{13}$H$_{10}$	166.225
Benzoyl	—H → —COC$_6$H$_5$	C$_7$H$_4$O	104.109
Benzyl	—H → —CH$_2$C$_6$H$_5$	C$_7$H$_6$	90.126
p-Bromophenacyl	—H → —CH$_2$COC$_6$H$_4$Br	C$_8$H$_5$OBr	197.032
Carbobenzoxy	—H → —COOCH$_2$C$_6$H$_5$	C$_8$H$_6$O$_2$	134.136
Cathyl (carbethoxyl)	—H → —COOC$_2$H$_5$	C$_3$H$_4$O$_2$	72.064
Chloroacetyl	—H → —COCH$_2$Cl	C$_2$HOCl	76.483
Dimedon	=O → =(C$_8$H$_{11}$O$_2$)$_2$ − O	C$_{16}$H$_{22}$O$_3$	262.352
2,4-Dinitrobenzenesulfenyl	—H → —SC$_6$H$_3$(NO$_2$)$_2$	C$_6$H$_2$O$_4$N$_2$S	198.158
2,4-Dinitrobenzal	=H$_2$ → =CHC$_6$H$_3$(NO$_2$)$_2$	C$_7$H$_2$O$_4$N$_2$	178.105
3,5-Dinitrobenzoyl	—H → —COC$_6$H$_3$(NO$_2$)$_2$	C$_7$H$_2$O$_5$N$_2$	194.104
2,4-Dinitrophenyl	—H → —C$_6$H$_3$(NO$_2$)$_2$	C$_6$H$_2$O$_4$N$_2$	166.094
2,4-Dinitrophenylhydrazone	=O → =NNHC$_6$H$_3$(NO$_2$)$_2$	C$_6$H$_4$O$_3$N$_4$	180.124
Ethanol solvate	X + C$_2$H$_5$OH	C$_2$H$_6$O	46.095
Ethyl	—H → —C$_2$H$_5$	C$_2$H$_4$	28.054
Ethylenehemithioketal	=O → (ring: S—CH$_2$ / O—CH$_2$)	C$_2$H$_4$S	60.1182
Ethyleneketal	=O → (ring: O—CH$_2$ / O—CH$_2$)	C$_2$H$_4$O	44.054
Ethylenethioketal	=O → (ring: S—CH$_2$ / S—CH$_2$)	C$_2$H$_4$S$_2$ − O	76.183
Formyl	—H → —CHO	CO	28.011
Hydrate	X + H$_2$O	H$_2$O	18.015
Hydrazone	=O → =NNH$_2$	N$_2$H$_2$ − O	14.030
Mesyl	—H → —SO$_2$CH$_3$	CH$_2$O$_2$S	78.090
Methanol solvate	X + CH$_3$OH	CH$_4$O	32.042
Methiodide	X + CH$_3$I	CH$_3$I	141.940
Methyl	—H → —CH$_3$	CH$_2$	14.027
α-Naphthylurethan	—H → —CONHC$_{10}$H$_7$	C$_{11}$H$_7$ON	169.185
p-Nitrobenzyl	—H → —CH$_2$C$_6$H$_4$NO$_2$	C$_7$H$_5$O$_2$N	135.123
p-Nitrocarbobenzoxy	—H → —COOCH$_2$C$_6$H$_4$NO$_2$	C$_8$H$_5$O$_4$N	179.133
Oxime	=O → =NOH	HN	15.015
Phenylhydrazone	=O → =NNHC$_6$H$_5$	C$_6$H$_6$N$_2$ − O	90.129
Phenylosazone (of α-ketol)	—CO — ... — CHOH → —C=NNHC$_6$H$_5$ — ... — C=NNHC$_6$H$_5$	C$_{12}$H$_{10}$N$_4$ − O$_2$	178.242
Phenylurethan	—H → —CONHC$_6$H$_5$	C$_7$H$_5$ON	119.124
Picrate (salt or complex)	X + C$_6$H$_2$(NO$_2$)$_3$OH	C$_6$H$_3$O$_7$N$_3$	229.107
Pyrano	—H → (ring: —CH·O·CH$_2$ / CH$_2$CH$_2$CH$_2$)	C$_5$H$_8$O	84.119
Quinoxaline	—COCO— → —C—C— (‖ ‖ NC$_6$H$_4$N)	C$_6$H$_4$N$_2$ − O$_2$	72.113
Semicarbazone	=O → =NNHCONH$_2$	CH$_3$N$_3$	57.055
Tosyl	—H → —SO$_2$C$_6$H$_4$CH$_3$	C$_7$H$_6$O$_2$S	154.189
Trifluoroacetyl	—H → —COCF$_3$	C$_2$OF$_3$ − H	96.009
s-Trinitrobenzene complex	X + C$_6$H$_3$(NO$_2$)$_2$	C$_6$H$_3$O$_6$N$_3$	213.107
2,4,7-Trinitrofluorenone complex	X + C$_{12}$H$_5$CO(NO$_2$)$_3$	C$_{13}$H$_5$O$_7$N$_3$	315.201
Trityl	—H → —C(C$_6$H$_5$)$_3$	C$_{19}$H$_{14}$	242.323

MULTIPLES OF ELEMENTS' WEIGHTS AND THEIR LOGARITHMS

Element	Weight	Log
C	12.0112	1.07958
C$_2$	24.0223	1.38061
C$_3$	36.0335	1.55671
C$_4$	48.0446	1.68164
C$_5$	60.0557	1.77855
C$_6$	72.0669	1.85774
C$_7$	84.0780	1.92468
C$_8$	96.0892	1.98267
C$_9$	108.100	2.03383
C$_{10}$	120.111	2.07958
C$_{11}$	132.123	2.12098
C$_{12}$	144.134	2.15877
C$_{13}$	156.145	2.19353
C$_{14}$	168.156	2.22571
C$_{15}$	180.167	2.25568
C$_{16}$	192.178	2.28370
C$_{17}$	204.190	2.31003
C$_{18}$	216.201	2.33486
C$_{19}$	228.212	2.35834
C$_{20}$	240.223	2.38061
C$_{21}$	252.234	2.40180
C$_{22}$	264.245	2.42201
C$_{23}$	276.256	2.44131
C$_{24}$	288.268	2.45980
C$_{25}$	300.279	2.47752
C$_{26}$	312.290	2.49456
C$_{27}$	324.301	2.51095
C$_{28}$	336.312	2.52674
C$_{29}$	348.323	2.54198
C$_{30}$	360.334	2.55671
C$_{31}$	372.346	2.57095
C$_{32}$	384.357	2.58473
C$_{33}$	396.368	2.59810
C$_{34}$	408.379	2.61106
C$_{35}$	420.390	2.62365
C$_{36}$	432.401	2.63589
C$_{37}$	444.413	2.64779
C$_{38}$	456.424	2.65937
C$_{39}$	468.435	2.67065
C$_{40}$	480.446	2.68164
C$_{41}$	492.457	2.69237
C$_{42}$	504.468	2.70283
C$_{43}$	516.479	2.71305
C$_{44}$	528.491	2.72304
C$_{45}$	540.502	2.73280
C$_{46}$	552.513	2.74234
C$_{47}$	564.524	2.75168
C$_{48}$	576.535	2.76083
C$_{49}$	588.546	2.76978
C$_{50}$	600.558	2.77855
H	1.00797	0.003448
H$_2$	2.01594	0.304478
H$_3$	3.02391	0.480569
H$_4$	4.03188	0.605508
H$_5$	5.03985	0.702418
H$_6$	6.04782	0.781599

Element	Weight	Log
H$_7$	7.05579	0.848546
H$_8$	8.06376	0.906538
H$_9$	9.07173	0.95769
H$_{10}$	10.0797	1.00345
H$_{11}$	11.0877	1.04484
H$_{12}$	12.0956	1.08263
H$_{13}$	13.1036	1.11739
H$_{14}$	14.1116	1.14958
H$_{15}$	15.1196	1.17954
H$_{16}$	16.1275	1.20757
H$_{17}$	17.1355	1.23390
H$_{18}$	18.1435	1.25872
H$_{19}$	19.1514	1.28220
H$_{20}$	20.1594	1.30448
H$_{21}$	21.1674	1.32567
H$_{22}$	22.1753	1.34587
H$_{23}$	23.1833	1.36518
H$_{24}$	24.1913	1.38366
H$_{25}$	25.1993	1.40139
H$_{26}$	26.2072	1.41842
H$_{27}$	27.2152	1.43481
H$_{28}$	28.2232	1.45061
H$_{29}$	29.2311	1.46585
H$_{30}$	30.2391	1.48057
H$_{31}$	31.2471	1.49481
H$_{32}$	32.2550	1.50860
H$_{33}$	33.2630	1.52196
H$_{34}$	34.2710	1.53493
H$_{35}$	35.2790	1.54752
H$_{36}$	36.2869	1.55975
H$_{37}$	37.2949	1.57165
H$_{38}$	38.3029	1.58323
H$_{39}$	39.3108	1.59451
H$_{40}$	40.3188	1.60551
H$_{41}$	41.3268	1.61623
H$_{42}$	42.3347	1.62670
H$_{43}$	43.3427	1.63692
H$_{44}$	44.3507	1.64690
H$_{45}$	45.3587	1.65666
H$_{46}$	46.3666	1.66621
H$_{47}$	47.3746	1.67555
H$_{48}$	48.3826	1.68469
H$_{49}$	49.3905	1.69364
H$_{50}$	50.3985	1.70242
H$_{51}$	51.4065	1.71102
H$_{52}$	52.4144	1.71945
H$_{53}$	53.4224	1.72772
H$_{54}$	54.4304	1.73584
H$_{55}$	55.4384	1.74381
H$_{56}$	56.4463	1.75164
H$_{57}$	57.4543	1.75932
H$_{58}$	58.4623	1.76688
H$_{59}$	59.4702	1.77430
H$_{60}$	60.4782	1.78160
H$_{61}$	61.4862	1.78878
H$_{62}$	62.4941	1.79584
H$_{63}$	63.5021	1.80279

Element	Weight	Log
H$_{64}$	64.5101	1.80963
H$_{65}$	65.5181	1.81636
O	15.9994	1.20410
O$_2$	31.9988	1.50513
O$_3$	47.9982	1.68122
O$_4$	63.9976	1.80616
O$_5$	79.9970	1.90307
O$_6$	95.9964	1.98225
O$_7$	111.996	2.04920
O$_8$	127.995	2.10719
O$_9$	143.995	2.15835
O$_{10}$	159.994	2.20410
N	14.0067	1.14634
N$_2$	28.0134	1.44737
N$_3$	42.0201	1.62346
N$_4$	56.0268	1.74840
N$_5$	70.0335	1.84531
N$_6$	84.0402	1.92449
S	32.064	1.50602
S$_2$	64.128	1.80705
S$_3$	96.192	1.98314
S$_4$	128.256	2.10808
S$_5$	160.320	2.20499
S$_6$	192.384	2.28417
F	18.9984	1.27872
F$_2$	37.9968	1.57975
F$_3$	56.9952	1.75584
F$_4$	75.9936	1.88078
F$_5$	94.992	1.97769
F$_6$	113.99	2.05687
F$_7$	132.989	2.12382
F$_8$	151.987	2.18181
F$_9$	170.986	2.23296
F$_{10}$	189.984	2.27872
Cl	35.453	1.54966
Cl$_2$	70.906	1.85069
Cl$_3$	106.359	2.02678
Cl$_4$	141.812	2.15172
Cl$_5$	177.265	2.24863
Br	79.904	1.90257
Br$_2$	159.808	2.2036
Br$_3$	239.712	2.37969
Br$_4$	319.616	2.50463
Br$_5$	399.52	2.60154
I	126.904	2.10348
I$_2$	253.809	2.40451
I$_3$	380.713	2.5806
OCH$_3$	31.0345	1.49185
(OCH$_3$)$_2$	62.0689	1.79288
(OCH$_3$)$_3$	93.1034	1.96897
(OCH$_3$)$_4$	124.138	2.09391
(OCH$_3$)$_5$	155.172	2.19082
(OCH$_3$)$_6$	186.207	2.27
(OCH$_3$)$_7$	217.241	2.33695

Element	Weight	Log
(OCH$_3$)$_8$	248.276	2.3949
(OCH$_3$)$_9$	279.31	2.4460
(OCH$_3$)$_{10}$	310.345	2.4918
OC$_2$H$_5$	45.0616	1.6538
(OC$_2$H$_5$)$_2$	90.1231	1.9548
(OC$_2$H$_5$)$_3$	135.185	2.1309
(OC$_2$H$_5$)$_4$	180.246	2.2558
(OC$_2$H$_5$)$_5$	225.308	2.3527
(OC$_2$H$_5$)$_6$	270.369	2.4319
(OC$_2$H$_5$)$_7$	315.431	2.4989
(OC$_2$H$_5$)$_8$	360.492	2.5569
OCOCH$_3$	59.045	1.7711
(OCOCH$_3$)$_2$	118.090	2.0722
(OCOCH$_3$)$_3$	177.135	2.2483
(OCOCH$_3$)$_4$	236.180	2.3732
(OCOCH$_3$)$_5$	295.225	2.4701
(OCOCH$_3$)$_6$	354.270	2.5493
(OCOCH$_3$)$_7$	413.315	2.6162
(OCOCH$_3$)$_8$	472.360	2.6742
(OCOCH$_3$)$_9$	531.405	2.7254
(OCOCH$_3$)$_{10}$	590.450	2.7711
(H$_2$O)$_{1/2}$	9.00767	0.9546
H$_2$O	18.0153	1.2556
(H$_2$O)$_{1 1/2}$	27.0230	1.4317
(H$_2$O)$_2$	36.0307	1.5566
(H$_2$O)$_3$	54.0460	1.7327
(H$_2$O)$_4$	72.0614	1.8577
(H$_2$O)$_5$	90.0767	1.9546
(H$_2$O)$_6$	108.092	2.0338
P	30.9738	1.491
P$_2$	61.9476	1.7920
P$_3$	92.9214	1.9681
P$_4$	123.895	2.0930
P$_5$	154.869	2.1899
P$_6$	185.843	2.2691
Na	22.9898	1.3615
Na$_2$	45.9796	1.6625
Na$_3$	68.9694	1.8386
K	39.102	1.5922
K$_2$	78.204	1.8932
K$_3$	117.306	2.0693
Ag	107.868	2.0329
Ag$_2$	215.736	2.3339
Cu	63.546	1.8030
Cu$_2$	127.092	2.1041
Cr	51.996	1.7159
Hg	200.59	2.3023
Pb	207.19	2.3163
Pt	195.09	2.2902
Pt$_2$	390.18	2.5912
Se	78.96	1.8974
Th	204.37	2.3104

Log C/CO$_2$ = 0.436029   Log H$_2$/H$_2$O = 0.048836